Clear Cell Renal Cell Carcinoma 2022–2023

Clear Cell Renal Cell Carcinoma 2022–2023

Editors

José I. López
Claudia Manini

MDPI

Basel • Beijing • Wuhan • Barcelona • Belgrade • Novi Sad • Cluj • Manchester

Editors

José I. López
Biobizkaia Health Research
Institute
Barakaldo
Spain

Claudia Manini
San Giovanni Bosco Hospital
Turin
Italy

Editorial Office
MDPI
St. Alban-Anlage 66
4052 Basel, Switzerland

This is a reprint of articles from the Special Issue published online in the open access journal *Cancers* (ISSN 2072-6694) (available at: https://www.mdpi.com/journal/cancers/special_issues/A2O199N7M9).

For citation purposes, cite each article independently as indicated on the article page online and as indicated below:

Lastname, A.A.; Lastname, B.B. Article Title. *Journal Name* **Year**, *Volume Number*, Page Range.

ISBN 978-3-7258-0475-7 (Hbk)
ISBN 978-3-7258-0476-4 (PDF)
doi.org/10.3390/books978-3-7258-0476-4

Cover image courtesy of José I. López and Claudia Manini

Contents

About the Editors

José I. López

José I. López is a Research Advisor at the Biomarker in Cancer Unit of the Biobizkaia Health Research Institute. He graduated from the Faculty of Medicine of the University of the Basque Country, Leioa, Spain, and trained in pathology at the Hospital Universitario 12 de Octubre, Madrid, Spain. He received his PhD degree from the Universidad Complutense of Madrid, Spain. Dr. López has worked as a pathologist for more than 30 years in several hospitals in Spain, and he has specialized in uropathology, a topic on which he has published more than 200 peer-reviewed articles and reviews. Dr. López is interested in translational uropathology in general, and in renal cancer in particular, and collaborates with several international research groups, working to unveil the genomic landscape of urological cancer. Intratumor heterogeneity, tumor sampling, tumor microenvironments, tumor ecology, game theory, cell motility, immunotherapy, and the basic mechanisms of carcinogenesis are his main areas of interest.

Claudia Manini

Claudia Manini is Head of the Department of Pathology at the San Giovanni Bosco Hospital in Turin, Italy. She graduated from the Faculty of Medicine and Surgery at the University of Turin, Turin, Italy, and completed a postgraduate degree in surgical pathology at the same institution . Dr. Manini has worked as a pathologist for more than 25 years in several hospitals in Italy, developing her expertise in diagnostic uropathology, neuropathology, and gynecological pathology. She is also affiliated with the Department of Public Health and Pediatric Sciences at the University of Turin. Her main research interest is translational pathology.

cancers

Editorial

Clear Cell Renal Cell Carcinoma: A Test Bench for Investigating Tumor Complexity

Claudia Manini [1,2], Estíbaliz López-Fernández [3,4], Gorka Larrinaga [5,6,7] and José I. López [7,*]

1 Department of Pathology, San Giovanni Bosco Hospital, ASL Città di Torino, 10154 Turin, Italy; claudia.manini@aslcittaditorino.it
2 Department of Sciences of Public Health and Pediatrics, University of Turin, 10124 Turin, Italy
3 FISABIO Foundation, 46020 Valencia, Spain; estibaliz.lopez@universidadeuropea.es
4 Faculty of Health Sciences, European University of Valencia, 46023 Valencia, Spain
5 Department of Nursing, Faculty of Medicine and Nursing, University of the Basque Country (UPV/EHU), 48940 Leioa, Spain; gorka.larrinaga@ehu.eus
6 Department of Physiology, Faculty of Medicine and Nursing, University of the Basque Country (UPV/EHU), 48940 Leioa, Spain
7 Biobizkaia Health Research Institute, 48903 Barakaldo, Spain
* Correspondence: joseignacio.lopez@biocrucesbizkaia.org or jilpath@gmail.com

Clear cell renal cell carcinoma (CCRCC), by far the most common renal cancer subtype, is an aggressive tumor variant, serving in recent years as a prolific test bench in cancer research. Many of the recent advances in our knowledge about intratumor heterogeneity, tumor evolution, eco-oncology, metastatic competence, and therapeutic resistance have been achieved by analyzing this neoplasm [1–3]. The varied spectrum of papers collected in this Special Issue, composed of five articles, six reviews, one systematic review, and one commentary, confirms this fact.

Here, basic researchers and clinicians from Italy, Japan, USA, Spain, South Africa, Sweden, Romania, and Germany present their findings from very different perspectives, i.e., the role of the complement system in the immune background of tumors [4], imaging-based biomarker identification [5], the relevance of the immune microenvironment [6,7], the influence of clock genes and the circadian rhythm in tumor therapy [8], the morphological/molecular characteristics of cystic CCRCCs [9], the single-cell RNA-sequencing signature of primary and metastatic neoplasms [10], an extensive review of tumor biomarkers in CCRCC [11], the nitric oxide cycle-related pathways of this tumor [12], the current trends and complications of partial and radical nephrectomy in CCRCC [13], the characteristics of the intratumor immune heterogeneity in non-metastatic tumors [14], the importance of NPC1 targeting in CCRCC [15], and a comparison between the efficacy of immune checkpoint inhibitors (ICI) and tyrosine kinase inhibitors (TKI)/everolimus in the adjuvant therapy of CCRCC [16].

Panebianco et al. [4] review the role of the complement system (CS) in fostering the growth and progression of CCRCC through its interaction with the tumor microenvironment. The authors detail first the physiology of the CS, including the canonical and non-canonical pathways of complement activation and its negative regulators. Then, they analyze the role of the CS in cancer in general and in CCRCC in particular. Finally, the CS as a possible target in cancer therapy is considered. They conclude that CS may represent a predictive marker in the evaluation of immune checkpoint inhibitors-based therapies.

Posada Calderon et al. [5] focus on the difficulties of distinguishing renal cell carcinoma from other renal diseases based on current techniques of imaging. Magnetic resonance imaging (MRI) and positron emission tomography (PET)/CT using different radiolabeled molecules such as ^{18}F-fluorodeoxyglucose, ^{124}I-cG250, radiolabeled prostate-specific membrane antigen, and ^{11}C-acetate, together with a computational approach to CT images, are unveiling some data; however, in the opinion of the authors, these data still require

Citation: Manini, C.; López-Fernández, E.; Larrinaga, G.; López, J.I. Clear Cell Renal Cell Carcinoma: A Test Bench for Investigating Tumor Complexity. *Cancers* **2024**, *16*, 829. https://doi.org/10.3390/cancers16040829

Received: 7 February 2024
Accepted: 17 February 2024
Published: 19 February 2024

standardization and external validation before their integration into clinical practice. In this sense, recent research proposes two new feature selection strategies in the interpretation of radiomics studies to predict molecular and clinical targets in CCRCC [17].

In their first contribution to this Special Issue, Shapiro et al. [7] review the complexity of the immune microenvironment in renal cell carcinomas with special focus on CCRCC. The authors contextualize the issue starting with the immune response to cancer in general and the cells involved in the tumor immune microenvironment. Then, they follow with the characteristics of the immune environment in renal cancer, paying special attention to the different lymphocyte subsets involved and to tumor-associated macrophages. The role of tumor neoantigens and the genomic correlations with the immune response, as well as the predictive and prognostic ability of the immune microenvironment in CCRCC survival, have been also considered in their review. They conclude by emphasizing the critical importance of the tumor immune microenvironment in the understanding of the still hidden mechanisms operating in this neoplasm.

CCRCC exhibits resistance to standard chemotherapy, and the objective of current first-line therapies is to control tumor angiogenesis via tyrosine TKIs and to inhibit immune checkpoints through ICIs [18], which indicates the importance of the tumor microenvironment (TME) in the intricate biology of the CCRCC ecosystem. The interplay between cancer-associated fibroblasts (CAF), a predominant constituent of tumor stroma, and immune cells in the TME emerges as a decisive factor in CCRCC progression and in therapeutic responsiveness [19–21]. Through cytokine secretion, CAFs have the capability to induce the differentiation of M2 macrophages and attenuate the cytotoxic capacity of CD8 cells, fostering an immunosuppressive milieu [19,22]. In addition, CAFs' remodeling actions on the extracellular matrix impede the arrival of tumor-killing immune cells [22]. CAFs exhibit abundant expression of fibroblast activation protein-α (FAP) [23], a key player in these biological phenomena [22]. Both mRNA and immunohistochemical FAP expression in CAFs correlate with poor prognosis and limited response to TKIs and ICIs [20,21,24]. Current evidence suggests that FAP is a promising biomarker and a target for histopathological imaging, advanced radio-diagnostics, and emerging therapeutic modalities [25]. As a recent example of diagnostic usefulness, FAP+ CAFs specifically concentrate at the point of stromal invasion in colonic biopsies with uncertain invasion, thus helping in diagnosing adenocarcinoma microinvasion [26].

In a very interesting article, Fujita et al. [6] analyze the association of the tumor immune microenvironment in the primary tumor with the intervals of metastases. For this purpose, the authors considered synchronous (metastases within 3 months) and metachronous (metastases after 3 months) CCRCC in a series of 568 patients. They found that PD-L1 expression in tumor-infiltrating immune cells and immunophenotypes of the primary tumor were different according to the time recurrence, with the increased PD-L1 expression and the inflamed phenotypes being associated with shorter recurrences and tumor aggressiveness. These findings agree with previous experiences with inflamed [27] and PD-L1+ [28] CCRCC.

The circadian rhythm is involved in the regulation of cellular differentiation and physiology. Santoni et al. [8] revisit the influence of the altered expression of clock genes on the onset of cancer. The authors first consider the role of clock genes in cancer focusing specifically on renal cell carcinoma tumorigenesis and prognosis. They also analyze the circadian variations in cytokines and chemokines and their influence on the efficacy of immunotherapy and targeted therapy. They conclude that the role of the circadian clock genes in patients with renal cell carcinoma deserves further investigation since they represent potential therapeutic targets.

Pini et al. [9] review the morphological and molecular characteristics of cystic clear cell renal cell carcinomas. They point specifically to CCRCC with cystic changes, multilocular CCRCC, and clear cell papillary renal cell carcinomas. This short review provides useful criteria for pathologists when making a differential diagnosis in daily practice. The review includes a complete list of references on this topic.

A review of the recent advances in single-cell RNA-sequencing of CCRCC (primary and metastatic) [10] shows the importance of new tools in unveiling cancer biology. The authors revisit the origin of tumor cells in CCRCC, the transcriptomic identity of metastasizing cells, the role of the tumor microenvironment, immune and non-immune, in cancer progression, and the treatment opportunities of metastatic and non-metastatic CCRCCs.

Since CCRCC is a paradigm of intratumor heterogeneity, the selection of the best treatment modality in every CCRCC patient is a difficult task. In a systematic review, Dani et al. [11] present valuable insights into the spectrum of biomarkers available in predicting treatment response, prognosis, and therapeutic monitoring in patients with metastatic disease. The authors revisit retrospective studies and analyze prospective options in immunotherapy, VEGF-TKIs, and mTOR treatment strategies.

Ene et al. [12] analyze the role of the nitric oxide (NO) in the therapy of CCRCC. The authors review fundamental aspects such as the disruption of NO homeostasis, the dysregulation ureagenic cycle, the upregulation of glutamine, the cellular depletion of arginine, hyperammonemia, the reduction in branched-chain amino acids, the inactivation of VHL and the accumulation of HIFs, and the endogenous inhibition of NO synthesis.

Surgical trends and complications in partial and radical nephrectomy are reviewed by Pyrgidis et al. [13], using the GeRmAn Nationwide Inpatient Data provided by the Research Data Center of the Federal Bureau of Statistics of Germany between 2005 and 2021. The authors compare the perioperative morbidity, mortality, hospital stay, and costs between patients who underwent partial and radical nephrectomies. They show that the number of partial nephrectomies increased in this period. The authors detect an increment in comorbidities and risk factors in patients selected for radical nephrectomy. For example, statistically significant differences were detected in several clinical complications, such as transfusion, sepsis, acute respiratory failure, acute kidney disease, thromboembolism, ileus, surgical wound infection, 30-day mortality, and intensive care unit admission, between the two groups.

The second contribution of Shapiro et al. [14] points to the immune cell infiltration heterogeneity in non-metastatic CCRCC. The authors find that an increased number of CD8 cells within the tumor is associated with a decreased likelihood of progression to metastatic disease. In addition, they find that CD8 cells situated in close proximity to tumor cells are a sign of non-metastatic evolution in these patients. These findings strengthen the role of CD8 cells as a prognostic biomarker in CCRCC.

Proliferating cancer cells have greater requirements for cholesterol than non-tumor cells, and Fazliyeva et al. [15] show in their study that CCRCC cells have redundant mechanisms of cholesterol acquisition; for example, all major lipoproteins have comparable ability to support tumor cell growth and are equally effective in counteracting the antitumor activity of TKIs. Interestingly, the endolysosomal cholesterol transport regulated by the Niemann–Pick type C1 (NPC1) protein is a therapeutic target because this is a point where lipoproteins-derived cholesterol trafficking routes converge and may be simultaneously targeted in CCRCC patients.

Ossato et al. [16] aim in their study to compare the efficacy of mTOR, TKI, and CI inhibition therapies in metastatic CCRCC using a reconstruction of individual patient data from Kaplan–Meier curves. This novel approach allowed the authors to conduct all indirect head-to-head comparisons between these agents in a context in which no 'real' comparative trials have been conducted.

To conclude, this Special Issue comprises the recent experience of basic researchers and clinicians in several key issues of CCRCC pathogenesis, evolution, and treatment modalities that currently impact patient prognosis. Once more, the convenience of promoting multidisciplinary and translational approaches in cancer research is highlighted using CCRCC as a test bench.

Author Contributions: Conceptualization: C.M. and J.I.L.; formal analysis, C.M., E.L.-F. and G.L.; writing—original draft preparation, C.M.; supervision, J.I.L. All authors have read and agreed to the published version of the manuscript.

Funding: This research received no external funding.

Institutional Review Board Statement: Not applicable.

Informed Consent Statement: Not applicable.

Data Availability Statement: Not applicable.

Conflicts of Interest: The authors declare no conflicts of interest.

References

1. Noble, R.; Burley, J.T.; Le Sueur, C.; Hochberg, M.E. When, why and how tumour clonal diversity predicts survival. *Evol. Appl.* **2020**, *13*, 1558–1568. [CrossRef]
2. Tippu, Z.; Au, L.; Turajlic, S. Evolution of renal cell carcinoma. *Eur. Urol. Focus* **2021**, *7*, 148–151. [CrossRef]
3. Pallikonda, H.A.; Turajlic, S. Predicting cancer evolution for patient benefit: Renal cell carcinoma paradigm. *Biochim. Biophys. Acta Rev. Cancer* **2022**, *1877*, 188759. [CrossRef]
4. Panebianco, M.; Ciccarese, C.; Strusi, A.; Beccia, V.; Carbone, C.; Agostini, A.; Piro, G.; Tortora, G.; Iacovelli, R. The role of the complement in clear cell renal cell carcinoma (ccRCC). What future prospects are there for its use in clinical practice? *Cancers* **2024**, *16*, 490. [CrossRef]
5. Posada Calderon, L.; Eismann, L.; Reese, S.W.; Reznik, E.; Ari Hakimi, A. Advances in imaging-based biomarkers in renal cell carcinoma: A critical analysis of the current literature. *Cancers* **2023**, *15*, 354. [CrossRef]
6. Fujita, K.; Kimura, G.; Tsuzuki, T.; Kato, T.; Banno, E.; Kazama, A.; Yamashita, R.; Matsushita, Y.; Ishii, D.; Fukawa, T.; et al. The association of tumor immune microenvironment of the primary lesion with time to metastasis in patients with renal cell carcinoma: A retrospective analysis. *Cancers* **2022**, *14*, 5258. [CrossRef] [PubMed]
7. Saphiro, D.D.; Dolan, B.; Laklouk, I.A.; Rassi, S.; Lozar, T.; Emamekhoo, H.; Wentland, A.L.; Lubner, M.G.; Abel, E.J. Understanding the tumor immune microenvironment in renal cell carcinoma. *Cancers* **2023**, *15*, 2500. [CrossRef]
8. Santoni, M.; Molina-Cerrillo, J.; Santoni, G.; Lam, E.T.; Massari, F.; Mollica, V.; Mazzaschi, G.; Rapoport, B.L.; Grande, E.; Buti, S. Role of clock genes and circadian rhythm in renal cell carcinoma: Recent evidence and therapeutic consequences. *Cancers* **2023**, *15*, 408. [CrossRef] [PubMed]
9. Pini, G.M.; Lucianò, R.; Colecchia, M. Cystic clear cell renal cell carcinoma: A morphological and molecular reappraisal. *Cancers* **2023**, *15*, 3352. [CrossRef]
10. Alchahin, A.M.; Tsea, I.; Baryawno, N. Recent advances in single-cell RNA-sequencing of primary and metastatic clear cell renal cell carcinoma. *Cancers* **2023**, *15*, 4734. [CrossRef]
11. Dani, K.A.; Rich, J.M.; Kumar, S.S.; Cen, R.; Duddalwar, V.A.; D'Souza, A. Comprehensive systematic review of biomarkers in metastatic renal cell carcinoma: Predictors, prognostics, and therapeutic monitoring. *Cancers* **2023**, *15*, 4934. [CrossRef] [PubMed]
12. Ene, C.D.; Tampa, M.; Georgescu, S.R.; Matei, C.; Leulescu, I.M.T.; Dogaru, C.I.; Penescu, M.N.; Nicolae, I. Disturbances in nitric oxide cycle and related molecular pathways in clear cell renal cell carcinoma. *Cancers* **2023**, *15*, 5797. [CrossRef] [PubMed]
13. Pyrgidis, N.; Schulz, G.B.; Stief, C.; Blajan, I.; Ivanova, T.; Graser, A.; Staehler, M. Surgical trends and complications in partial and radical nephrectomy: Results from the GRAND study. *Cancers* **2024**, *16*, 97. [CrossRef]
14. Saphiro, D.D.; Lozar, T.; Cheng, L.; Xie, E.; Laklouk, I.; Lee, M.H.; Huang, W.; Jarrard, D.F.; Allen, G.O.; Hu, R.; et al. Non-metastatic clear cell renal cell carcinoma immune cell infiltration heterogeneity and prognostic ability in patients following surgery. *Cancers* **2024**, *16*, 478. [CrossRef]
15. Fazliyeva, R.; Makhov, P.; Uzzo, R.G.; Kolenko, V.M. Targeting NPC1 in renal cell carcinoma. *Cancers* **2024**, *16*, 517. [CrossRef] [PubMed]
16. Ossato, A.; Gasperoni, L.; Del Bono, L.; Messori, A.; Damuzzo, V. Efficacy of immune checkpoint inhibitors vs tyrosine kinase inhibitors/everolimus in adjuvant renal cell carcinoma: Indirect comparison of disease-free survival. *Cancers* **2024**, *16*, 557. [CrossRef] [PubMed]
17. Orton, M.R.; Hann, E.; Doran, S.J.; Shepherd, S.T.C.; Dafydd, D.A.; Spencer, C.E.; López, J.I.; Albarrán-Artahona, V.; Comito, F.; Warren, H.; et al. Interpretability of radiomics models is improved when using feature group selection strategies for predicting molecular and clinical targets in clear-cell renal cell carcinoma: Insights from the TRACERx Renal study. *Cancer Imaging* **2023**, *23*, 76. [CrossRef] [PubMed]
18. Astore, S.; Baciarello, G.; Cerbone, L.; Calabrò, F. Primary and acquired resistance to first-line therapy for clear cell renal cell carcinoma. *Cancer Drug Resist.* **2023**, *6*, 517–546. [CrossRef]
19. Hou, C.M.; Qu, X.M.; Zhang, J.; Ding, T.T.; Han, W.; Ji, G.C.; Zhong, Z.H.; Chen, H.; Zhang, F. Fibroblast activation proteins-α suppress tumor immunity by regulating T cells and tumor-associated macrophages. *Exp. Mol. Pathol.* **2018**, *104*, 29–37. [CrossRef]
20. Warli, S.M.; Putrantyo, I.I.; Laksmi, L.I. Correlation between tumor-associated collagen signature and fibroblast activation protein expression with prognosis of clear cell renal cell carcinoma patient. *World J. Oncol.* **2023**, *14*, 145–149. [CrossRef]

21. Davidson, G.; Helleux, A.; Vano, Y.A.; Lindner, V.; Fattori, A.; Cerciat, M.; Elaidi, R.T.; Verkarre, V.; Sun, C.M.; Chevreau, C.; et al. Mesenchymal-like tumor cells and myofibroblastic cancer-associated fibroblasts are associated with progression and immunotherapy response of clear cell renal cell carcinoma. *Cancer Res.* **2023**, *83*, 2952–2969. [CrossRef]
22. Xu, Y.; Li, W.; Lin, S.; Liu, B.; Wu, P.; Li, L. Fibroblast diversity and plasticity in the tumor microenvironment: Roles in immunity and relevant therapies. *Cell Commun. Signal.* **2023**, *21*, 234. [CrossRef]
23. Cords, L.; Tietscher, S.; Anzeneder, T.; Langwieder, C.; Rees, M.; de Souza, N.; Bodenmiller, B. Cancer-associated fibroblast classification in single-cell and spatial proteomics data. *Nat. Commun.* **2023**, *14*, 4294. [CrossRef] [PubMed]
24. Solano-Iturri, J.D.; Errarte, P.; Etxezarraga, M.C.; Echevarria, E.; Angulo, J.C.; López, J.I.; Larrinaga, G. Altered tissue and plasma levels of fibroblast activation protein-α (FAP) in renal tumors. *Cancers* **2020**, *12*, 3393. [CrossRef] [PubMed]
25. Shahvali, S.; Rahiman, N.; Jaafari, M.R.; Arabi, L. Targeting fibroblast activation protein (FAP): Advances in CAR-T cell, antibody, and vaccine in cancer immunotherapy. *Drug Deliv. Transl. Res.* **2023**, *13*, 2041–2056. [CrossRef] [PubMed]
26. Tarín-Nieto, A.; Solano-Iturri, J.D.; Arrieta-Aguirre, I.; Valdivia, A.; Etxezarraga, M.C.; Loizate, A.; López, J.I. Fibroblast activation protein-α (FAP) identifies stromal invasion in colorectal neoplasia. *Am. J. Surg. Pathol.* **2023**, *47*, 1027–1033. [CrossRef] [PubMed]
27. Brugarolas, J.; Rajaram, S.; Christie, A.; Kapur, P. The evolution of angiogenic and inflamed tumors: The renal cancer paradigm. *Cancer Cell* **2020**, *38*, 771–773. [CrossRef] [PubMed]
28. Larrinaga, G.; Solano-Iturri, J.D.; Errarte, P.; Unda, M.; Loizaga-Iriarte, A.; Pérez-Fernández, A.; Echevarría, E.; Asumendi, A.; Manini, C.; Angulo, J.C.; et al. Soluble PD-L1 is an independent prognostic factor in clear cell renal cell carcinoma. *Cancers* **2021**, *13*, 667. [CrossRef]

cancers MDPI

Review

The Role of the Complement in Clear Cell Renal Carcinoma (ccRCC)—What Future Prospects Are There for Its Use in Clinical Practice?

Martina Panebianco [1], Chiara Ciccarese [1], Alessandro Strusi [2], Viria Beccia [2], Carmine Carbone [1], Antonio Agostini [1], Geny Piro [1], Giampaolo Tortora [1,2,†] and Roberto Iacovelli [1,2,*,†]

1 Medical Oncology, Department of Medical and Surgical Sciences, Fondazione Policlinico Universitario Agostino Gemelli IRCCS, 00168 Rome, Italy; martina.panebianco@aslroma4.it (M.P.); chiara.ciccarese@policlinicogemelli.it (C.C.); carmine.carbone@policlinicogemelli.it (C.C.); agoan27@gmail.com (A.A.); geny.piro@policlinicogemelli.it (G.P.); giampaolo.tortora@policlinicogemelli.it (G.T.)
2 Medical Oncology, Department of Translational Medicine and Surgery, Catholic University of the Sacred Heart, 00168 Rome, Italy; ale.strusi94@gmail.com (A.S.); viria.beccia@gmail.com (V.B.)
* Correspondence: roberto.iacovelli@policlinicogemelli.it
† These authors share the last co-authorship.

Simple Summary: In recent years the first-line treatment of advanced renal cancer cell was implemented by new combination strategy. However, despite numerous research efforts, the choice of the type of treatment is still entrusted to clinical parameters. In this review we deepened the role of the complement system (CS) as a prognostic and predictive marker in kidney cancer. In particular, we described the physiology of the CS, its interaction with the tumor microenvironment and its role in oncogenesis and tumor progression. Based on the data reported in the literature, we concluded that the CS has a negative prognostic role in this pathology and its predictive role of response to immune-checkpoint inhibitors should be tested in prospective studies in order to optimize personalization of treatment for our patients, with the aim of reducing toxicity.

Abstract: In recent years, the first-line available therapeutic options for metastatic renal cell carcinoma (mRCC) have radically changed with the introduction into clinical practice of new immune checkpoint inhibitor (ICI)-based combinations. Many efforts are focusing on identifying novel prognostic and predictive markers in this setting. The complement system (CS) plays a central role in promoting the growth and progression of mRCC. In particular, mRCC has been defined as an "aggressive complement tumor", which encompasses a group of malignancies with poor prognosie and highly expressed complement components. Several preclinical and retrospective studies have demonstrated the negative prognostic role of the complement in mRCC; however, there is little evidence on its possible role as a predictor of the response to ICIs. The purpose of this review is to explore more deeply the physio-pathological role of the complement in the development of RCC and its possible future use in clinical practice as a prognostic and predictive factor.

Keywords: mRCC; complement; classical pathway; "aggressive complement" tumor; biomarkers; immunotherapy

Citation: Panebianco, M.; Ciccarese, C.; Strusi, A.; Beccia, V.; Carbone, C.; Agostini, A.; Piro, G.; Tortora, G.; Iacovelli, R. The Role of the Complement in Clear Cell Renal Carcinoma (ccRCC)—What Future Prospects Are There for Its Use in Clinical Practice? *Cancers* **2024**, *16*, 490. https://doi.org/10.3390/cancers16030490

Academic Editors: José I. López and Claudia Manini

Received: 15 December 2023
Revised: 12 January 2024
Accepted: 18 January 2024
Published: 23 January 2024

1. Introduction

In recent years, we have witnessed a change in the implementation of first-line treatment options for metastatic renal cell carcinoma (mRCC), with the introduction into clinical practice of two different immune checkpoint inhibitor (ICI)-based combinations, not directly compared to each other: the dual ICI regimen of ipilimumab + nivolumab (Ipi/Nivo) [1] or combinations of an ICI plus a VEGFR tyrosine kinase inhibitor (ICI + TKI) [2–5]. All the pivotal studies leading to the approval of ICI-based combinations stratified patients according to the International Metastatic RCC Database Consortium (IMDC) prognostic risk score, which relies only on clinical features [6]. Therefore, one of the main goals of the translational research aims at identifying new prognostic and predictive biomarkers of mRCC in tumor tissue, in order to select at diagnosis the patients who would benefit more from ICI rather than TKI. The identification of new biomarkers is essential for a better characterization of this disease and a tailored treatment selection [7–13]. One of the aspects that research has deepened in terms of RCC is the role of the complement system (CS) as a modulator of the tumor microenvironment (TME) and a possible prognostic and predictive factor in RCC. In fact, there is considerable evidence that the CS could play a key role in RCC progression and metastatization [14,15]. Roumenia et al. identified RCC as part of the so-called "aggressive complement tumors", which are characterized by a high expression of the complement components of the classical and alternative pathways and are associated with a worse prognosis [16]. Moreover, some preclinical studies evaluated the predictive role of the CS in mRCC as an immune-sensitive marker [17,18].

The purpose of this review is to elucidate the role of the complement in the TME, particularly in RCC, and to evaluate its possible use as a prognostic and predictive factor of response to ICI-based combinations.

2. The Physiology of the Complement System

The complement was first discovered in 1996 by Jules Bordet. The scientist did not observe bacterial lysis when serum containing antibodies against bacteria, preheated to a temperature of 56 °C, was incubated with the pathogens. Therefore, since it was already known that antibodies are heat-resistant, he hypothesized the presence of an additional heat-labile serum component that could assist the action of the antibody-mediated lysis, which was named the "complement" [19]. The complement system (CS) involves about 50 constituents, such as pattern recognition molecules (PRM), protein components, proteases, regulators, and cell surface receptors [20], which are produced mostly by the liver [21]. At the beginning, the CS was considered part of innate immunity due to the important role it plays in opsonization, chemotaxis, the lysis of pathogens, and inflammation. However, new evidence shows that the CS acts as a bridge between our innate and adaptive immunity, as it is able to increase the antibody response, promote the T cell response, eliminate self-reactive B cells, and enhance immunologic memory. In addition, the CS maintains cell homeostasis by eliminating cellular debris and immune complexes [22]. The activation of the complement occurs according to a mechanism called a "cascade" with sequential activation of the various components that circulate in inactive form (zymogens). The CS can be activated in canonical and non-canonical manners [14].

2.1. Canonical Pathways of Complement Activation

There are three canonical pathways of complement activation: classical, alternative, and lectin-driven. All three canonical pathways lead first to the formation of C3 convertase, which, in turn, cleaves C3 into two proteins (C3a and C3b). C3b is part of the C5 convertase that cleaves the C5 protein into C5a and C5b. C5b binds C6 and C7, forming the C5b,6,7 complex. C7, which is a hydrophobic protein, allows the insertion of the C5b,6,7 complex into the lipid double layer of the microbial membrane, where it becomes a receptor for C8 (complex C5b–8). The last step in the formation of the membrane attack complex (MAC) is the binding of the complex C5b–8 with C9. Then, the complex C5b–9 polymerizes and forms a pore in the bacterial membrane, which ultimately leads to the lysis of the cell (Figure 1).

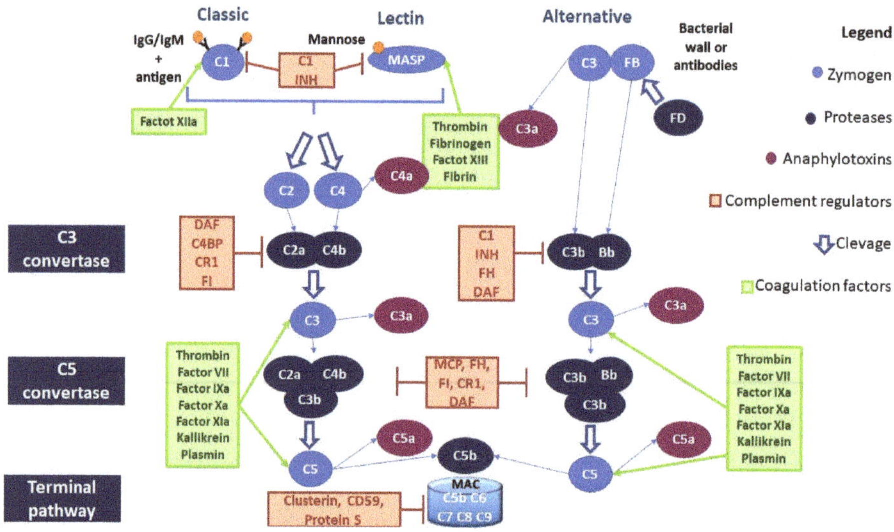

Figure 1. Canonical pathways of the complement activation and interaction of the coagulation system with the CS. Abbreviations. C1 INH, C1 inhibitor; DAF, decay-accelerating factor, or CD55, cluster of differentiation 55; C4BP, C4, binding protein; MCP, membrane cofactor protein, or CD46, cluster of differentiation 46; FH, factor H; FD, factor D; FI, factor I; MAC, membrane attack complex.

2.1.1. Classical Pathway

Classical pathway (CP) activation starts with the binding of the C1 fraction to the constant immunoglobulin domains IgG and IgM complexed to the antigen. The C1 fraction is a protein complex composed of C1q, C1r, and C1s subunits: C1q is assigned to the antibody binding, while C1r and C1s have protease activity. The binding of C1q to the immunoglobulins' Fc region leads to the activation of C1r, which cleaves and activates C1s. C1s cleaves the next component of the complement cascade, the C4 fraction, generating C4a and C4b fragments: the first is released in the fluid phase, while the second will bind on the microbial surface. Another important component is the C2 fraction, which is cleaved by C1s after the binding with the C4b fragment, generating soluble C2b and C2a. C2a remains physically associated with the C4b fragment bound to the microbial surface. The complex C4b2a represents the C3 convertase of the CP [23] (Figure 1).

2.1.2. Lectin Pathway

The lectin pathway (LP) is antibody-independent. It is activated by the interaction of microbial polysaccharides with circulating lectins such as mannose-binding lectin (MBL). MBL interacts with the associated serine protease (MASP), whose members are MASP-1, MASP-2, and MASP-3. MASP proteins have a homologous structure to that of the C1r and C1s proteases of the C1 fraction and perform very similar functions. The protein MASP-1 can form a tetrameric complex with MASP-2 similar to that formed by the subunits C1r and C1s; specifically, MASP-2 cleaves C4 and C2. The events resulting from this reaction are identical to those occurring in the classical pathway [24] (Figure 1).

2.1.3. Alternative Pathway

The alternative pathway (AP) is activated by direct C3 binding to the bacterial wall or antibodies. C3 proteolysis forms C3b, which covalently binds to the microbial surface. This bond induces exposure to an additional binding site for factor B (FB). FB, which binds to C3b, is cleaved by a plasma serine protease called factor D (FD). The cleavage generates a small fragment called Ba and a large fragment called Bb that remains linked to C3b. The

C3bBb complex represents the C3 convertase of the AP. C3b fragments, generated by the cleavage of C3, can bind to the surface of the microbe or bind to the C3 convertase itself, leading to the formation of a complex containing a Bb fragment and two C3b fragments (C5 convertase) [25] (Figure 1).

2.1.4. Anaphylatoxins

C3a, C5a, and C4a represent the anaphylatoxins produced by the activation of the CS. C4a has been shown to play a possible role in the activity of macrophages and monocytes, but further studies are needed to discover the receptors and functions of this anaphylatoxin [26–28]. C3a and C5a exert their important inflammatory role through the C3aR and C5aR receptors expressed on the cells of the immune system, such as non-myeloid cells [29]. In particular, C3a and C5a have the following functions: (a) contribute to the activation of macrophages, eosinophils, and neutrophils by stimulating the production of reactive oxygen species (ROS); (b) promote chemotaxis; (c) stimulate the release of histamine from mast cells; (d) and promote vasodilation [29–33]. The role of the C5CL2 receptor, which binds both C5a and C3a, is still unclear [34].

Anaphylatoxins also play a fundamental role in adaptive immunity. The CR2 receptor, expressed by B cells, contributes to the activation of B-lymphocytes after interaction with C3d-opsonized pathogens [35–37]. C3a and C5a are able to modulate the differentiation of T cells through the production of specific cytokines from the T cells and antigen-presenting cells (APC) [37,38]. Furthermore, local C3 synthesis by the dendritic cells is necessary to induce a Th1 response [39].

2.2. Non-Canonical Pathways of Complement Activation

2.2.1. Complement Activation through Coagulation Factors

Several studies have shown the possibility of activating the CS through proteases belonging to the coagulation system, and vice versa. Moreover, the CS may have both procoagulant (MAC) and anticoagulant action (C5a) [40] (Figure 1). For example, as reported in Table 1, factor XII is able to activate C1r and MASP1; thrombin leads to the activation of MASP1, MASP2, C3, and C4, whereas it inhibits the CS pathway by activating DAF [41–45]. Table 1 summarizes the interactions between the complement and coagulation systems.

Table 1. The relationship between the complement and coagulation systems.

Factor	Substrate	Action of the Complement/Coagulation Factor on the Substrate: Activated (+)/Inactivated (−)	References
Factor XIIa	C1r	+	[37]
Factor XIII, Fibrinogen, Fibrin, Thrombin	MASP1	+	[38,40,42,43]
Fibrin, Thrombin	MASP2	+	[43,46,47]
Thrombin, Factor VII, Factor IXa, Factor Xa, Factor Xia, Kallikrein, Plasmin	C3 and C5	+	[40,48–53]
Platelets	C3	+	[50,51,54,55]
Thrombin	DAF	+ (− complement activity)	[41]
MAC, C5a	Tissue Factor	+	[56,57]
Complement	Heparin	-	[58]

2.2.2. Local Complement Production

Although the majority of complement components are produced by the liver, new evidence has suggested also the local production of C3 and C5. T cells are among the first discovered to be C3 producers. C3 is expressed in the lysosomes and the endoplasmic reticulum of resting T lymphocytes. T-cell-expressed protease cathepsin L (CTSL) processes C3 into biologically active C3a and C3b. "Tonic" intracellular C3a generation is required for homeostatic T cell survival through mTOR activation. This intracellular CS is defined as a "complosome" (Figure 2A).

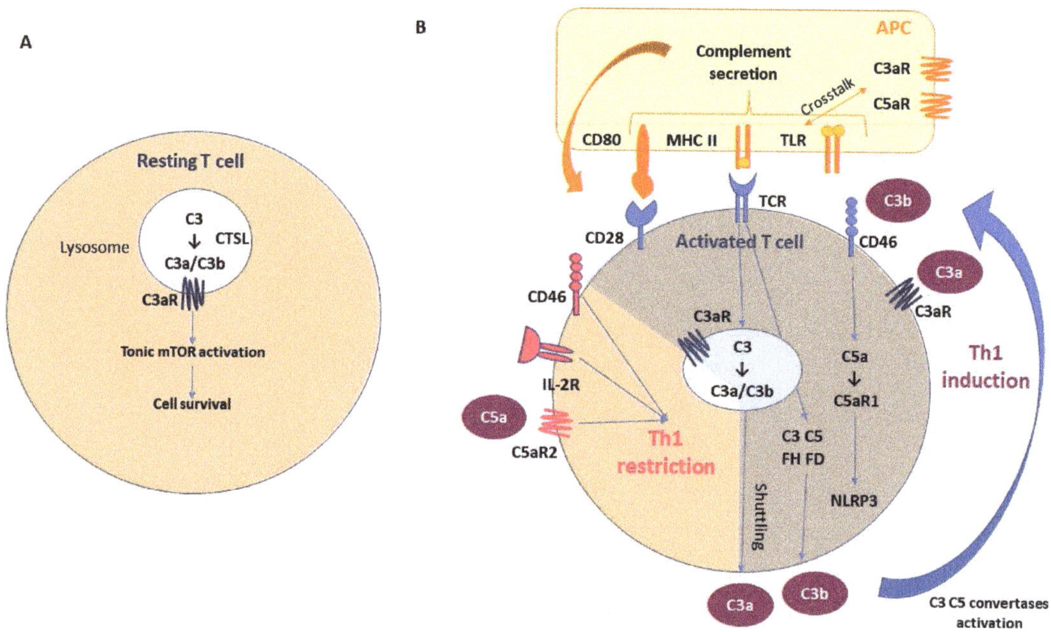

Figure 2. Local complement activation: the fundamental role of "complosome". (**A**) T-cell-expressed protease cathepsin L (CTSL) processes C3 into biologically active C3a and C3b. "Tonic" intracellular C3a generation is required for homeostatic T cell survival through mTOR activation. This intracellular CS is defined as "complosome". (**B**) As a result of T cell activation, the activated intracellular complement and complement receptors are secreted by T cells; C3a and C3b bind to C3aR and CD46 (membrane cofactor protein, MCP), respectively, and promote protective T Helper1 (Th1) differentiation. The activation of T cell receptors (TCRs) promotes also intracellular C3, C5, FH, FD secretion. Moreover, intracellular C5 activation (induced by CD46) and stimulation of C5a receptor 1 (C5aR1) promote pyrin-domain-containing protein 3 (NLRP3) inflammasome assembly, caspase-1-dependent interleukin-1β secretion, and induction of Th1 functional activation. In the early stages, CD46 promotes differentiation and expansion of T cells into Th1 effector lymphocytes. However, in the second phase, CD46, in association with the IL-2 receptor (IL-2R), leads to a Th1 restriction, promoting the secretion of IL-10. This is associated with the expression of C5aR2 via surface-shuttled C5a/C5a-desArg, which probably has a role in assisting this inhibitory signal. In addition, a synergistic effect has been observed between Toll-like receptor (TLR) and complement receptors expressed on the antigen-presenting cells (APCs). Abbreviations. APC, antigen-presenting cells; CTSL, T-cell-expressed protease cathepsin L; MHC II, major histocompatibility complex class II; TLR, Toll-like receptor; C3aR, C3a receptor; C5aR, C5a receptor; IL-2R, interleukin-2 receptor; NLRP3, NLR family pyrin-domain-containing 3.

As a result of T cell activation, the activated intracellular complement and complement receptors are secreted by the T cells; C3a and C3b bind to C3aR and CD46 (membrane cofactor protein, MCP), respectively, and promote protective T Helper1 (Th1) differentiation [27,59,60]. The activation of the T cell receptors (TCRs) promotes also intracellular C3, C5, FH, and FD secretion. Moreover, intracellular C5 activation (induced by CD46) and stimulation of C5a receptor 1 (C5aR1) promote pyrin-domain-containing protein 3 (NLRP3) inflammasome assembly, caspase-1-dependent interleukin-1β secretion, and the induction of Th1 functional activation [61,62]. In the early stages, CD46 promotes the differentiation and expansion of T cells into Th1 effector lymphocytes. However, in the second phase, CD46, in association with the IL-2 receptor (IL-2R), leads to a Th1 restriction, promoting the secretion of IL-10. This is associated with the expression of the C5aR2 via surface-shuttled C5a/C5a-desArg, which probably has a role in assisting this inhibitory signal [60,63]. In addition, a synergistic effect has been observed between Toll-like receptor (TLR) and the complement receptors expressed on the antigen-presenting cells (APCs) [64]. These mechanisms emphasize the importance of paracrine and autocrine complement secretion in the T-cell/APC interaction (Figure 2B), with the possibility of complement intracellular storage not being restricted to the immune cells.

2.3. Complement Negative Regulators

Several complement regulators (either soluble or membrane-bound) prevent aberrant activation of the CS (Figure 1). In particular, the negative regulators of the CS include:

(a) C1 inhibitor (C1 INH) is a competitive inhibitor of the C1r-C1s complex and MASP2 and interferes with the C3b–factor B interaction, blocking the activation of all three canonical pathways of CS;

(b) C3 and C5 convertase inhibitors: C4b-binding protein (C4BP) controls the activation at the C4 level of the CP and LP; factor H (FH) competes with FB to C3b-bond; complement receptor 1 (CR1) is capable of binding both C3b and C4b, displacing the link with Bb and C2b, and acts as a co-factor for factor I (FI), which degrades the fragment C3b; membrane cofactor protein (MCP) is capable of binding both C3b and C4b, displacing the link with Bb and C2b, respectively; and decay-accelerating factor (DAF) accelerates the decay of C3 convertases.

(c) MAC formation inhibitors: CD59 prevents the binding of the C9 fraction; S protein binds the C5b,6,7 complex, inhibiting insertion into the membranes; clusterin binds to C7 and the β-subunit of C8 and C9, inhibiting the correct complex assembly [65–67].

3. Cancer and Complement

The first evidence of an intrinsic interplay between the complement and cancer was demonstrated with the introduction of rituximab into lymphoma treatment. Rituximab is a chimeric (murine/human) monoclonal antibody that targets the CD20 receptor, expressed by the B cells. In particular, rituximab activates the classical pathway of the CS through the binding of its Fc portion to the C1q component. The activation of the complement results in the opsonization and lysis of cancer cells and in the recruitment of immune cells with phagocytic properties (i.e., macrophages and neutrophils) [14,68]. The complement has a dichotomous role in the tumor microenviroment (TME): in fact, the physiological complement concentrations activate the immune system's response to cancer cells, whereas an aberrant and chronic activation of the CS has been related to an immunosuppressed TME [69]. Moreover, cancer cells evade the CS's activity by hyper-expressing negative complement regulators such as CD35, CD46, CD55, and CD59 [70,71].

Complement proteins are produced in the TME by tumor-infiltrating immune cells and cancer cells and have access to the TME through vessels [72,73]. Complement activation is a response to tumor neoantigens and it has been shown to promote growth, progression, invasiveness, and tumor metastasization [74]. C5a is involved in the recruitment of myeloid-derived suppressor cells (MDSCs), resulting in suppressed antitumor immune responses [75]. The MDSCs inhibited the effector T cells, promoted angiogenesis, and were associated with an increased expression of immunomodulators such as programmed cell death 1 ligand 1 (PD-L1), arginase 1 (ARG1), cytotoxic T-lymphocyte antigen 4 (CTLA4), interleukin 10 (IL-10), IL-6, and lymphocyte-activation gene 3 (LAG3) in a murine model of lung cancer [76]. In addition, C5a promotes the extrusion of neutrophil extracellular traps (NETs), which are fibers of extracellular DNA released from the neutrophils. In particular, the nuclear protein high-mobility group box 1 (HMGB1) seems to be an endogenous stimulus produced by tumor cells that mediates the induction of NETosis in C5a-stimulated polymorphonuclear MDSCs (PMN-MDSCs). Furthermore, C5a facilitates the migration of these cells into the tumor, reducing the surface expression of β1 and β3 integrins and upregulating CD11b, a mediator of leukocyte extravasation [77].

Similar to C5a, C3a is involved in neutrophil recruitment [78]. Specifically, C3aR-dependent NETosis conduces coagulation and N2 polarization, promoting tumorigenesis. This shows a novel link between coagulation, neutrophilia, and complement activation [79]. Both C3a and C5a recruit tumor-associated macrophages (TAMs) and determine their differentiation into an M2-like phenotype, driving the immunosuppression of the T cells and tumor promotion [80,81]. M2 TAMs produce vascular endothelial growth factor (VEGF), fibroblast growth factor (FGF) chemokines, IL-17, IL-23, TGF-β, and other growth factors that determine the stimulation of vascular endothelial cell proliferation and the release of metalloproteases (MMPs), favoring tumor neovascularization, invasiveness, and progression [82–84].

A new C1q+ TAM subpopulation was recently identified in the TME. The chemokine CXCL-10, secreted by the C1Q+ TAM, activates and recruits CD8+ and CD4+ T cells, especially Th1 cells, binding to the receptor CXCR3 expressed by these immune cells. C1q+ TAM, favors T cell exhaustion. Moreover, C1q seems to perform the following functions: (a) activating an extracellular complement cascade and promoting chronic inflammation; (b) interacting with the endothelial cells promoting neoangiogenesis; (c) regulating macrophage polarization through autocrine action, interacting with LAIR1 (Figure 3).

Fibroblasts are also able to secrete C1q-containing exosomes and influence M2-like macrophage differentiation [85]. Autocrine stimulation of C3aR and C5aR1, expressed in the tumor cells, promotes tumor proliferation, triggering the phosphoinositide 3-kinase (PI3K)–AKT pathway [73]. Similarly, the C5b–C9 deposited on the cancer cells activates signal transduction pathways and induces cell cycle progression [86,87].

With regard to invasiveness and metastatization, both C3a and C5a play a role: C3a inhibits the expression of E-caderin, while C5a, through a C5aR1-mediated signal, stimulates the production of MMPs [88,89]. C3a activates the WNT–β-catenin pathway and enhances tumor cell proliferation, migration, and stemness in a mouse model with cutaneous squamous cell carcinoma [90].

Long pentraxin (PTX3) has been defined as a possible link between inflammation and cancer. PTX3 in physiological conditions acts as a functional ancestor of antibodies. It has an opsonic function via the Fc receptors, activates and regulates the complement cascade, regulates inflammation, and has a role in vascular biology [91–96]. Particularly, PTX3 binds C1q, activating the CP of the CS [93,96,97], modulates the LP by interacting with M-ficolins and MBL [98], and interacts with FH, modulating negatively AP activation through the recruitment of FH in PTX3-opsonized cells [99]. Therefore, PTX3 can be considered an oncosuppressor gene able to block complement-induced tumorigenesis. In fact, in knockout mice lacking the PTX3 gene, major susceptibility to mesenchymal and epithelial carcinogenesis; complement C3 deposition and higher C5a circulating levels, macrophage infiltration, cytokine production, and angiogenesis were observed. Therefore, this preclinical model suggests evidence of the central role of PTX3, as driven by complement activated macrophages, in counteracting tumor progression and of its loss potentially contributing to TME immunosuppression [100]. Moreover, loss of PTX3 was associated with genetic instability, contributing to malignant transformation [101].

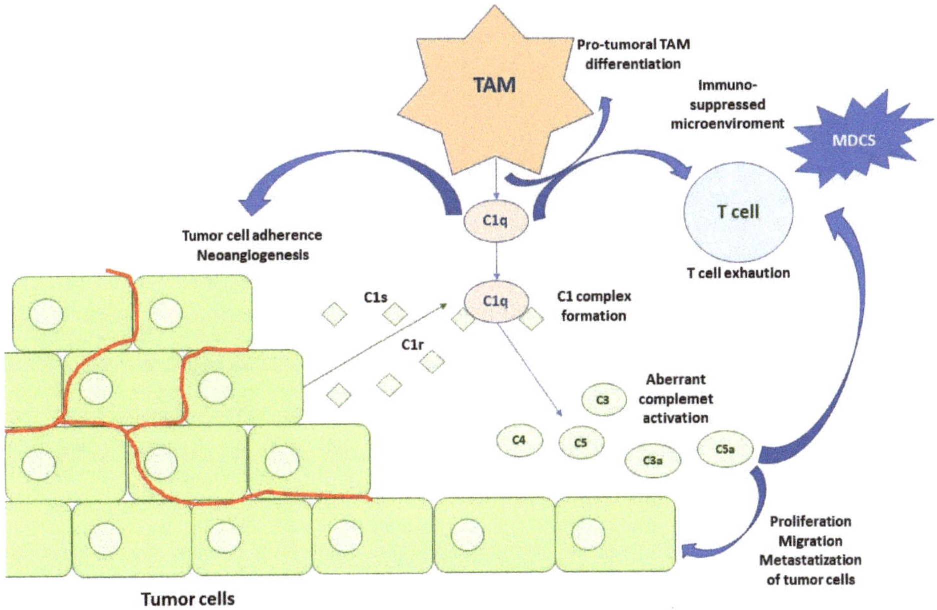

Figure 3. Activation of the CS in an RCC microenvironment. The first step is represented by C1q TAM secretion that contributes to pro-tumoral TAM phenotypes, T cell exhaustion, adhesion of tumor cells to the extracellular matrix, and neoangiogenesis. RCC cancer cells produce C1r and C1s, leading to an active C1 complex formation. This, in association with IgG deposits on tumor cells, promotes the classical pathway of the CS. Tumor cells also secrete the subsequent complement components, leading to C4 and C3 activation fragment deposition (C4b, C4d and C3b, iC3b, C3d). Moreover, anaphylatoxins C3a and C5a produced by the activation of the classical pathway modulate tumor cells and TME. Abbreviations: TAMs, tumor-associated macrophages; mDCs, myeloid dendritic cells.

4. The Prognostic Role of the Complement in Renal Cell Carcinoma

Based on the complement component gene expression analysis from The Cancer Genome Atlas (TCGA)'s database, four groups of neoplasms were characterized, including the so-called "aggressive complement" tumors. This group includes kidney renal clear cell carcinoma (KIRC), lung squamous carcinoma (LUSC), uveal melanoma (UVM), lower-grade glioma (LGG), and glioblastoma (GBM). The "aggressive complement" cancers highly express factors of both the classical and alternative pathways, which are significantly correlated with a worse prognosis [16]. Two important studies investigated the prognostic role of the C1q, C1s, C3, and C4 components in RCC [102,103]. Roumenia et al. conducted a study aimed at understanding the activation mechanisms of the CS in RCC and its impact on patients' clinical outcome. This study included an analysis of cell culture, mouse models, three retrospective cohorts of patients (with a total of 303 patients), and a prospective cohort of 7 patients, affected by stages I–IV ccRCC. Immunohistochemistry analyses of the C1q expression were conducted. Firstly, tumors were scored into three groups according to the percentage of C1q non-neoplastic cells: score 0 (weak): <1%; score 1 (intermediate): 1–30%; and score 2 (strong): >30%. Later, given the worse prognosis of patients with a score of 2 compared to those with scores 0–1, all subsequent evaluations were performed separating tumors into C1q high (score 2) and C1q low (scores 0 to 1) staining. The same score was used for the histological coloration criteria of cytoplasmic C4/C3 and C4d/C3d deposits on the tumor cell membranes. The analysis was first performed on 106 patients with stages I–IV ccRCC (Cohort 1), which revealed that the expression of C1q was a significant negative prognostic factor in the overall population, but specifically in patients with advanced stage disease (III–IV stages) for both overall survival (OS) ($p < 0.002$) and progression-free survival (PFS) ($p < 0.004$). These findings were then validated using cohorts 2 and 3. Immunofluorescence analysis showed that macrophages were the major cell type producing C1q in ccRCC tumors (80% of the C1q+-infiltrating cells were CD68+CD163+ macrophages), whereas the tumor cells stained negative for cytoplasmic C1q. Moreover, the M5 macrophage subtype was the principal C1q productor, associated with exhausted T cells. In fact, a positive correlation was found between the C1q+ cell density and PD-1 ($p = 0.012$), as well as LAG3 ($p = 0.0008$) in the 102 patients of cohort 1. A correlation between the C1q gene expression and exhausted T cell markers was confirmed in ccRCC tumors in publicly available transcriptomic data from the TCGA database ($n = 537$). The co-localization of IgG deposits and C1q on the surface of the tumor cells (30% of cases) and the co-localization of C1q with C4d staining (about 50% of samples) demonstrated in situ CP activation in RCC. Patients with a high density of C4-producing tumor cells and high C4d deposits had a significantly worse OS ($p = 0.007$) and PFS ($p = 0.013$). C1q, C4, and C4d were independent prognostic factors in both univariate and multivariate analyses, as opposed to C3 and C3d. Moreover, the ablation of C1q, C4, and C3 in mice was associated with decreased tumor growth. Impaired vascularization and downregulated VEGF-C gene expression were observed in a C1q knockout murine model [102]. This study confirmed the intra-tumoral activation of the CS, thanks to the in loco production of C1q by the macrophages and given the tumor cells' secretion of C3 and C4. Of note, C1q seemed to have a role in the angiogenesis and contributed to an immunosuppressed TME [102]. This study demonstrated for the first time the negative prognostic role of the complement components in RCC; moreover, it used cellular and animal models to reconstruct the intra-tumoral biological pathways through which complement activation occurs in RCC. This was fundamental to understanding the biological rationale behind the negative prognostic value of the CS, as well as its possible predictive role in identifying patients more likely to respond to ICIs.

The second study was conducted on three cohorts of patient affected by any stage of RCC [103]. Cohort 1 was a retrospective series of 82 patients, while cohorts 2 and 3 were prospective and enrolled 26 and 92 patients, respectively. In these two prospective cohorts, higher plasma C4d levels had a negative prognostic impact in terms of PFS, albeit not a statistically significant one. A trend between high levels of C4d tissue deposition and increased plasma C4d was observed in cohort 3. Moreover, co-localization of C1q with C1s and C4d with C1s indicated the CP of the CS as the main source of C4d deposition. Furthermore, combining the C4d deposits and C4d plasma values, it was possible to distinguish patients with a poor prognosis (high/high) from those with an intermediate (low/high) and good prognosis (low/low), respectively. Based on immunohistochemical sample examination of cohort 1, patients in the high-C1s group (C1s expressed in >30% of tumor cells) had a significantly decreased OS ($p = 0.006$) and PFS ($p = 0.004$) compared with those in the low-C1s group (C1s expression < 30%). High levels of C1s expression were correlated with TNM and metastatic status and they were associated with a worse prognosis for early disease stages (I and II), but not for advanced RCC (stage III and IV) in the ccRCC TCGA cohort. No correlation was observed between C1s expression and Fuhrman grade. These results were confirmed also in cohorts 1 and 3. Moreover, the prognostic value of C1s was independent regardless of other complement activation markers, indicating the possible cascade-independent role of C1s. C1s is involved also in tumor cell proliferation. In fact, silencing of C1S in Caki-1 and A498 cells induced a decrease in the proliferation capacity. The number of C1s+ cells within the tumor correlated with increased CD8+ T cell infiltration and PD-1 expression as detected using IHC in cohort 1, but this was not observed for C4d deposits or plasma C4d and C3, underling the potentially complement-independent role of C1s in the modulation of the TME [104]. Based on these results, the authors hypothesized a noncanonical intracellular function of C1s. In particular, three nonexclusive processes were suggested: (i) tumor cells secreted C1r, C1s, C3, C4, and C5, which, in association with C1q macrophage production, led to the canonical complement cascade activation; (ii) C1s modified the tumor cell transcription program and phenotype and promoted their proliferation and survival; (iii) a complex synergic interaction between the C1q-producing tumor cells and T cells was observed. In fact, IFNγ stimulated the production of C1s. The C1s-expressing tumor cells failed to activate CD4 and CD8 T cells. In contrast, C1s depletion in the tumor cells resulted in the aberrant expression of MHC class I and alleviated the inability of the tumor cells to activate the CD8 T cells [103]. Therefore, this study suggested that C1s not only took part in the activation of the classical complement pathway cascade but it also promoted tumor cell proliferation and would appear to have a role in conditioning tumor immunogenicity (higher T cell infiltrate and PD-L1 expression). In summary, this second study confirmed the negative prognostic value of the CS and in particular of C1s, in line with what was previously reported by Roumenine et al. Moreover, it evaluated for the first time the complement as a circulating biomarker, showing a correlation between plasmatic CS values and C4d tissue expression [103].

Based on the results of the previous two studies, a mechanism of classical pathway activation of the complement in "aggressive complement" tumors has been proposed (Figure 3). The first step is represented by C1q TAM secretion, which contributes to pro-tumoral TAM phenotypes, T cell exhaustion, the adhesion of tumor cells to the extracellular matrix, and neoangiogenesis. RCC cancer cells produce C1r and C1s, leading to an active C1 complex formation. This, in association with the IgG deposits on the tumor cells, promotes the classical pathway of the CS. Tumor cells also secrete the subsequent complement components, leading to C4 and C3 activation fragment deposition (C4b, C4d, C3b, iC3b, C3d). Moreover, the anaphylatoxins C3a and C5a produced by the activation of the classical pathway modulate the tumor cells and TME [16].

The role of other complement components in ccRCC was explored more deeply. An Asian study conducted on 272 RCC patients who underwent nephrectomy demonstrated the negative prognostic impact of high C5a histochemical expression in term of disease-free survival (DFS) ($p = 0.079$) and OS ($p = 0.011$), with a statistical significance for OS observed only for stages III and IV ($p < 0.001$) [59].

A more recent study confirmed the negative prognostic significance of a high expression of C5a in mRCC. This trial included 231 patients that received a TKI (sorafenib or sunitinib as first-line treatment). All tissue samples analyzed (mainly from nephrectomy and only a minority from metastases) were obtained prior to the start of TKI treatment. High immunohistochemical expression of C5a was significantly correlated with a worse OS and PFS ($p = 0.0199$ and $p = 0.0138$, respectively). C5a expression was significantly correlated with the MMP9 ($p = 0.000$), vimentin ($p = 0.000$), tumor PD-L1 ($p = 0.001$), stromal PD-L1 ($p = 0.002$), PD-1 ($p = 0.003$), and Ki67 ($p = 0.000$) expression levels. Additionally, high levels of C5a expression were strongly correlated with resistance to TKI ($p < 0.001$). Multivariate analysis identified C5a expression as an independent prognostic factor for mRCC patient outcomes [105].

Another study clarified the prognostic role of C3, analyzing three gene expression datasets (GSE36895 contained data from 29 ccRCC tissue samples and 23 tumor adjacent tissue samples; GSE53757 from 72 ccRCC samples and 72 tumor-adjacent samples; GSE66272 from 26 ccRCC samples and 26 tumor-adjacent samples) and a cohort of RCC patients in the TCGA database. Survival analysis was conducted only for this last cohort. In particular, C3, FN1, and C3AR1 were all more highly expressed in the RCC tissues ($n = 539$) than healthy renal tissues ($n = 72$) ($p < 0.001$). However, the protein levels of C3 and FN1, but not of C3AR1, were significantly higher in the RCC tissues than in normal kidney tissues. The RCC patients with high C3 or FN1 expression had a poorer OS (all $p < 0.05$). High C3 expression was also associated with significantly worse relapse-free survival (RFS). C3AR1 had no prognostic value in terms of both OS and RFS [106]. These studies were concordant and supported the negative prognostic significance of the CS in RCC.

The role of FH, which is part of the AP, was also explored in RCC. From histochemical tissue examination, the expression of FH was heterogeneous since it was possible to recognize both extracellular (mb-FH) and intracellular (int-FH) deposits. The co-localization of mb-FH with IgG suggested complement activation. Tumor cells were the main cells staining positive for FH. This study examined four cohorts of patients, three of which included ccRCC patients. Retrospective cohorts 1 and 2 enrolled a total of 224 patients and cohort 4 prospectively analyzed 61 patients. Histological analysis was conducted on cohorts 1 and 2, while cohort 4 was considered for plasma analysis. The Mb-FH deposits did not impact survival, whereas a high int-FH density was significantly correlated with a worse outcome. FH-silenced ccRCC cell lines were characterized by an alteration in proliferation, the cell cycle (arrested at G0–G1), morphology, viability, and migration. Int-FH but not mb-FH seemed to be essential for the tumor cell phenotype. Therefore, int-FH could play a role as a regulator of complement activation and impact the survival outcomes [107].

PTX3 was evaluated as a prognostic and predictive factor in a prospective cohort of 168 RCC patients. Both in cell lines and tissue samples, PTX3 was more highly expressed in cancer cells than in healthy ones. Moreover, PTX3 was co-localized with C1q, CD59, C3aR, and C5R. At time of diagnosis, the PTX3 serum levels were higher in the patients with RCC compared to healthy controls ($p < 0.001$) and significantly correlated with a high Fuhrman grade (G3–4 vs. G1–2 $p < 0.01$), lymph node involvement (N1 vs. N0 $p = 0.0008$), and visceral metastases (M1 vs. M0 $p < 0.001$). Patients with low baseline PTX3 levels (<165.0 pg/mL) showed a significantly higher 10-year OS rate as compared with ccRCC patients with high PTX3 serum levels (73.7% vs. 48.4%, $p = 0.002$) [108].

Thus, all studies examined (Table 2) were concordat on the negative prognostic value of different components of the CS, although a prospective, large, and more homogeneous cohort of patients is required for their validation, and further analyses are highly expected for assessing their potential role as predictors of response at the immunotherapy stage.

Table 2. Negative prognostic components of the complement system investigated in RCC.

Study	Negative Prognostic Complement Factor Evaluated	Laboratory Investigation	Patients (n)	Type of Study [R: Retrospective; P: Prospective]	Stage (%)	Survival Outcome
L.T. Roumenina et al., Cancer Immunol. Res., 2019 [102]	C1q C3 C4 C4d	IHC	Cohort 1: 106	R	I (40) II (6) III (41) IV (14)	**C1q** **OP: PFS $p = 0.008$, OS $p = 0.0016$** Stage I–II: PFS $p = 0.711$, OS $p = 0.256$ **Stage III–IV: PFS $p = 0.00356$, OS $p = 0.00198$** **C4c OP: PFS $p = 0.0235$, OS $p = 0.0299$** **C4d OP: PFS $p = 0.013$, OS $p = 0.007$** **C3 OP: PFS $p = 0.0349$, OS $p = 0.07$**
			Cohort 2: 154	R	I (40) II (5) III (54) IV (2)	**C1q:** Stage I–II: PFS $p = 0.527$, **Stage III–IV: PFS $p = 0.0109$**
			Cohort 3: 43	R	IV (100)	**C1q: PFS $p = 0.00276$, OS $p = 0.00126$** **C4d: PFS $p = 0.0176$**

Table 2. *Cont.*

Study	Negative Prognostic Complement Factor Evaluated	Laboratory Investigation	Patients (*n*)	Type of Study [R: Retrospective; P: Prospective]	Stage (%)	Survival Outcome
M.V. Daugan et al. Cancer Immunol. Res., 2021 [103]	C1s C4d deposits Plasma C4d	IHC IHC Plasma	Cohort 1: 82	R	I (40) II (6) III (39) IV (15)	**C4d deposits OP: PFS $p = 0.00176$**
			Cohort 2: 26	P	/	Plasma C4d OP: PFS $p = 0.09$
			Cohort 3:92 (longer FU)	P	I (54) II (8) III (17) IV (18)	**Plasma C4d OP: PFS $p = 0.0125$**
Wei Xi et al. Scientific Reports, 2016 [59]	C5a	IHC	272	R	I (62) II (8) III (24) IV (7)	OP: **OS $p = 0.011$**, DFS $p = 0.079$ Stage I–II: OS $p = 0.845$ Stage III–IV: OS $p < 0.001$
C. Yang et al. IJBM, 2023 [105]	C5a	IHC	231		IV	**OS $p = 0.0199$, PFS $p = 0.0138$**
Dong et al., BMC 2021 [106]	C3 C3AR1	Transcriptomics analysis	532 (TCGA-KIRC dataset)	R		C3 OP: OS $p = 0.0004$, RFS $p = 0.007$ C3AR1 OP: OS $p = 0.204$, RFS $p = 0.323$
Daugan et al. Cancer Immunol. Res., 2021 [107]	mb-FH int-FH	IHC	Cohort 1: 133	R	I (0) II (24) III (64) IV (9)	OP mb-FH: DFS $p = 0.14$ **OP int-FH: DFS $p = 0.004$**
			Cohort 2:91	R	I (5) II (22) III (48) IV (8)	OP mb-FH: PFS $p = 0.226$, OS $p = 0.627$ **OP int-FH: PFS $p = 0.0274$, OS $p = 0.0727$**
Netti et al. Aging (Albany NY). 2020 [108]	PTX3	Plasma	Cohort 1: 168	R	pT1 (62) pT2 (14) pT3 (21) pT4 (3) pN+ (20) cM+ (18)	**10-yr OS rate: 73.7% ↓PTX3 vs. 48.4% ↑PTX3, $p = 0.002$**

Abbreviations: IHC, immunohistochemistry; OP, overall population; OS, overall survival; PFS, progression-free survival; C4d, C4 deposits: RFS, relapse-free survival; DFS, disease-free survival. Up arrows: "high level"; down arrows: "low level".

5. The Complement System as a Possible Therapeutic Target

Rees et al. measured the plasma serum levels of some complement components, including C1q, C3, C5, FB, FD, FH, FI, C3c, sCD59, and s5b-9 (terminal complement complex—TCC) in a prospective sample of 24 mRCC patients prior to the initiation of immunotherapy (nivolumab monotherapy or a combination of ipilimumab and nivolumab). The concentration of plasma complement proteins was correlated with the time to next treatment (TNT). Patients with FH and FD levels below the cutoff had a worse response to ICIs, while low levels of both FI and TCC were associated with a better response and longer TNT. The authors also proposed an algorithm in order to select ICI-sensitive patients: low levels of TCC were predictive of the ICI response. In the presence of high TCC values, high C5 levels determined a good response to ICIs [17]. This study also evaluated a murine RCC model resistant to ICIs. Mice treated with C3aR1 (SB290157) and C5aR1 (PMX53) inhibitors had reduced tumor growth. The correlation between tumor growth and complement activation was not evaluated. Genetic deficiencies and pharmacological blockade of the complement receptors improved TIL function (increased production of IFN-γ). In conclusion, the complement appears to act as an additional checkpoint in RCC [17]. Although this prospective study enrolled limited series of mRCC patients and did not lead to robust conclusions, it is of great interest and suggests further prospective investigation to better clarify the predictive role of the CS in immunotherapy so as to guide treatment selection in the near future.

Another preclinical study evaluated a double block of both C5a and PD-1 in a syngeneic mouse model based on the subcutaneous growth of 393P cells that constitute a KRAS-driven lung adenocarcinoma. The combination of anti-C5a and anti-PD-1 monoclonal antibodies resulted in a significant reduction in tumor growth, showing the synergistic effect of the combined inhibition of both immune checkpoints and the complement, warranting additional studies for evaluating this potential synergistic therapeutic strategy [18].

6. Discussion

Our review further explored the central role of the CS and in particular the activation of the CP in RCC. Chronic imbalanced activation of the CS (especially the CP) negatively modulates the activity of the immune system, leading to an immunosuppressed TME. This favors tumor growth and progression [109]. A mechanism of complement activity proposed in RCC considers the C1q-secreting TAMs central in the first step of intra-tumor complement activation [16,102]. Then, the complex interaction between the tumor cells secreting the other components of the complement (C1r, C1s, C3, and C4), the MDSCs, and exhausted T lymphocytes leads to proliferation, increased tumor invasiveness, and neoangiogenesis, which ultimately results in tumor progression. In addition, there is new evidence regarding non-canonical mechanisms that involve the CS and that make the scenario even more complex [102–104,107].

All the studies reported in our review, although retrospective, agree on the negative prognostic role of the CS components' expression. Specifically, high levels of complement factors, tested in both tissue and plasma, are correlated with a worse outcome in RCC patients, mainly in the advanced stages (III–IV) [59,102,103,106–108]. However, these data need be validated prospectively.

Furthermore, the role of the CS as a predictor of the response to ICIs is of great interest, warranting further investigation. The only prospective study that evaluated the predictive role of the CS involved a limited number of patients with mRCC (24 total) but could be a first exploratory study for the realization of a prospective clinical validation trial. In fact, this study identified 11 plasma-detectable complement proteins (C1q, C3, C5, FB, FD, FH, FI, C3c, sCD59, and s5b-9), which were quantified prior to treatment with ICIs (nivolumab or ipilimumab/nivolumab). The study hypothesized an algorithm for selecting those patients more likely to respond to ICIs based on the plasma values of TCC and C5a [17]. The above findings show that the complement could play a crucial role in determining an immunosuppressed TME.

Therefore, high levels of CS components could identify highly immunosuppressed tumors, which therefore could be more responsive to ICIs. In this setting, ICIs would restore the proper activity of the immune system against cancer cells. This evidence demonstrates that high complement levels could be predictive of ICI response and this could be used as a positive predictive factor in clinical practice. Indeed, as already stated before, several studies have demonstrated a correlation between high levels of complement factors and the expression of molecules such as PD-L1 and LAG3. This is representative of an exhausted TME, which is therefore more susceptible to the action of ICIs [102]. In fact, exhausted CD8+ TILs with a mild expression of PD-1 could be re-activated by a PD-1/PD-L1 blockade [110–112]. However, the role of the CS in the development of treatment resistance to ICIs in mRCC has yet to be clarified, as there are conflicting data. Probably, the complex interaction between the TME and tumor cells has a fundamental role in the response to immunotherapy and anti-angiogenic agents [27,113].

It is important to notice that most of these studies evaluated only complement activation in tumor tissue. It would be interesting to evaluate the role of the CS as a dynamic biomarker of response/resistance in the course of immunotherapy by evaluating plasma levels. In fact, having a dynamic marker that could be easily evaluated in plasma during systemic therapy could be fundamental in selecting those patients who would benefit most from immunotherapy rather than anti-VEGFR-based treatments.

The CS represents only one of the pathways studied in order to identify new prognostic and predictive biomarkers in mRCC patients. In recent years, given the advent of ICI-based combinations as a first-line therapy for mRCC, the research has focused on the definition of biological features, including new immune-modulatory and angiogenesis gene expression signatures (GESs) that could play a role as predictors of response to guide clinicians in the selection of the most active combination. Therefore, exploratory analyses from four main randomized prospective trials (JAVELIN Renal 101, IMmotion151, IMmotion150, and CheckMate 9ER) of ICI-based combinations in treatment-naïve mRCC patients tried to identify potential biomarkers of response, however, without any definitive conclusion. Motzer et al. investigated GESs in histological samples from the phase 3 JAVELIN Renal 101 trial (evaluating avelumab plus axitinib as a first-line therapy for mRCC patients compared to sunitinib monotherapy). This analysis demonstrated that the expression of immune signatures, which included natural-killer-cell-related transcripts, as well as chemokine- and cytotoxic-T-cell-related elements, was associated with an improved PFS with ICI-based treatment compared to anti-VEGFR monotherapy [114]. Similarly, an exploratory analysis of integrated RNA sequencing (RNA-seq) and targeted somatic variant analysis from tumor samples of the IMmotion151 study (comparing atezolizumab plus bevacizumab to sunitinib) defined seven clusters of RCC associated with a different response to ICI-based combinations and VEGFR-TKI alone. The T-effector/proliferative cluster (number 4), proliferative cluster (number 5), and snoRNA cluster (number 7) responded better to immunotherapy. Moreover, regardless of the clusters, among the patients responding to immunotherapy, genes associated with the proliferation and immune pathways were more expressed. On the other hand, the patients responding to sunitinib showed a major expression of genes associated with VEGF signaling (hypoxia) [9]. A translational exploratory analysis from IMmotion150 was conducted with the aim of defining molecular biomarkers potentially correlated with the clinical outcome in each treatment group (atezolizumab vs. atezolizumab + bevacizumab vs. sunitinib). This analysis indicated that tumor mutation and neoantigen burdens were not associated with PFS, whereas angiogenesis, the T-effector/IFN-γ response, and myeloid inflammatory gene expression signatures were strongly and differentially associated with PFS within and across the treatments [10]. During the 2023 ASCO Genitourinary Symposium, Choueiri presented the results of a post hoc molecular analysis on pre-treatment tumor samples from the CheckMate 9ER trial. The study compared the combination of nivolumab and cabozantinib with sunitinib in a first-line setting for mRCC patients. The analysis identified a high angio-GES was associated with a longer PFS compared to a medium and low angio-GES in patients receiving nivolumab + cabozantinib; however, the predictive value of all seven GESs was not confirmed using Cox PH models. In addition, no single gene with predictive value was identified [8].

Further prospective studies are needed to confirm and deepen the role of the CS in RCC. It would be interesting, based on the study of Reese et al. [17], to design clinical studies aimed at evaluating the efficacy of combining ICIs with a complement inhibitor.

7. Conclusions

In actuality, the treatment selection in first-line mRCC therapy is currently based on the IMDC's prognostic factors, which rely only on clinical data and do not reflect the molecular heterogeneity of RCC tumors. Preliminary gene sequencing studies have paved the way for the identification of those tumors more sensitive to ICIs rather than to anti-VEGFR agents. However, further studies are highly expected to identify prognostic and predictive factors that can also be applied in clinical practice. In particular, the goal would be to identify an easily detectable, reproducible, and highly specific molecular profile able to select the best treatment option for each patient for a tailored personalized treatment approach. In fact, identifying patients who could benefit from immunotherapy alone could lead to more personalized and less toxic treatment strategies, such as the possibility of intermittent use of VEGFR-TKIs while maintaining immunotherapy in those patients with a disease response after initial induction with the ICI-TKI combo [115].

In conclusion, the CS, in consideration of its role in modulating immunogenicity and angiogenesis in RCC, could represent a dynamic marker with a predictive value of the response to ICI-based therapies. Further prospective investigations are highly awaited.

Author Contributions: Conceptualization, R.I., M.P. and C.C. (Chiara Ciccarese); methodology, A.A., C.C. (Chiara Ciccarese) and G.P.; software, M.P.; validation, R.I., C.C. (Carmine Carbone) and G.T.; formal analysis, C.C. (Chiara Ciccarese); investigation, M.P.; resources, V.B.; data curation, A.S.; writing—original draft preparation, M.P.; writing—review and editing, M.P. and C.C. (Carmine Carbone); visualization, M.P.; supervision, G.T.; project administration, R.I.; funding acquisition, R.I. All authors have read and agreed to the published version of the manuscript.

Funding: This research received no external funding.

Conflicts of Interest: The authors declare no conflict of interest.

References

1. Albiges, L.; Tannir, N.M.; Burotto, M.; McDermott, D.; Plimack, E.R.; Barthélémy, P.; Porta, C.; Powles, T.; Donskov, F.; George, S.; et al. Nivolumab plus ipilimumab versus sunitinib for first-line treatment of advanced renal cell carcinoma: Extended 4-year follow-up of the phase III CheckMate 214 trial. *ESMO Open* **2020**, *5*, e001079. [CrossRef] [PubMed]
2. Powles, T.; Plimack, E.R.; Soulières, D.; Waddell, T.; Stus, V.; Gafanov, R.; Nosov, D.; Pouliot, F.; Melichar, B.; Vynnychenko, I.; et al. Pembrolizumab plus axitinib versus sunitinib monotherapy as first-line treatment of advanced renal cell carcinoma (KEYNOTE-426): Extended follow-up from a randomised, open-label, phase 3 trial. *Lancet Oncol.* **2020**, *21*, 1563–1573. [CrossRef] [PubMed]
3. Choueiri, T.K.; Powles, T.; Burotto, M.; Escudier, B.; Bourlon, M.T.; Zurawski, B.; Oyervides Juárez, V.M.; Hsieh, J.J.; Basso, U.; Shah, A.Y.; et al. Nivolumab plus Cabozantinib versus Sunitinib for Advanced Renal-Cell Carcinoma. *N. Engl. J. Med.* **2021**, *384*, 829–841. [CrossRef]
4. Motzer, R.; Alekseev, B.; Rha, S.-Y.; Porta, C.; Eto, M.; Powles, T.; Grünwald, V.; Hutson, T.E.; Kopyltsov, E.; Méndez-Vidal, M.J.; et al. Lenvatinib plus Pembrolizumab or Everolimus for Advanced Renal Cell Carcinoma. *N. Engl. J. Med.* **2021**, *384*, 1289–1300. [CrossRef] [PubMed]
5. Motzer, R.J.; Penkov, K.; Haanen, J.; Rini, B.; Albiges, L.; Campbell, M.T.; Venugopal, B.; Kollmannsberger, C.; Negrier, S.; Uemura, M.; et al. Avelumab plus axitinib versus sunitinib for advanced renal-cell carcinoma. *N. Engl. J. Med.* **2019**, *380*, 1103–1115. [CrossRef] [PubMed]
6. Motzer, R.J.; Mazumdar, M.; Bacik, J.; Russo, P.; Berg, W.J.; Metz, E.M. Effect of cytokine therapy on survival for patients with advanced renal cell carcinoma. *J. Clin. Oncol. Off. J. Am. Soc. Clin. Oncol.* **2000**, *18*, 1928–1935. [CrossRef]
7. Ricketts, C.J.; De Cubas, A.A.; Fan, H.; Smith, C.C.; Lang, M.; Reznik, E.; Bowlby, R.; Gibb, E.A.; Akbani, R.; Beroukhim, R.; et al. The Cancer Genome Atlas Comprehensive Molecular Characterization of Renal Cell Carcinoma. *Cell Rep.* **2018**, *23*, 3698. [CrossRef] [PubMed]
8. Choueiri, T.K.; Motzer, R.J.; Powles, T.; Burotto, M.; Apolo, A.B.; Escudier, B.; Tomita, Y.; McDermott, D.F.; Braun, D.A.; Han, C.; et al. Biomarker analysis from the phase 3 CheckMate 9ER trial of nivolumab + cabozantinib v sunitinib for advanced renal cell carcinoma (aRCC). *J. Clin. Oncol.* **2023**, *41* (Suppl. S6), 608. [CrossRef]
9. Motzer, R.J.; Banchereau, R.; Hamidi, H.; Powles, T.; McDermott, D.; Atkins, M.B.; Escudier, B.; Liu, L.-F.; Leng, N.; Abbas, A.R.; et al. Molecular Subsets in Renal Cancer Determine Outcome to Checkpoint and Angiogenesis Blockade. *Cancer Cell* **2020**, *38*, 803–817.e4. [CrossRef]

10. McDermott, D.F.; Huseni, M.A.; Atkins, M.B.; Motzer, R.J.; Rini, B.I.; Escudier, B.; Fong, L.; Joseph, R.W.; Pal, S.K.; Reeves, J.A.; et al. Clinical activity and molecular correlates of response to atezolizumab alone or in combination with bevacizumab versus sunitinib in renal cell carcinoma. *Nat. Med.* **2018**, *24*, 749–757. [CrossRef]
11. Ciccarese, C.; Buti, S.; Roberto, M.; Calabro, F.; Masini, C.; Massari, F.; Cannella, M.A.; Mazzaschi, G.; Astore, S.; Di Girolamo, S.; et al. Evaluation of PBRM1, PD-L1, CD31, and CD4/CD8 ratio as a predictive signature of response to VEGFR-TKI–based therapy in patients with metastatic renal cell carcinoma (mRCC) with IMDC intermediate prognosis: Results from the APAChE-I Study. *J. Clin. Oncol.* **2023**, *41* (Suppl. S6), 714. [CrossRef]
12. Bimbatti, D.; Ciccarese, C.; Fantinel, E.; Sava, T.; Massari, F.; Bisogno, I.; Romano, M.; Porcaro, A.; Brunelli, M.; Martignoni, G.; et al. Predictive role of changes in the tumor burden and International Metastatic Renal Cell Carcinoma Database Consortium class during active surveillance for metastatic renal cell carcinoma. *Urol. Oncol.* **2018**, *36*, e13–e526.e18. [CrossRef] [PubMed]
13. Iacovelli, R.; Lanoy, E.; Albiges, L.; Escudier, B. Tumour burden is an independent prognostic factor in metastatic renal cell carcinoma. *BJU Int.* **2012**, *110*, 1747–1753. [CrossRef]
14. Reis, E.S.; Mastellos, D.C.; Ricklin, D.; Mantovani, A.; Lambris, J.D. Complement in cancer: Untangling an intricate relationship. *Nat. Rev. Immunol.* **2018**, *18*, 5–18. [CrossRef] [PubMed]
15. Magrini, E.; Minute, L.; Dambra, M.; Garlanda, C. Complement activation in cancer: Effects on tumor-associated myeloid cells and immunosuppression. *Semin. Immunol.* **2022**, *60*, 101642. [CrossRef]
16. Roumenina, L.T.; Daugan, M.V.; Petitprez, F.; Sautès-Fridman, C.; Fridman, W.H. Context-dependent roles of complement in cancer. *Nat. Rev. Cancer* **2019**, *19*, 698–715. [CrossRef]
17. Reese, B.; Silwal, A.; Daugherity, E.; Daugherity, M.; Arabi, M.; Daly, P.; Paterson, Y.; Woolford, L.; Christie, A.; Elias, R.; et al. Complement as Prognostic Biomarker and Potential Therapeutic Target in Renal Cell Carcinoma. *J. Immunol.* **2020**, *205*, 3218–3229. [CrossRef]
18. Ajona, D.; Ortiz-Espinosa, S.; Moreno, H.; Lozano, T.; Pajares, M.J.; Agorreta, J.; Bértolo, C.; Lasarte, J.J.; Vicent, S.; Hoehlig, K.; et al. A Combined PD-1/C5a Blockade Synergistically Protects against Lung Cancer Growth and Metastasis. *Cancer Discov.* **2017**, *7*, 694–703. [CrossRef]
19. Kolev, M.; Le Friec, G.; Kemper, C. The role of complement in $CD4^+$ T cell homeostasis and effector functions. *Semin. Immunol.* **2013**, *25*, 12–19. [CrossRef]
20. Ricklin, D.; Lambris, J.D. Complement in immune and inflammatory disorders: Pathophysiological mechanisms. *J. Immunol.* **2013**, *190*, 3831–3838. [CrossRef]
21. Lubbers, R.; van Essen, M.F.; van Kooten, C.; Trouw, L.A. Production of complement components by cells of the immune system. *Clin. Exp. Immunol.* **2017**, *188*, 183–194. [CrossRef]
22. Ricklin, D.; Lambris, J.D. Complement-targeted therapeutics. *Nat. Biotechnol.* **2007**, *25*, 1265–1275. [CrossRef]
23. Porter, R.R.; Reid, K.B. Activation of the complement system by antibody-antigen complexes: The classical pathway. *Adv. Protein Chem.* **1979**, *33*, 1–71. [CrossRef]
24. Garred, P.; Genster, N.; Pilely, K.; Bayarri-Olmos, R.; Rosbjerg, A.; Ma, Y.J.; Skjoedt, M.-O. A journey through the lectin pathway of complement-MBL and beyond. *Immunol. Rev.* **2016**, *274*, 74–97. [CrossRef] [PubMed]
25. Cooper, P.D. Complement and cancer: Activation of the alternative pathway as a theoretical base for immunotherapy. *Adv. Immun. Cancer Ther.* **1985**, *1*, 125–166. [CrossRef]
26. Tsuruta, T.; Yamamoto, T.; Matsubara, S.; Nagasawa, S.; Tanase, S.; Tanaka, J.; Takagi, K.; Kambara, T. Novel function of C4a anaphylatoxin. Release from monocytes of protein which inhibits monocyte chemotaxis. *Am. J. Pathol.* **1993**, *142*, 1848–1857.
27. Zhao, Y.; Xu, H.; Yu, W.; Xie, B.-D. Complement anaphylatoxin C4a inhibits C5a-induced neointima formation following arterial injury. *Mol. Med. Rep.* **2014**, *10*, 45–52. [CrossRef] [PubMed]
28. Barnum, S.R. C4a: An Anaphylatoxin in Name Only. *J. Innate Immun.* **2015**, *7*, 333–339. [CrossRef] [PubMed]
29. Klos, A.; Tenner, A.J.; Johswich, K.-O.; Ager, R.R.; Reis, E.S.; Köhl, J. The role of the anaphylatoxins in health and disease. *Mol. Immunol.* **2009**, *46*, 2753–2766. [CrossRef] [PubMed]
30. El-Lati, S.G.; Dahinden, C.A.; Church, M.K. Complement peptides C3a- and C5a-induced mediator release from dissociated human skin mast cells. *J. Investig. Dermatol.* **1994**, *102*, 803–806. [CrossRef]
31. Elsner, J.; Oppermann, M.; Czech, W.; Kapp, A. C3a activates the respiratory burst in human polymorphonuclear neutrophilic leukocytes via pertussis toxin-sensitive G-proteins. *Blood* **1994**, *83*, 3324–3331. [CrossRef] [PubMed]
32. Elsner, J.; Oppermann, M.; Czech, W.; Dobos, G.; Schöpf, E.; Norgauer, J.; Kapp, A. C3a activates reactive oxygen radical species production and intracellular calcium transients in human eosinophils. *Eur. J. Immunol.* **1994**, *24*, 518–522. [CrossRef] [PubMed]
33. Murakami, Y.; Imamichi, T.; Nagasawa, S. Characterization of C3a anaphylatoxin receptor on guinea-pig macrophages. *Immunology* **1993**, *79*, 633–638. [PubMed]
34. Chen, N.-J.; Mirtsos, C.; Suh, D.; Lu, Y.-C.; Lin, W.-J.; McKerlie, C.; Lee, T.; Baribault, H.; Tian, H.; Yeh, W.-C. C5L2 is critical for the biological activities of the anaphylatoxins C5a and C3a. *Nature* **2007**, *446*, 203–207. [CrossRef]
35. Klaus, G.G.; Humphrey, J.H. A re-evaluation of the role of C3 in B-cell activation. *Immunol. Today* **1986**, *7*, 163–165. [CrossRef] [PubMed]
36. Dempsey, P.W.; Allison, M.E.; Akkaraju, S.; Goodnow, C.C.; Fearon, D.T. C3d of complement as a molecular adjuvant: Bridging innate and acquired immunity. *Science* **1996**, *271*, 348–350. [CrossRef]

37. Fang, Y.; Xu, C.; Fu, Y.X.; Holers, V.M.; Molina, H. Expression of complement receptors 1 and 2 on follicular dendritic cells is necessary for the generation of a strong antigen-specific IgG response. *J. Immunol.* **1998**, *160*, 5273–5279. [CrossRef]
38. Strainic, M.G.; Liu, J.; Huang, D.; An, F.; Lalli, P.N.; Muqim, N.; Shapiro, V.S.; Dubyak, G.R.; Heeger, P.S.; Medof, M.E. Locally produced complement fragments C5a and C3a provide both costimulatory and survival signals to naive CD4+ T cells. *Immunity* **2008**, *28*, 425–435. [CrossRef]
39. Peng, Q.; Li, K.; Patel, H.; Sacks, S.H.; Zhou, W. Dendritic cell synthesis of C3 is required for full T cell activation and development of a Th1 phenotype. *J. Immunol.* **2006**, *176*, 3330–3341. [CrossRef]
40. Amara, U.; Flierl, M.A.; Rittirsch, D.; Klos, A.; Chen, H.; Acker, B.; Brückner, U.B.; Nilsson, B.; Gebhard, F.; Lambris, J.D.; et al. Molecular intercommunication between the complement and coagulation systems. *J. Immunol.* **2010**, *185*, 5628–5636. [CrossRef]
41. Ghebrehiwet, B.; Silverberg, M.; Kaplan, A.P. Activation of the classical pathway of complement by Hageman factor fragment. *J. Exp. Med.* **1981**, *153*, 665–676. [CrossRef] [PubMed]
42. Gulla, K.C.; Gupta, K.; Krarup, A.; Gal, P.; Schwaeble, W.J.; Sim, R.B.; O'Connor, C.D.; Hajela, K. Activation of mannan-binding lectin-associated serine proteases leads to generation of a fibrin clot. *Immunology* **2010**, *129*, 482–495. [CrossRef]
43. Krarup, A.; Wallis, R.; Presanis, J.S.; Gál, P.; Sim, R.B. Simultaneous activation of complement and coagulation by MBL-associated serine protease 2. *PLoS ONE* **2007**, *2*, e623. [CrossRef]
44. Hess, K.; Ajjan, R.; Phoenix, F.; Dobó, J.; Gál, P.; Schroeder, V. Effects of MASP-1 of the complement system on activation of coagulation factors and plasma clot formation. *PLoS ONE* **2012**, *7*, e35690. [CrossRef] [PubMed]
45. Lidington, E.A.; Haskard, D.O.; Mason, J.C. Induction of decay-accelerating factor by thrombin through a protease-activated receptor 1 and protein kinase C-dependent pathway protects vascular endothelial cells from complement-mediated injury. *Blood* **2000**, *96*, 2784–2792. [CrossRef] [PubMed]
46. Morita, T.; Kato, H.; Iwanaga, S.; Takada, K.; Kimura, T. New fluorogenic substrates for alpha-thrombin, factor Xa, kallikreins, and urokinase. *J. Biochem.* **1977**, *82*, 1495–1498. [CrossRef] [PubMed]
47. Presanis, J.S.; Hajela, K.; Ambrus, G.; Gál, P.; Sim, R.B. Differential substrate and inhibitor profiles for human MASP-1 and MASP-2. *Mol. Immunol.* **2004**, *40*, 921–929. [CrossRef]
48. Nangaku, M.; Couser, W.G. Mechanisms of immune-deposit formation and the mediation of immune renal injury. *Clin. Exp. Nephrol.* **2005**, *9*, 183–191. [CrossRef]
49. Markiewski, M.M.; Nilsson, B.; Ekdahl, K.N.; Mollnes, T.E.; Lambris, J.D. Complement and coagulation: Strangers or partners in crime? *Trends Immunol.* **2007**, *28*, 184–192. [CrossRef]
50. Amara, U.; Rittirsch, D.; Flierl, M.; Bruckner, U.; Klos, A.; Gebhard, F.; Lambris, J.D.; Huber-Lang, M. Interaction between the coagulation and complement system. *Adv. Exp. Med. Biol.* **2008**, *632*, 71–79. [CrossRef]
51. Wiggins, R.C.; Giclas, P.C.; Henson, P.M. Chemotactic activity generated from the fifth component of complement by plasma kallikrein of the rabbit. *J. Exp. Med.* **1981**, *153*, 1391–1404. [CrossRef]
52. Frade, R.; Rodrigues-Lima, F.; Huang, S.; Xie, K.; Guillaume, N.; Bar-Eli, M. Procathepsin-L, a proteinase that cleaves human C3 (the third component of complement), confers high tumorigenic and metastatic properties to human melanoma cells. *Cancer Res.* **1998**, *58*, 2733–2736. [CrossRef] [PubMed]
53. Huber-Lang, M.; Sarma, J.V.; Zetoune, F.S.; Rittirsch, D.; Neff, T.A.; McGuire, S.R.; Lambris, J.D.; Warner, R.L.; Flierl, M.A.; Hoesel, L.M.; et al. Generation of C5a in the absence of C3: A new complement activation pathway. *Nat. Med.* **2006**, *12*, 682–687. [CrossRef] [PubMed]
54. Nilsson-Ekdahl, K.; Nilsson, B. Phosphorylation of C3 by a casein kinase released from activated human platelets increases opsonization of immune complexes and binding to complement receptor type 1. *Eur. J. Immunol.* **2001**, *31*, 1047–1054. [CrossRef] [PubMed]
55. Ekdahl, K.N.; Nilsson, B. Phosphorylation of complement component C3 and C3 fragments by a human platelet protein kinase. Inhibition of factor I-mediated cleavage of C3b. *J. Immunol.* **1995**, *154*, 6502–6510. [CrossRef]
56. Tedesco, F.; Pausa, M.; Nardon, E.; Introna, M.; Mantovani, A.; Dobrina, A. The cytolytically inactive terminal complement complex activates endothelial cells to express adhesion molecules and tissue factor procoagulant activity. *J. Exp. Med.* **1997**, *185*, 1619–1627. [CrossRef]
57. Ikeda, K.; Nagasawa, K.; Horiuchi, T.; Tsuru, T.; Nishizaka, H.; Niho, Y. C5a induces tissue factor activity on endothelial cells. *Thromb. Haemost.* **1997**, *77*, 394–398. [CrossRef]
58. Weiler, J.M.; Edens, R.E.; Linhardt, R.J.; Kapelanski, D.P. Heparin and modified heparin inhibit complement activation in vivo. *J. Immunol.* **1992**, *148*, 3210–3215. [CrossRef]
59. Xi, W.; Liu, L.; Wang, J.; Xia, Y.; Bai, Q.; Long, Q.; Wang, Y.; Xu, J.; Guo, J. High Level of Anaphylatoxin C5a Predicts Poor Clinical Outcome in Patients with Clear Cell Renal Cell Carcinoma. *Sci. Rep.* **2016**, *6*, 29177. [CrossRef]
60. Liszewski, M.K.; Kolev, M.; Le Friec, G.; Leung, M.; Bertram, P.G.; Fara, A.F.; Subias, M.; Pickering, M.C.; Drouet, C.; Meri, S.; et al. Intracellular complement activation sustains T cell homeostasis and mediates effector differentiation. *Immunity* **2013**, *39*, 1143–1157. [CrossRef]
61. Arbore, G.; West, E.E.; Spolski, R.; Robertson, A.A.B.; Klos, A.; Rheinheimer, C.; Dutow, P.; Woodruff, T.M.; Yu, Z.X.; O'Neill, L.A.; et al. T helper 1 immunity requires complement-driven NLRP3 inflammasome activity in $CD4^+$ T cells. *Science* **2016**, *352*, aad1210. [CrossRef] [PubMed]

62. Kunz, N.; Kemper, C. Complement Has Brains-Do Intracellular Complement and Immunometabolism Cooperate in Tissue Homeostasis and Behavior? *Front. Immunol.* **2021**, *12*, 629986. [CrossRef] [PubMed]
63. Cardone, J.; Le Friec, G.; Vantourout, P.; Roberts, A.; Fuchs, A.; Jackson, I.; Suddason, T.; Lord, G.; Atkinson, J.P.; Cope, A.; et al. Complement regulator CD46 temporally regulates cytokine production by conventional and unconventional T cells. *Nat. Immunol.* **2010**, *11*, 862–871. [CrossRef] [PubMed]
64. Song, W.-C. Crosstalk between complement and toll-like receptors. *Toxicol. Pathol.* **2012**, *40*, 174–182. [CrossRef] [PubMed]
65. Sarma, J.V.; Ward, P.A. The complement system. *Cell Tissue Res.* **2011**, *343*, 227–235. [CrossRef]
66. Poppelaars, F.; Faria, B.; Gaya da Costa, M.; Franssen, C.F.M.; van Son, W.J.; Berger, S.P.; Daha, M.R.; Seelen, M.A. The Complement System in Dialysis: A Forgotten Story? *Front. Immunol.* **2018**, *9*, 71. [CrossRef]
67. Merle, N.S.; Noe, R.; Halbwachs-Mecarelli, L.; Fremeaux-Bacchi, V.; Roumenina, L.T. Complement System Part II: Role in Immunity. *Front. Immunol.* **2015**, *6*, 257. [CrossRef]
68. Smith, M.R. Rituximab (monoclonal anti-CD20 antibody): Mechanisms of action and resistance. *Oncogene* **2003**, *22*, 7359–7368. [CrossRef]
69. Gunn, L.; Ding, C.; Liu, M.; Ma, Y.; Qi, C.; Cai, Y.; Hu, X.; Aggarwal, D.; Zhang, H.-G.; Yan, J. Opposing roles for complement component C5a in tumor progression and the tumor microenvironment. *J. Immunol.* **2012**, *189*, 2985–2994. [CrossRef]
70. Gorter, A.; Meri, S. Immune evasion of tumor cells using membrane-bound complement regulatory proteins. *Immunol. Today* **1999**, *20*, 576–582. [CrossRef]
71. Schmidt, C.Q.; Lambris, J.D.; Ricklin, D. Protection of host cells by complement regulators. *Immunol. Rev.* **2016**, *274*, 152–171. [CrossRef]
72. Piao, C.; Cai, L.; Qiu, S.; Jia, L.; Song, W.; Du, J. Complement 5a Enhances Hepatic Metastases of Colon Cancer via Monocyte Chemoattractant Protein-1-mediated Inflammatory Cell Infiltration. *J. Biol. Chem.* **2015**, *290*, 10667–10676. [CrossRef]
73. Cho, M.S.; Vasquez, H.G.; Rupaimoole, R.; Pradeep, S.; Wu, S.; Zand, B.; Han, H.-D.; Rodriguez-Aguayo, C.; Bottsford-Miller, J.; Huang, J.; et al. Autocrine effects of tumor-derived complement. *Cell Rep.* **2014**, *6*, 1085–1095. [CrossRef]
74. Köhl, J. Self, non-self, and danger: A complementary view. *Adv. Exp. Med. Biol.* **2006**, *586*, 71–94. [CrossRef] [PubMed]
75. Markiewski, M.M.; DeAngelis, R.A.; Benencia, F.; Ricklin-Lichtsteiner, S.K.; Koutoulaki, A.; Gerard, C.; Coukos, G.; Lambris, J.D. Modulation of the antitumor immune response by complement. *Nat. Immunol.* **2008**, *9*, 1225–1235. [CrossRef]
76. Corrales, L.; Ajona, D.; Rafail, S.; Lasarte, J.J.; Riezu-Boj, J.I.; Lambris, J.D.; Rouzaut, A.; Pajares, M.J.; Montuenga, L.M.; Pio, R. Anaphylatoxin C5a creates a favorable microenvironment for lung cancer progression. *J. Immunol.* **2012**, *189*, 4674–4683. [CrossRef] [PubMed]
77. Ortiz-Espinosa, S.; Morales, X.; Senent, Y.; Alignani, D.; Tavira, B.; Macaya, I.; Ruiz, B.; Moreno, H.; Remírez, A.; Sainz, C.; et al. Complement C5a induces the formation of neutrophil extracellular traps by myeloid-derived suppressor cells to promote metastasis. *Cancer Lett.* **2022**, *529*, 70–84. [CrossRef] [PubMed]
78. Nabizadeh, J.A.; Manthey, H.D.; Steyn, F.J.; Chen, W.; Widiapradja, A.; Md Akhir, F.N.; Boyle, G.M.; Taylor, S.M.; Woodruff, T.M.; Rolfe, B.E. The Complement C3a Receptor Contributes to Melanoma Tumorigenesis by Inhibiting Neutrophil and CD4+ T Cell Responses. *J. Immunol.* **2016**, *196*, 4783–4792. [CrossRef]
79. Guglietta, S.; Chiavelli, A.; Zagato, E.; Krieg, C.; Gandini, S.; Ravenda, P.S.; Bazolli, B.; Lu, B.; Penna, G.; Rescigno, M. Coagulation induced by C3aR-dependent NETosis drives protumorigenic neutrophils during small intestinal tumorigenesis. *Nat. Commun.* **2016**, *7*, 11037. [CrossRef] [PubMed]
80. Khan, M.A.; Assiri, A.M.; Broering, D.C. Complement and macrophage crosstalk during process of angiogenesis in tumor progression. *J. Biomed. Sci.* **2015**, *22*, 58. [CrossRef]
81. Piao, C.; Zhang, W.-M.; Li, T.-T.; Zhang, C.-C.; Qiu, S.; Liu, Y.; Liu, S.; Jin, M.; Jia, L.-X.; Song, W.-C.; et al. Complement 5a stimulates macrophage polarization and contributes to tumor metastases of colon cancer. *Exp. Cell Res.* **2018**, *366*, 127–138. [CrossRef]
82. Fu, L.-Q.; Du, W.-L.; Cai, M.-H.; Yao, J.-Y.; Zhao, Y.-Y.; Mou, X.-Z. The roles of tumor-associated macrophages in tumor angiogenesis and metastasis. *Cell. Immunol.* **2020**, *353*, 104119. [CrossRef] [PubMed]
83. Ribatti, D.; Nico, B.; Crivellato, E.; Vacca, A. Macrophages and tumor angiogenesis. *Leukemia* **2007**, *21*, 2085–2089. [CrossRef] [PubMed]
84. Lamagna, C.; Aurrand-Lions, M.; Imhof, B.A. Dual role of macrophages in tumor growth and angiogenesis. *J. Leukoc. Biol.* **2006**, *80*, 705–713. [CrossRef]
85. Revel, M.; Sautès-Fridman, C.; Fridman, W.-H.; Roumenina, L.T. C1q+ macrophages: Passengers or drivers of cancer progression. *Trends Cancer* **2022**, *8*, 517–526. [CrossRef] [PubMed]
86. Towner, L.D.; Wheat, R.A.; Hughes, T.R.; Morgan, B.P. Complement Membrane Attack and Tumorigenesis: A Systems Biology Approach. *J. Biol. Chem.* **2016**, *291*, 14927–14938. [CrossRef]
87. Vlaicu, S.I.; Tegla, C.A.; Cudrici, C.D.; Danoff, J.; Madani, H.; Sugarman, A.; Niculescu, F.; Mircea, P.A.; Rus, V.; Rus, H. Role of C5b-9 complement complex and response gene to complement-32 (RGC-32) in cancer. *Immunol. Res.* **2013**, *56*, 109–121. [CrossRef]
88. Nitta, H.; Murakami, Y.; Wada, Y.; Eto, M.; Baba, H.; Imamura, T. Cancer cells release anaphylatoxin C5a from C5 by serine protease to enhance invasiveness. *Oncol. Rep.* **2014**, *32*, 1715–1719. [CrossRef]
89. Cho, M.S.; Rupaimoole, R.; Choi, H.-J.; Noh, K.; Chen, J.; Hu, Q.; Sood, A.K.; Afshar-Kharghan, V. Complement Component 3 Is Regulated by TWIST1 and Mediates Epithelial-Mesenchymal Transition. *J. Immunol.* **2016**, *196*, 1412–1418. [CrossRef]

90. Fan, Z.; Qin, J.; Wang, D.; Geng, S. Complement C3a promotes proliferation, migration and stemness in cutaneous squamous cell carcinoma. *J. Cell. Mol. Med.* **2019**, *23*, 3097–3107. [CrossRef]
91. Casula, M.; Montecucco, F.; Bonaventura, A.; Liberale, L.; Vecchié, A.; Dallegri, F.; Carbone, F. Update on the role of Pentraxin 3 in atherosclerosis and cardiovascular diseases. *Vascul. Pharmacol.* **2017**, *99*, 1–12. [CrossRef]
92. Presta, M.; Foglio, E.; Churruca Schuind, A.; Ronca, R. Long Pentraxin-3 Modulates the Angiogenic Activity of Fibroblast Growth Factor-2. *Front. Immunol.* **2018**, *9*, 2327. [CrossRef] [PubMed]
93. Daigo, K.; Inforzato, A.; Barajon, I.; Garlanda, C.; Bottazzi, B.; Meri, S.; Mantovani, A. Pentraxins in the activation and regulation of innate immunity. *Immunol. Rev.* **2016**, *274*, 202–217. [CrossRef]
94. Bottazzi, B.; Garlanda, C.; Cotena, A.; Moalli, F.; Jaillon, S.; Deban, L.; Mantovani, A. The long pentraxin PTX3 as a prototypic humoral pattern recognition receptor: Interplay with cellular innate immunity. *Immunol. Rev.* **2009**, *227*, 9–18. [CrossRef]
95. Souza, D.G.; Amaral, F.A.; Fagundes, C.T.; Coelho, F.M.; Arantes, R.M.E.; Sousa, L.P.; Matzuk, M.M.; Garlanda, C.; Mantovani, A.; Dias, A.A.; et al. The long pentraxin PTX3 is crucial for tissue inflammation after intestinal ischemia and reperfusion in mice. *Am. J. Pathol.* **2009**, *174*, 1309–1318. [CrossRef] [PubMed]
96. Inforzato, A.; Bottazzi, B.; Garlanda, C.; Valentino, S.; Mantovani, A. Pentraxins in humoral innate immunity. *Adv. Exp. Med. Biol.* **2012**, *946*, 1–20. [CrossRef] [PubMed]
97. Daigo, K.; Hamakubo, T. Host-protective effect of circulating pentraxin 3 (PTX3) and complex formation with neutrophil extracellular traps. *Front. Immunol.* **2012**, *3*, 378. [CrossRef]
98. Gout, E.; Moriscot, C.; Doni, A.; Dumestre-Pérard, C.; Lacroix, M.; Pérard, J.; Schoehn, G.; Mantovani, A.; Arlaud, G.J.; Thielens, N.M. M-ficolin interacts with the long pentraxin PTX3: A novel case of cross-talk between soluble pattern-recognition molecules. *J. Immunol.* **2011**, *186*, 5815–5822. [CrossRef]
99. Deban, L.; Jarva, H.; Lehtinen, M.J.; Bottazzi, B.; Bastone, A.; Doni, A.; Jokiranta, T.S.; Mantovani, A.; Meri, S. Binding of the long pentraxin PTX3 to factor H: Interacting domains and function in the regulation of complement activation. *J. Immunol.* **2008**, *181*, 8433–8440. [CrossRef]
100. Bonavita, E.; Gentile, S.; Rubino, M.; Maina, V.; Papait, R.; Kunderfranco, P.; Greco, C.; Feruglio, F.; Molgora, M.; Laface, I.; et al. PTX3 is an extrinsic oncosuppressor regulating complement-dependent inflammation in cancer. *Cell* **2015**, *160*, 700–714. [CrossRef]
101. Olivieri, F.; Albertini, M.C.; Orciani, M.; Ceka, A.; Cricca, M.; Procopio, A.D.; Bonafè, M. DNA damage response (DDR) and senescence: Shuttled inflamma-miRNAs on the stage of inflamm-aging. *Oncotarget* **2015**, *6*, 35509–35521. [CrossRef] [PubMed]
102. Roumenina, L.T.; Daugan, M.V.; Noé, R.; Petitprez, F.; Vano, Y.A.; Sanchez-Salas, R.; Becht, E.; Meilleroux, J.; Le Clec'h, B.; Giraldo, N.A.; et al. Tumor Cells Hijack Macrophage-Produced Complement C1q to Promote Tumor Growth. *Cancer Immunol. Res.* **2019**, *7*, 1091–1105. [CrossRef] [PubMed]
103. Daugan, M.V.; Revel, M.; Russick, J.; Dragon-Durey, M.-A.; Gaboriaud, C.; Robe-Rybkine, T.; Poillerat, V.; Grunenwald, A.; Lacroix, G.; Bougouin, A.; et al. Complement C1s and C4d as Prognostic Biomarkers in Renal Cancer: Emergence of Noncanonical Functions of C1s. *Cancer Immunol. Res.* **2021**, *9*, 891–908. [CrossRef] [PubMed]
104. Magrini, E.; Garlanda, C. Noncanonical Functions of C1s Complement Its Canonical Functions in Renal Cancer. *Cancer Immunol. Res.* **2021**, *9*, 855. [CrossRef]
105. Yang, C.; Yang, F.; Chen, X.; Li, Y.; Hu, X.; Guo, J.; Yao, J. Overexpression of complement C5a indicates poor survival and therapeutic response in metastatic renal cell carcinoma. *Int. J. Biol. Markers* **2023**, *38*, 124–132. [CrossRef] [PubMed]
106. Dong, Y.; Ma, W.-M.; Yang, W.; Hao, L.; Zhang, S.-Q.; Fang, K.; Hu, C.-H.; Zhang, Q.-J.; Shi, Z.-D.; Zhang, W.; et al. Identification of C3 and FN1 as potential biomarkers associated with progression and prognosis for clear cell renal cell carcinoma. *BMC Cancer* **2021**, *21*, 1135. [CrossRef] [PubMed]
107. Daugan, M.V.; Revel, M.; Thouenon, R.; Dragon-Durey, M.-A.; Robe-Rybkine, T.; Torset, C.; Merle, N.S.; Noé, R.; Verkarre, V.; Oudard, S.M.; et al. Intracellular Factor H Drives Tumor Progression Independently of the Complement Cascade. *Cancer Immunol. Res.* **2021**, *9*, 909–925. [CrossRef]
108. Netti, G.S.; Lucarelli, G.; Spadaccino, F.; Castellano, G.; Gigante, M.; Divella, C.; Rocchetti, M.T.; Rascio, F.; Mancini, V.; Stallone, G.; et al. PTX3 modulates the immunoflogosis in tumor microenvironment and is a prognostic factor for patients with clear cell renal cell carcinoma. *Aging* **2020**, *12*, 7585–7602. [CrossRef]
109. Ajona, D.; Ortiz-Espinosa, S.; Pio, R. Complement anaphylatoxins C3a and C5a: Emerging roles in cancer progression and treatment. *Semin. Cell Dev. Biol.* **2019**, *85*, 153–163. [CrossRef]
110. Lin, E.; Liu, X.; Liu, Y.; Zhang, Z.; Xie, L.; Tian, K.; Liu, J.; Yu, Y. Roles of the Dynamic Tumor Immune Microenvironment in the Individualized Treatment of Advanced Clear Cell Renal Cell Carcinoma. *Front. Immunol.* **2021**, *12*, 653358. [CrossRef]
111. Lasorsa, F.; di Meo, N.A.; Rutigliano, M.; Milella, M.; Ferro, M.; Pandolfo, S.D.; Crocetto, F.; Tataru, O.S.; Autorino, R.; Battaglia, M.; et al. Immune Checkpoint Inhibitors in Renal Cell Carcinoma: Molecular Basis and Rationale for Their Use in Clinical Practice. *Biomedicines* **2023**, *11*, 1071. [CrossRef] [PubMed]
112. O'Donnell, J.S.; Long, G.V.; Scolyer, R.A.; Teng, M.W.L.; Smyth, M.J. Resistance to PD1/PDL1 checkpoint inhibition. *Cancer Treat. Rev.* **2017**, *52*, 71–81. [CrossRef] [PubMed]
113. Lasorsa, F.; Rutigliano, M.; Milella, M.; Ferro, M.; Pandolfo, S.D.; Crocetto, F.; Tataru, O.S.; Autorino, R.; Battaglia, M.; Ditonno, P.; et al. Cellular and Molecular Players in the Tumor Microenvironment of Renal Cell Carcinoma. *J. Clin. Med.* **2023**, *12*, 3888. [CrossRef] [PubMed]

114. Motzer, R.J.; Robbins, P.B.; Powles, T.; Albiges, L.; Haanen, J.B.; Larkin, J.; Mu, X.J.; Ching, K.A.; Uemura, M.; Pal, S.K.; et al. Avelumab plus axitinib versus sunitinib in advanced renal cell carcinoma: Biomarker analysis of the phase 3 JAVELIN Renal 101 trial. *Nat. Med.* **2020**, *26*, 1733–1741. [CrossRef]
115. Iacovelli, R.; Atzori, F.; Basso, U.; Bersanelli, M.; Bonomi, L.; Bracarda, S.; Buttigliero, C.; Fantinel, E.; Calabro, F.; Chiuri, V.E.; et al. Phase II study of avelumab plus intermittent axitinib in previously untreated patients with metastatic renal cell carcinoma (Tide-A study). *J. Clin. Oncol.* **2020**, *38* (Suppl. S6), TPS762. [CrossRef]

cancers

MDPI

Review

Advances in Imaging-Based Biomarkers in Renal Cell Carcinoma: A Critical Analysis of the Current Literature

Lina Posada Calderon [1,†], Lennert Eismann [1,†], Stephen W. Reese [1], Ed Reznik [2] and Abraham Ari Hakimi [1,*]

1 Urology Service, Department of Surgery, Memorial Sloan Kettering Cancer Center, New York, NY 10065, USA
2 Computational Oncology, Department of Epidemiology & Biostatistics, Memorial Sloan Kettering Cancer Center, New York, NY 10065, USA
* Correspondence: hakimia@mskcc.org
† These authors contributed equally to this work.

Simple Summary: Current imaging techniques do not reliably distinguish renal cell carcinoma from other renal diseases. This review summarizes recent advances in other imaging methods for the diagnosis and monitoring of potential kidney tumors. Magnetic resonance imaging (MRI), positron emission tomography (PET)/CT using various radiolabeled molecules to detect specific cancer-associated features, and computational extraction of data from CT images have all proven useful for various purposes, but more research is needed to verify their reliability.

Abstract: Cross-sectional imaging is the standard diagnostic tool to determine underlying biology in renal masses, which is crucial for subsequent treatment. Currently, standard CT imaging is limited in its ability to differentiate benign from malignant disease. Therefore, various modalities have been investigated to identify imaging-based parameters to improve the noninvasive diagnosis of renal masses and renal cell carcinoma (RCC) subtypes. MRI was reported to predict grading of RCC and to identify RCC subtypes, and has been shown in a small cohort to predict the response to targeted therapy. Dynamic imaging is promising for the staging and diagnosis of RCC. PET/CT radiotracers, such as ^{18}F-fluorodeoxyglucose (FDG), ^{124}I-cG250, radiolabeled prostate-specific membrane antigen (PSMA), and ^{11}C-acetate, have been reported to improve the identification of histology, grading, detection of metastasis, and assessment of response to systemic therapy, and to predict oncological outcomes. Moreover, ^{99}Tc-sestamibi and SPECT scans have shown promising results in distinguishing low-grade RCC from benign lesions. Radiomics has been used to further characterize renal masses based on semantic and textural analyses. In preliminary studies, integrated machine learning algorithms using radiomics proved to be more accurate in distinguishing benign from malignant renal masses compared to radiologists' interpretations. Radiomics and radiogenomics are used to complement risk classification models to predict oncological outcomes. Imaging-based biomarkers hold strong potential in RCC, but require standardization and external validation before integration into clinical routines.

Keywords: imaging; renal cell carcinoma; biomarker; renal mass; radiomics; radiogenomics

Citation: Posada Calderon, L.; Eismann, L.; Reese, S.W.; Reznik, E.; Hakimi, A.A. Advances in Imaging-Based Biomarkers in Renal Cell Carcinoma: A Critical Analysis of the Current Literature. *Cancers* **2023**, *15*, 354. https://doi.org/10.3390/cancers15020354

Academic Editors: José I. López and Claudia Manini

Received: 5 December 2022
Revised: 31 December 2022
Accepted: 3 January 2023
Published: 5 January 2023

1. Introduction

Renal cell carcinoma (RCC) has an incidence of 12 per 100,000 in North America, and a peak incidence at the age of 60–70 years [1]. RCC incidence continues to rise, with an estimated 79,000 new cases and 13,920 deaths from RCC in 2022 in the United States alone [2]. The most common histologic subtype of renal cell carcinoma is clear-cell renal cell carcinoma (ccRCC), with five-year survival rates declining by stage of the disease. In the industrial world, the incidence of localized RCC continues to rise, with almost 70% of tumors being detected incidentally [3,4], secondary to increased utilization of abdominal imaging.

Current challenges in the treatment of renal cell carcinoma include diagnostic uncertainty, which leads to both under- and overtreatment of the disease. Conventional cross-sectional imaging techniques do not allow for the discrimination of malignant tumor sub-types, nor can they differentiate between benign lesions. Additionally, current imaging gives us opaque insight into patients with metastatic disease and its response to and progression with therapeutics. In recent years, research has focused on improved imaging techniques to enhance diagnostic precision and prognosis in patients with renal tumors. Advancements in the field have been multi-factorial, from enhancements in current ultrasound and cross-sectional imaging technologies, nuclear medicine studies, and the field of radiomics, which infers renal mass insights from radiologic data. In this review, we summarize and analyze the future of imaging modalities, and the advancements in radiomics and radiogenomics as they develop new ways of diagnosing and distinguishing renal cell masses (Table 1).

Table 1. Summary of imaging-based parameters.

Imaging Technique/Model	Description	Advantages	Disadvantages
MRI			
Multiparametric MRI	DWI uses water particle movement to identify tumor-like tissue, which has slower movement of water particles, and calculated an apparent diffusion coefficient (ADC)	Can be used to calculate likelihood of cancer vs. non cancer Can predict Fuhrman grade with a of 78% and 86% sensitivity and specificity, respectively.	Studies use a variety of non-standardized parameters for MRI that have not been validated in a larger population setting Poor-lipid AMLs remain a challenge to distinguish from chromophobe RCC and oncocytoma
Perfusion MRI (DCE, DSC, ASL)	Works by assessing perfusion at the micropapillary level, calculating changes in signal before and after contrast (DCE and DSC) or detecting water protons in blood (ASL)	Different histologic subtypes of RCC have different perfusion coefficients. ASL MRI can be used to predict response to treatment with sunitinib and pazopanib, with responders having higher baseline tumor perfusion	
PET-CT			
18F-FDG	FDG binds to metabolically active tissue, signaling cancer activity. From a meta-analysis, pooled sensitivity to detect renal lesions is 62% and specificity 88%.	Proposed surrogate for tumor aggressiveness, with maximum SUV of lesions in patients with advanced RCC is independently associated with overall survival, also related to higher Fuhrman grade, higher stage and sarcomatoid features.	Limited applicability in RCC due to physiologic uptake in renal parenchyma. Limited also by practicality, cost, and variable results across multiple studies.
Girentuximab, Xr-Girentuximab to CA-IX	CA-IX is a protein that is overexpressed in VHL-mutated pathways and expressed in 95–100% of ccRCC. Average sensitivity and specificity of 86.2% and 85.9%, respectively for identifying ccRCC.	Studies are validated with surgical pathology. Multiple ongoing studies for different molecules that target CA-IX. Recently zirconium girentuximab showed promising sensitivity and specificity of 86% and 86% in identifying ccRCC.	Long half life time of girentuximab, where injection needs to be administered 2–6 days prior to imaging. Logistics and timing of molecule remain main barriers

Table 1. *Cont.*

Imaging Technique/Model	Description	Advantages	Disadvantages
Tc-MIBI SPECT/CT	^{99}Tc-sestamibi accumulates in cells with high mitochondrial content and low multidrug resistance (MDR) pump expression, which are characteristic of renal oncocytoma. Sensitivity of 87.5% and a specificity of 95.2% in differentiating oncocytomas and HOCTs	Widespread and usability of ^{99}Tc-sestamibi SPECT/CT and high concordance of imaging findings with pathology, results are promising in the identification of oncocytomas	Other benign pathologies such as chronic sclerosis, fibroma, hydatid cyst and angiomyolipoma don't have any uptake.
PSMA/PET	PSMA is a cell surface protein that is expressed in prostatic tissue and also in neovasculature of some cancers, including RCC, specifically clear cell histology	Increased sensitivity of PSMA PET/CT in detecting distant metastasis, with sensitivities of 89–95%, compared to 67–78% with conventional CT scan Can predict presence of adverse histopathological characteristics (necrosis, sarcomatoid an rhabdoid features)	Evaluation of primary lesions is limited, and studies have small sample sizes. Non ccRCC masses have a low PSMA uptake. Use may be limited to metastatic clear cell histology.
C-acetate PET	^{11}C-acetate is actively incorporated into tumor cells and is integrated in cellular lipid structures. Has high uptake rates in papillary and ccRCC.	Better sensitivity rates than FDG PET. Using dual tracer c-acetate and FDG PET, AML was differentiated from RCC with sensitivity and specificity of 94% and 98%, respectively.	Evidence based on small sample size studies Differentiating AML and RCC would require dual complex imaging techniques and there is no added information on histology.
Radiomics	Objective and detailed analysis of imaging characteristics analyzed via quantitative methods and statistical models. Specific morphological characteristics, texture analysis and intensity of different parameters within the tumor can be standardized and integrated into algorithms and AI models to predict tumor malignancy, histology, grade and molecular characteristics. Convolutional Neural network (CNN) is a deep learning algorithm that processes pixel and clinical data to create models to predict malignancy of renal masses.	Reported AUC of 0.87 for differentiating benign versus malignant renal masses Radiomic models have reported to be superior to conventional radiological interpretation of images in distinguishing histologic subtypes and presence of sarcomatoid features Investigated as a biomarker for response to therapy, and texture analysis was found to be an independent factor associated with time to progression in patients with metastatic RCC being treated with TKI	Lack of generalizability and clinical application. There is few external validity and reproducibility of studies because of insufficient access to cades and images that serve for the creation of the models. Most studies are compared to surgical specimens, which implies a selection bias. Intra-tumoral heterogeneity may not be accounted for as there are only few areas of the tumor that are used for imaging analysis and creation of models

Table 1. *Cont.*

Imaging Technique/Model	Description	Advantages	Disadvantages
Radiogenomics	Genetic pathways express with different phenotypic imaging characteristics, so radiogenomic is the integration of radiomics with genetic tumoral data and molecular signatures.	Reported genetic associations with imaging characteristics for *VHL*, *KDMC5*, *BAP1*, and *MUC4*. Radiogenomic Risk Score (RSS) was developed to identify CT imaging features that are correlated to genetic signatures that have shown to predict oncological outcomes. This risk score was shown to correlate with progression free survival and response to treatment	Known genetic alterations in RCC have a very low prevalence, and are mostly applicable to ccRCC Given the complexity of molecular pathways, heterogeneity within the tumor and change in time, it is challenging to make direct correlations of gene and molecular pathways to specific imaging findings

MRI: magnetic resonance imaging; CT: computerized tomography; DWI: Diffusion-weighted imaging; DCE: dynamic contrast-enhanced; DSC: dynamic susceptibility contrast; ASL: arterial spin labeling; RCC: renal cell cancer; FDG: fluorodeoxy-glucose; SUV: standardized uptake value; CA-IX: carbonic-anhydrase XI; ccRCC: clear cell RCC; HOCTs: hybrid oncocytic/chromophobe tumors; AI: artificial intelligence; AUC: area under the curve; TKI: Tyrosin kinase inhibitors.

2. Magnetic Resonance Imaging (MRI)

Multiparametric MRI (mpMRI) allows for the evaluation of anatomic as well as functional characteristics of renal masses [5]. Specifically, diffusion MRI and perfusion MRI have been studied as imaging tools to aid in differentiating tumor histology or subtype, and assessing the response to treatment [6]. MRI has been proposed as an alternative to computed tomography (CT), which is limited in its ability to identify benign lesions such as fat-poor angiomyolipomas (AMLs) and oncocytomas [7,8]. Diffusion-weighted imaging (DWI) quantifies the mobility of protons that are associated with water (Brownian motion). Tissue that is highly cellular, such as tumors, restricts water molecules' movement, and thus appears as a high-intensity signal on DWI, and has a low apparent diffusion coefficient (ADC) [9]. A systematic review and meta-analysis including four studies that used DWI to differentiate between malignant and non-malignant lesions showed DWI to have 86% sensitivity and 78% specificity. In this meta-analysis, DWI used to differentiate high-grade and low-grade RCCs had an area under the curve (AUC) of 83%, reflecting moderately accurate test performance. However, there were no standardized criteria to compare radiological findings to different imaging modalities or pathological specimens [10]. These values are comparable to those associated with CT scans, where the sensitivity has been reported to be 88% and the specificity 75% [11]. Using diffusion MRI, parenchymal wash index, and ADC ratio were correlated with clear-cell RCC Fuhrman grade, with a pooled sensitivity and specificity of DWI to differentiate between high and low grades of 78% and 86%, respectively [12,13].

Perfusion MRI, which assesses tissue perfusion at the micropapillary level, offers the possibility of improving performance characteristics. There are three main types of perfusion MRI: dynamic contrast-enhanced (DCE), dynamic susceptibility contrast (DSC), and arterial spin labeling (ASL) [6,14]. DCE and DSC calculate changes in signal intensity before and after intravenous gadolinium contrast injection, which measures perfusion parameters. ASL does not require intravenous contrast, and measures perfusion by detecting water protons in the blood [14]. Using ASL perfusion MRI, Lanzman et al. compared pre-operative MRI perfusions of 42 patients with various types of renal masses. The RCC histology was associated with different mean perfusion levels, with papillary RCC having lower perfusion levels than all other RCC types, and oncocytomas having significantly higher perfusion levels than RCCs [15].

Differentiating fat-poor AML and RCC based solely on imaging is a known challenge. In a systematic review and meta-analysis by Wilson et al., MRI was found to be 83%

sensitive and 90% specific for the detection of fat-poor AMLs, with an AUC of 0.93 [16]. In a retrospective study of 109 renal masses, Kay et al. developed an MRI diagnostic algorithm comprising 11 MR imaging features to determine the most likely histology of a renal mass. They found a sensitivity and specificity of 85% and 76%, respectively, for predicting clear cell histology, and 80% and 94%, respectively, for predicting papillary histology. Their algorithm, however, was weak in predicting chromophobe, oncocytoma, and fat-poor AML histologies [17]. Using this algorithm, Canvasser and colleagues developed a clear cell likelihood scale of 1 (less likely) to 5 (most likely), and found a sensitivity of 78% and a specificity of 90% for scores of 4 and 5 [18]. The clear cell likelihood scale was evaluated in a larger retrospective cohort of 454 renal masses, and the authors found a 93% positive predictive value for a score of 5, and a sensitivity and specificity of predicting clear-cell RCC of 91% and 56%, respectively, for scores of 4 and 5 [19]. Although these scales do not provide insight into tumor aggressiveness, they may be used to help select treatment for small renal masses, and to determine candidates for surveillance [5].

There is a growing interest in using mpMRI, not only to predict renal mass histology and behavior, but also to assess response to therapy. In a prospective mixed cohort of treatment-naïve and exposed patients, Tsai et al. evaluated changes in tumor ASL MRI perfusion as a measure of response to sunitinib and pazopanib treatment for metastatic RCC. Perfusion on MRI imaging, as evaluated by objective response rate, was compared among 6 responders and 11 non-responders at multiple time points during treatment and up to disease progression. Responders had a higher baseline tumor perfusion than non-responders (404 mL/100 g/min vs. 199 mL/100 g/min; $p = 0.02$), suggesting this could aid in identifying responders to therapy with tyrosine kinase inhibitors [20]. In a prospective, randomized, double-blinded trial that compared sorafenib and placebo, DCE MRI was also evaluated as a pharmacodynamic biomarker of response to sorafenib in metastatic RCC. Of the 44 patients with two available MRIs for comparison, two DCE parameters (area under the contrast concentration versus time curve 90 s after contrast injection [IAUC90], and volume transfer constant of contrast agent [Ktrans]) were evaluated. Although patients with high baseline Ktrans had better progression-free survival (PFS) compared to patients with low baseline Ktrans (log-rank $p = 0.027$), there was no significant association between change in IAUC90 and Ktrans with PFS [21].

In summary, while multiple studies have evaluated the use of MRI to predict the histology and grade of renal masses, and to assess response to treatment in metastatic RCC, they were generally small studies that used a variety of non-standardized mpMRI metrics [14]. Future studies are needed to validate the use of these metrics and demonstrate their usefulness in clinical scenarios.

3. Contrast-Enhanced Ultrasound

Ultrasound is a widely used diagnostic tool, and in many settings is the first modality used to evaluate renal pathologies. Focal lesions, hydronephrosis, and vascular pathologies can be identified, whereas benign lesions cannot be reliably distinguished from malignancies by conventional ultrasound [22]. Therefore, contrast-enhanced ultrasound (CEUS) has been proposed to visualize RCC characteristics. The contrast agent used for ultrasound is based on microbubbles, and amplifies the signal of microvascular structures [23]. CEUS has been shown to highly differentiate RCC from oncocytoma and angiomyolipoma [24,25]. In one study, combining CEUS parameters showed a 93% sensitivity and 100% specificity for renal malignancies [26]. Furthermore, a study of 85 patients with 93 renal masses showed that peak intensity and time to peak intensity in CEUS differed between clear-cell RCC, chromophobe RC and papillary RCC [27]. Additional studies have shown specific enhancement characteristics compared to clear-cell RCC [28]. CEUS has also been used in the diagnostic setting. For example, Lamuraglia et al., in 2006, showed that CEUS holds predictive value in metastatic RCC patients treated with the multi-kinase inhibitor sorafenib [29]. Similarly, Williams et al. reported significant changes in CEUS along with anti-angiogenic therapy of metastatic RCC, although CEUS parameters did not correlate with progression-free survival or best response rate to therapy [30]. Current research focuses on the assessment of CEUS to predict the response to immunotherapy in metastatic RCC (NCT05206942).

In summary, CEUS is a diagnostic modality that offers potential advantages in the characterization of renal masses, including enhanced diagnostic performance, characterization of renal mass histologic subtypes, its low cost, its low barrier to access, and the absence of ionizing radiation. Additional studies are needed in larger settings to validate these findings, including understanding performance characteristics in patients with different habitus.

4. Positron Emission Tomography–Computed Tomography (PET/CT)

Molecular or nuclear imaging studies rely on in vivo visualizations of biological processes at a cellular and molecular level, using radiopharmaceutical compounds that bind to a molecule of interest [5]. In RCC specifically, nuclear imaging allows for the identification of not only anatomic locations, but also for molecular pathways and processes that are associated with specific histologic features and tumor behavior. Multiple positron emission tomography (PET) radiotracers have been developed and studied as both prognostic and predictive biomarkers in RCC [6,31].

4.1. ^{18}F-Fluorodeoxy-Glucose (FDG) PET/CT

While ^{18}F-fluorodeoxy-glucose (FDG)-PET is the most common and well-known radiotracer used in other cancers, it has limited applicability in RCC due to its variable activity in primary and metastatic tumors, as well as physiologic uptake in normal renal parenchyma [5]. In a meta-analysis of 14 studies that assessed this modality in advanced RCC, the pooled sensitivity and specificity of FDG-PET/CT were 62% and 88%, respectively, for renal lesions, and 79% and 90%, respectively, for extrarenal lesions [32]. Despite variable uptake at the individual lesion level, the maximum standardized uptake value (SUVmax) of lesions in patients with advanced RCC has been independently associated with overall survival (OS) and PFS [33,34]. FDG-PET/CT activity has been proposed as a surrogate for tumor aggressiveness, as it has also been correlated with higher Fuhrman grade, TNM stage, and sarcomatoid features, and can aid in the prediction of progression and in clinical decision making [35–38]. Additionally, the detection of metastatic or recurrent sites was evaluated in a recent meta-analysis that included 14 studies [39]. The pooled sensitivity was described with 0.86, and specificity with 0.88 [39]. Accordingly, FDG-PET may be a useful re-staging tool for RCC, but current evidence is mostly based on retrospective studies, and lacks prospective investigations [39]. Hou et al. focused on the clinical value of FDG-PET in papillary RCC, and reported a similar sensitivity of 81% in the primary lesion and 100% in recurrent lesions [40]. These preliminary retrospective studies are promising, and need to be confirmed in larger prospective studies. Limitations to the use of FDG-PET/CT include practicality, cost, and variable results across multiple studies [41].

4.2. ^{124}I-cG250 (^{124}I-Girentuximab)

Girentuximab, formerly known as antibody cG250, is one of the most promising nuclear imaging methodologies in the characterization of solid renal masses [5]. It selectively binds to carbonic anhydrase IX (CA-IX), a protein that is overexpressed in VHL-mutated pathways in response to hypoxic conditions, and is expressed in 95 to 100% of clear-cell RCCs [6,42,43]. A multi-center phase III trial, the REDECT trial, evaluated the diagnostic efficacy of ^{124}I-girentuximab PET/CT and contrast-enhanced CT (CECT) in identifying clear-cell RCC in patients with indeterminate renal masses that were scheduled for surgical resection. Imaging was performed 2–6 days after intravenous administration of girentuximab, and prior to surgical resection. Imaging readings were classified as clear-cell RCC and non-clear-cell RCC, which were then compared to final surgical pathologies. In 195 patients that had imaging and pathology available for analysis, the average sensitivity and specificity were 86.2% and 85.9%, respectively, for girentuximab-PET/CT, and 75.5%

and 46.8%, respectively, for CECT. Furthermore, the inter-reader agreement was higher for girentuximab PET/CT [44]. Although the limitations of this study included a bias in patient selection, using only pre-surgical candidates, nevertheless, it provides the most accurate validation of pathology with imaging using PET/CT. Different radiotracers targeting CA-IX are currently being studied to improve clinical practice to reduce the long half-life of girentuximab. The molecule F-VM4-037, which is reported to have an 18-minute plasma half-life, has been studied in a phase II trial to allow same-day imaging [45]. Although the performance characteristics of this approach appear to be promising, logistics and timing remain ongoing barriers and, additionally, advancements in the technology will need to be validated in order for it to be used in clinical practice.

There is currently a prospective, open-label, multi-center phase III trial evaluating the performance characteristics of girentuximab (an anti-CAIX monoclonal antibody) labelled with 89Zr, to evaluate indeterminate renal masses to differentiate clear-cell RCC from other renal masses (ZIRCON Trial; NCT03849118). Its preliminary results were reported recently, and exceed the predetermined sensitivity and specificity study targets, with the imaging agent delivering 86% sensitivity and 87% specificity [46]. The phase I study showed in all ten cases a good toxicity profile, and was able to differentiate between clear-cell RCC and non-clear-cell RCC renal mass [47]. This technology is also being examined for diagnostic and therapeutic purposes in the STARLITE 2 Phase II study which evaluates the efficacy of Lu177 conjugated to girentuximab + nivolumab (anti-PD-1) systemic therapy. In a theranostic approach, Girentuximab could be labelled with 177Lu, a beta- emitter, that could induce single-strand DNA breaks into RCC cells. These agents are promising, and future research will focus on their incorporation into clinical practice.

4.3. Prostate-Specific Membrane Antigen (PSMA)–Targeted PET/CT

Prostate-specific membrane antigen (PSMA) is a cell surface protein that is overexpressed in prostate cancer, as well as in the neovasculature of some solid tumors, including RCC. PSMA-targeted imaging was first described in metastatic RCC by Demirci et al. in 2014 [48]. Small studies have reported the sensitivity of F-DCFPyL PSMA PET/CT in detecting distant metastases to range from 88.9% to 94.7%, compared with 66.7% to 78.0% for conventional CT scans [49–51]. For localized renal masses, Golan et al. found that the mean SUV_{max} of ^{68}Ga-PSMA-11 PET/CT was significantly higher in malignant as compared to benign lesions, and its washout coefficient K_2 was significantly lower in cancerous tissue [52]. Gao et al. reported that SUVmax of the same tracer could effectively differentiate high vs. low (WHO/SIUP grade I-II vs. III-IV) grade in 36 cases of clear-cell RCC. Furthermore, ^{68}Ga-PSMA-11 PET/CT could predict the presence of adverse histopathological characteristics, such as necrosis and sarcomatoid and rhabdoid features, with an AUC of 0.89 [53]. Both of these studies, however, have small sample sizes, and are not consistent with prior studies that show a high-background signal, limiting the evaluation of primary masses [54,55]. In general, most studies of PSMA-targeted PET/CT have included mostly clear-cell RCC, but the few non-clear-cell RCC lesions evaluated by this approach have shown lower uptake than surrounding renal parenchyma [55]. In particular, a meta-analysis described that PSMA PET/CT may also be suitable for chromophobe RCC, due to its relevant PSMA expression [56]. In contrast, only 13.6% of papillary RCC demonstrate a PSMA expression and therefore, FDG PET is the preferred dynamic imaging modality [56,57]. Based on the inconsistency of PSMA uptake in non-clear-cell RCC, PSMA PET is not appropriate for staging RCC subtypes other than clear-cell and chromophobe RCCs [56,58].

The specificity of PSMA PET/CT to patients with clear-cell RCC may limit its routine clinical use in the localized setting. However, this technique could potentially become useful in patients with metastatic disease as a way to measure treatment response or disease progression. Additional studies are warranted to validate these findings.

4.4. ^{11}C-Acetate PET-CT

The radiotracer 11C-acetate is actively incorporated into tumor cells and integrated into cellular lipid structures, and may be helpful in distinguishing between malignant and benign lesions [59]. In the imaging of RCC, ^{11}C-acetate has shown high uptake rates in clear-cell RCC, and even higher uptake rates in papillary RCC [60]. Additionally, in comparison to FDG-PET/CT, ^{11}C-acetate-PET/CT is reported to have better sensitivity for detecting RCC [61]. ^{11}C-acetate was evaluated as part of a dual-tracer technique with FDG PET/CT for the differentiation of AML from RCC; it was reported to have a sensitivity of 94% and a specificity of 98% [62]. In a case report, ^{11}C-acetate was reported to predict early response to sunitinib in metastatic RCC. In summary, ^{11}C-acetate is a promising radiotracer that may have the potential to be used to stage RCC, but evidence is based on small sample sizes. This tracer may also be relevant to the differentiation of AML from RCC as part of a complex dual imaging technique

5. Single Photon Emission-Computed Tomography (SPECT Scan)

^{99}Tc-Sestamibi SPECT/CT

As previously mentioned, a key limitation of several imaging modalities is their limited ability to distinguish benign from aggressive RCC tumors, such as clear-cell RCC and oncocytoma [63,64]. ^{99}Tc-sestamibi, a widely used nuclear imaging agent, offers the ability to differentiate these tumors based on their mitochondrial content. ^{99}Tc-sestamibi is a lipophilic cationic mitochondrial imaging agent that accumulates in cells with high mitochondrial content and low multidrug resistance (MDR) pump expression, which are characteristic of renal oncocytomas [5,65]. In contrast, clear-cell and chromophobe RCC masses have a higher MDR pump expression and low mitochondrial activity, although chRCC has generally higher mitochondrial activity than clear-cell RCC [65,66]. In a prospective study by Gorin et al., the use of preoperative ^{99}Tc-sestamibi single photon emission computed tomography (SPECT)/CT in detecting oncocytomas was assessed in 50 presurgical patients with T1 renal masses, and results were compared to final surgical pathologies. The authors found a sensitivity of 87.5% and a specificity of 95.2% in differentiating oncocytomas from hybrid oncocytic/chromophobe tumors [67]. Sistani et al. validated these findings and found that in 29 patients with 31 renal masses, all oncocytic lesions were positive on ^{99}Tc-sestamibi SPECT/CT, whereas uptake was low in chromophobe RCC and absent in other RCC subtypes [68]. In a 90-patient study by Asi et al., strong ^{99}Tc-sestamibi uptake was observed in 10 of 10 oncocytomas, while none was seen in most malignant lesions, except in 5 chromophobe RCC and 3 oncocytic papillary RCC masses. Other benign pathologies, such as chronic sclerosis, fibroma, hydatid cyst, and angiomyolipoma, also showed no uptake. The authors reported a positive predictive value of 60% and a negative predictive value of 91.3% in predicting benign pathologies. Additionally, they reported a relative uptake of 0.49 as an optimal cutoff to discriminate oncocytomas from other pathologies [69]. Given the already widespread use of ^{99}Tc-sestamibi SPECT/CT and the high concordance of imaging findings with pathologies, results support further evaluation of its use in the identification of oncocytomas and other benign renal lesions.

6. Radiomic and Radiogenomic Biomarkers

6.1. Radiomics

Radiomics and radiogenomics are two closely related fields with promising developments in characterizing cancers, predicting their behavior, and assessing treatment response. Radiomics consists of high-throughput extraction of quantitative data and the application of high-order statistical models to medical imaging to yield more objective and detailed analyses [5,70]. To make results more reproducible and interpretations more reliable, radiomic features quantifying morphological, intensity related, textural, and co-occurrence characteristics of CT, FDG-PET, and T1-weighted MRI features are being standardized [70]. These predefined quantitative radiomic features can be integrated into algorithms and artificial intelligence models to predict malignancy, tumor histology, tumor grade, and molecular characteristics [5].

Several radiomics models, most using texture analysis, have been designed and reported to be highly accurate in differentiating benign from malignant renal masses [71–73]. Varghese et al. evaluated 31 texture metrics of contrast-enhanced CT on 174 renal masses, and compared these findings with surgical pathologies. They found that six specific texture analysis features—entropy, entropy of fast-Fourier transform magnitude, mean, uniformity, information measure of correlation 2, and sum of averages—had high AUC values, with a mean AUC of 0.87 for differentiating benign versus malignant renal masses [72]. Furthermore, Uhlig et al. used machine learning algorithms to predict the malignancy of renal masses using 120 standardized radiomic features, and their diagnostic accuracies (base on surgical pathology) were compared with the those of blinded radiologists' assessments. The sensitivity and specificity of their models were higher than radiologists' diagnoses (0.88 vs. 0.80; $p = 0.045$, and 0.67 vs. 0.50; $p = 0.083$, respectively), with an AUC of 0.83 compared to 0.68 ($p = 0.47$) [73].

Other studies have also used convolutional neural network (CNN), a type of deep learning algorithm that processes images using pixel data recognition, in order to create models to differentiate malignant masses based on imaging [74–77]. In one of the largest series, Xi et al. developed a CNN model that included clinical and radiologic MRI data of 1162 renal lesions. This model was superior to radiological experts' interpretations, with an accuracy of 0.70 vs. 0.60 ($p = 0.053$), sensitivity of 0.92 vs. 0.80 ($p = 0.017$), and specificity of 0.41 vs. 0.35 ($p = 0.450$), respectively.

Using radiomics, various models have been reported to distinguish specific histologic features in renal masses. The accuracy of some models has exceeded conventional interpretation of radiologists, including those for CT to differentiate fat-poor AML from clear-cell RCC [78,79], oncocytomas from chromophobe RCC [80], and papillary type I from papillary type II RCC [81,82]. Similarly, such models have also been used to determine the presence of sarcomatoid features in clear-cell RCC, and to predict their nuclear grade [83–85].

Radiomics has also been investigated as a means of predicting responses to targeted therapy. In a retrospective analysis of 39 patients with 87 metastatic sites, Goh et al. compared the correlation of contrast-enhanced CT texture analysis parameters, at baseline and after two doses of tyrosine kinase inhibitors, with their progression with those of standard criteria. They reported texture analysis to be an independent factor associated with time to progression, supporting its potential to improve assessment and predict good response to therapy [86].

Although radiomics has promising utility for the diagnosis and assessment of renal masses and their response to therapy, its potential is limited by a lack of generalizability and clinical application. A recent systematic review and meta-analysis of 57 publications by Ursprung et al., of which 34 involved machine learning and artificial intelligence, reported that several similar characteristics have been investigated, but they have not been introduced into clinical practice because of a lack of external validation and reproducibility. This may be due to limited access to the codes and images that are used for the analysis and creation of the models [87].

One limitation to current radiomic analysis of renal masses is the fact that in most studies, the findings are compared to surgical pathologies, creating the possibility of selection bias. Additionally, some radiomics studies only analyzed single segments of the tumor, and therefore could not assess intratumoral heterogeneity, which may be an important factor in RCC [87]. This remains a promising field; however, prospective, randomized, and multicenter trials are required to accelerate its incorporation into clinical practice.

6.2. Radiogenomics

Radiogenomics is the integration of radiomics with genetic data and molecular signatures, based on the underlying principle that genetic alterations lead to distinct phenotypic imaging characteristics [5,6,88]. For clear-cell RCC specifically, the thorough analysis of genetic alterations with prognostic significance has led to an increased interest in the relationship between genomic signatures and imaging characteristics [88].

Karlo et al. analyzed the relationship of contrast-enhanced CT findings to genetic alterations in *VHL, PBRM1, SETD2, KDM5C,* and *BAP1* genes in 233 patients with clear-cell RCC. For *VHL,* they described a mutation frequency of 53.2% and imaging characteristics of well-defined tumor margins, nodular tumor enhancement, and gross appearance of intratumoral vascularity. *KDMC5* and *BAP1* had mutation frequencies of 6.9% and 6.0%, respectively, and were significantly associated with evidence of renal vein invasion. *PBRM1* mutations were observed in 28.8% of patients and, together with *VHL* mutations, were significantly more common among solid clear-cell RCC. *BAP1, KDMC5,* and *SETD2* mutations, with a frequency of 7.3%, were absent in multi-cystic clear-cell RCC [89]. In a similar study by Shinagare et al., *BAP1* mutations were significantly associated with ill-defined margins and the presence of calcifications, while *MUC4* mutations were associated with an exophytic growth pattern [90].

Radiogenomics has also been studied and integrated into models that can predict outcomes and response to treatment. Jamshidi et al. demonstrated how targeted, noninvasive, imaging-based surrogates of molecular assays (SOMA) can be constructed and used to determine outcomes in clear-cell RCC. They developed the Radiogenomic Risk Score (RRS), using a library of CT imaging features that have been correlated to genetic signatures shown to predict oncological outcomes. They followed 70 patients prospectively, classified RRS as high vs. low, and showed that the RSS predicts disease-specific survival, with a median survival of 40 months in patients with high RRS vs. 120 months in patients with low RRS ($p = 0.00024$) [91]. In another study, the ability of RRS to predict radiologic PFS was evaluated in patients with metastatic RCC undergoing presurgical treatment with bevacizumab in a phase II clinical trial. Patients with high RRSs on pretreatment CT scans had a median radiological PFS of 6 months vs. >25 months for patients with low RRSs ($p = 0.005$). Furthermore, overall survival differed significantly between the two cohorts: 25 months among high-RRS patients vs. >37 months among those with low RRSs ($p = 0.03$) [92]. These results must be interpreted carefully, as imaging characteristics may reflect tumor biology, and not necessarily the response to treatment.

The limitations of radiomics also apply to radiogenomics. In addition, the known genetic alterations in RCC have a very low prevalence, which may limit the utility of radiogenomics. Furthermore, image characteristics, as they relate to genetic alterations, may not be consistent in all phases of imaging, and given the complexity of molecular pathways, heterogeneity within the tumor, and change over time, it is challenging to make direct correlations to specific imaging findings. This field will continue to expand with the advancement of knowledge regarding the relationships among genetic alterations, molecular pathways, and prognosis and response to treatment [5,88].

7. Future Directions

There are currently multiple trials focusing on the imaging of RCC and indeterminate renal masses. An early stage study is currently investigating hyperpolarized ^{13}C pyruvate MRI to differentiate benign from malignant renal masses (NCT04687969). Due to the characteristic increased lactate production in malignant tissue, the conversion of hyperpolarized ^{13}C pyruvate to lactate can be visualized with MRI, and makes this noninvasive pathway specific technique a promising approach in visualizing renal masses [93]. Another trial is investigating the use of a machine learning algorithm in patients undergoing PET/MRI using [18F]-DCFPyL, a PSMA ligand, in solid tumors including renal masses to assess tumor aggressiveness (NCT04687969). Aggressiveness assessed in imaging will be compared to the final histopathology in 50 patients, and patients will undergo up to three scans to evaluate longitudinal differences. With respect to treatment response, Mittlmeier et al. reported a pilot study that revealed a potential approach to predict early response to tyrosine-kinase inhibitors in metastatic RCC using 18F-PSMA PET/CT [94]. The theranostics approach, involving molecular imaging and a subsequent targeted therapy using the same radiotracer, represents an exciting modality in clear-cell renal cell carcinoma. The natural association of neovascularization in clear-cell RCC and PSMA expression may allow for targeted therapy using 177Lu-PSMA in highly aggressive RCC [57,95]. Furthermore, a combination of PSMA-targeted therapy and immunotherapy may also be a promising approach [57].

8. Conclusions

Various imaging platforms are currently being studied that offer significant promise to better inform the diagnosis and prognosis of patients with RCC. Although most technologies described here are not yet ready for routine clinical use, advances in imaging will soon help clinicians make better informed management decisions.

Author Contributions: Both L.P.C. and L.E. contributed equally to the joint writing and preparation of the manuscript. Conceptualization, L.P.C., L.E. and S.W.R.; methodology, L.P.C., L.E. and S.W.R.; writing—original draft preparation, L.P.C., L.E. and S.W.R.; writing—review and editing, L.P.C., L.E., S.W.R., A.A.H. and E.R.; visualization, L.P.C. and S.W.R. supervision, A.A.H. and E.R. All authors have read and agreed to the published version of the manuscript.

Funding: A.A.H. and S.W.R. are supported by NIH MSK Cancer Center Support Grant. L.E. is supported by the Friedrich-Baur Foundation and a postdoctoral fellowship of Deutsche Forschungsgemeinschaft (German Research Foundation). A.A.H. and E.R. are supported by The Alan and Sandra Gerry Metastasis and Tumor Ecosystems Center (GMTEC) Grant. E.R. is supported by Cycle for Survival Equinox Innovation Award. All authors were supported by the National Institutes of Health Cancer Center Support Grant (P30 CA008748) to Memorial Sloan Kettering.

Institutional Review Board Statement: Not applicable.

Informed Consent Statement: Not applicable.

Data Availability Statement: Not applicable.

Conflicts of Interest: The funders had no role in the design of the study; in the collection, analyses, or interpretation of data; in the writing of the manuscript; or in the decision to publish the results.

References

1. Volpe, A.; Patard, J.J. Prognostic factors in renal cell carcinoma. *World J. Urol.* **2010**, *28*, 319–327. [CrossRef] [PubMed]
2. Cancer.Net. Cancer.Net: Kidney Cancer: Statistics. 2022. Available online: https://www.cancer.net/cancer-types/kidney-cancer/statistics (accessed on 1 January 2023).
3. Capitanio, U.; Bensalah, K.; Bex, A.; Boorjian, S.A.; Bray, F.; Coleman, J.; Gore, J.L.; Sun, M.; Wood, C.; Russo, P. Epidemiology of Renal Cell Carcinoma. *Eur. Urol.* **2019**, *75*, 74–84. [CrossRef] [PubMed]
4. Partin, A.W.; Dmochowski, R.R.; Kavoussi, L.R.; Peters, C.A. *Campbell-Walsh-Wein Urology*, 12th ed.; Chapter 57; Elsevier: Amsterdam, The Netherlands, 2020; ISBN 9780323546423.
5. Roussel, E.; Capitanio, U.; Kutikov, A.; Oosterwijk, E.; Pedrosa, I.; Rowe, S.P.; Gorin, M.A. Novel Imaging Methods for Renal Mass Characterization: A Collaborative Review. *Eur. Urol.* **2022**, *81*, 476–488. [CrossRef] [PubMed]
6. Farber, N.J.; Kim, C.J.; Modi, P.K.s.; Hon, J.D.; Sadimin, E.T.; Singer, E.A. Renal cell carcinoma: The search for a reliable biomarker. *Transl. Cancer Res.* **2017**, *6*, 620–632. [CrossRef] [PubMed]
7. Johnson, D.C.; Vukina, J.; Smith, A.B.; Meyer, A.M.; Wheeler, S.B.; Kuo, T.M.; Tan, H.J.; Woods, M.E.; Raynor, M.C.; Wallen, E.M.; et al. Preoperatively misclassified, surgically removed benign renal masses: A systematic review of surgical series and United States population level burden estimate. *J. Urol.* **2015**, *193*, 30–35. [CrossRef] [PubMed]
8. Sasaguri, K.; Takahashi, N. CT and MR imaging for solid renal mass characterization. *Eur. J. Radiol.* **2018**, *99*, 40–54. [CrossRef]
9. Gilet, A.G.; Kang, S.K.; Kim, D.; Chandarana, H. Advanced renal mass imaging: Diffusion and perfusion MRI. *Curr. Urol. Rep.* **2012**, *13*, 93–98. [CrossRef]
10. Kang, S.K.; Zhang, A.; Pandharipande, P.V.; Chandarana, H.; Braithwaite, R.S.; Littenberg, B. DWI for Renal Mass Characterization: Systematic Review and Meta-Analysis of Diagnostic Test Performance. *Am. J. Roentgenol.* **2015**, *205*, 317–324. [CrossRef]
11. Vogel, C.; Ziegelmüller, B.; Ljungberg, B.; Bensalah, K.; Bex, A.; Canfield, S.; Giles, R.H.; Hora, M.; Kuczyk, M.A.; Merseburger, A.S.; et al. Imaging in Suspected Renal-Cell Carcinoma: Systematic Review. *Clin. Genitourin. Cancer* **2019**, *17*, e345–e355. [CrossRef]
12. Cornelis, F.; Tricaud, E.; Lasserre, A.S.; Petitpierre, F.; Bernhard, J.C.; Le Bras, Y.; Yacoub, M.; Bouzgarrou, M.; Ravaud, A.; Grenier, N. Multiparametric magnetic resonance imaging for the differentiation of low and high grade clear cell renal carcinoma. *Eur. Radiol.* **2015**, *25*, 24–31. [CrossRef]

13. Woo, S.; Suh, C.H.; Kim, S.Y.; Cho, J.Y.; Kim, S.H. Diagnostic Performance of DWI for Differentiating High- From Low-Grade Clear Cell Renal Cell Carcinoma: A Systematic Review and Meta-Analysis. *AJR Am. J. Roentgenol.* **2017**, *209*, W374–W381. [CrossRef] [PubMed]
14. Wu, Y.; Kwon, Y.S.; Labib, M.; Foran, D.J.; Singer, E.A. Magnetic Resonance Imaging as a Biomarker for Renal Cell Carcinoma. *Dis. Mark.* **2015**, *2015*, 648495. [CrossRef] [PubMed]
15. Lanzman, R.S.; Robson, P.M.; Sun, M.R.; Patel, A.D.; Mentore, K.; Wagner, A.A.; Genega, E.M.; Rofsky, N.M.; Alsop, D.C.; Pedrosa, I. Arterial spin-labeling MR imaging of renal masses: Correlation with histopathologic findings. *Radiology* **2012**, *265*, 799–808. [CrossRef] [PubMed]
16. Wilson, M.P.; Patel, D.; Murad, M.H.; McInnes, M.D.F.; Katlariwala, P.; Low, G. Diagnostic Performance of MRI in the Detection of Renal Lipid-Poor Angiomyolipomas: A Systematic Review and Meta-Analysis. *Radiology* **2020**, *296*, 511–520. [CrossRef]
17. Kay, F.U.; Canvasser, N.E.; Xi, Y.; Pinho, D.F.; Costa, D.N.; Diaz de Leon, A.; Khatri, G.; Leyendecker, J.R.; Yokoo, T.; Lay, A.H.; et al. Diagnostic Performance and Interreader Agreement of a Standardized MR Imaging Approach in the Prediction of Small Renal Mass Histology. *Radiology* **2018**, *287*, 543–553. [CrossRef]
18. Canvasser, N.E.; Kay, F.U.; Xi, Y.; Pinho, D.F.; Costa, D.; de Leon, A.D.; Khatri, G.; Leyendecker, J.R.; Yokoo, T.; Lay, A.; et al. Diagnostic Accuracy of Multiparametric Magnetic Resonance Imaging to Identify Clear Cell Renal Cell Carcinoma in cT1a Renal Masses. *J. Urol.* **2017**, *198*, 780–786. [CrossRef]
19. Steinberg, R.L.; Rasmussen, R.G.; Johnson, B.A.; Ghandour, R.; De Leon, A.D.; Xi, Y.; Yokoo, T.; Kim, S.; Kapur, P.; Cadeddu, J.A.; et al. Prospective performance of clear cell likelihood scores (ccLS) in renal masses evaluated with multiparametric magnetic resonance imaging. *Eur. Radiol.* **2021**, *31*, 314–324. [CrossRef]
20. Tsai, L.L.; Bhatt, R.S.; Strob, M.F.; Jegede, O.A.; Sun, M.R.M.; Alsop, D.C.; Catalano, P.; McDermott, D.; Robson, P.M.; Atkins, M.B.; et al. Arterial Spin Labeled Perfusion MRI for the Evaluation of Response to Tyrosine Kinase Inhibition Therapy in Metastatic Renal Cell Carcinoma. *Radiology* **2021**, *298*, 332–340. [CrossRef]
21. Hahn, O.M.; Yang, C.; Medved, M.; Karczmar, G.; Kistner, E.; Karrison, T.; Manchen, E.; Mitchell, M.; Ratain, M.J.; Stadler, W.M. Dynamic contrast-enhanced magnetic resonance imaging pharmacodynamic biomarker study of sorafenib in metastatic renal carcinoma. *J. Clin. Oncol.* **2008**, *26*, 4572–4578. [CrossRef]
22. Sidhar, K.; McGahan, J.P.; Early, H.M.; Corwin, M.; Fananapazir, G.; Gerscovich, E.O. Renal Cell Carcinomas. *J. Ultrasound Med.* **2016**, *35*, 311–320. [CrossRef]
23. Sidhu, P.S.; Cantisani, V.; Dietrich, C.F.; Gilja, O.H.; Saftoiu, A.; Bartels, E.; Bertolotto, M.; Calliada, F.; Clevert, D.A.; Cosgrove, D.; et al. The EFSUMB Guidelines and Recommendations for the Clinical Practice of Contrast-Enhanced Ultrasound (CEUS) in Non-Hepatic Applications: Update 2017 (Long Version). *Ultraschall Med.* **2018**, *39*, e2–e44. [CrossRef] [PubMed]
24. Xu, Z.F.; Xu, H.X.; Xie, X.Y.; Liu, G.J.; Zheng, Y.L.; Lu, M.D. Renal cell carcinoma and renal angiomyolipoma: Differential diagnosis with real-time contrast-enhanced ultrasonography. *J. Ultrasound Med.* **2010**, *29*, 709–717. [CrossRef] [PubMed]
25. Barr, R.G.; Peterson, C.; Hindi, A. Evaluation of indeterminate renal masses with contrast-enhanced US: A diagnostic performance study. *Radiology* **2014**, *271*, 133–142. [CrossRef] [PubMed]
26. Tufano, A.; Drudi, F.M.; Angelini, F.; Polito, E.; Martino, M.; Granata, A.; Di Pierro, G.B.; Kutrolli, E.; Sampalmieri, M.; Canale, V.; et al. Contrast-Enhanced Ultrasound (CEUS) in the Evaluation of Renal Masses with Histopathological Validation-Results from a Prospective Single-Center Study. *Diagnostics* **2022**, *12*, 1209. [CrossRef]
27. Sun, D.; Wei, C.; Li, Y.; Lu, Q.; Zhang, W.; Hu, B. Contrast-Enhanced Ultrasonography with Quantitative Analysis allows Differentiation of Renal Tumor Histotypes. *Sci. Rep.* **2016**, *6*, 35081. [CrossRef]
28. Wei, S.; Tian, F.; Xia, Q.; Huang, P.; Zhang, Y.; Xia, Z.; Wu, M.; Yang, B. Contrast-enhanced ultrasound findings of adult renal cell carcinoma associated with Xp11.2 translocation/TFE3 gene fusion: Comparison with clear cell renal cell carcinoma and papillary renal cell carcinoma. *Cancer Imaging* **2019**, *20*, 1. [CrossRef]
29. Lamuraglia, M.; Escudier, B.; Chami, L.; Schwartz, B.; Leclère, J.; Roche, A.; Lassau, N. To predict progression-free survival and overall survival in metastatic renal cancer treated with sorafenib: Pilot study using dynamic contrast-enhanced Doppler ultrasound. *Eur. J. Cancer* **2006**, *42*, 2472–2479. [CrossRef]
30. Williams, R.; Hudson, J.M.; Lloyd, B.A.; Sureshkumar, A.R.; Lueck, G.; Milot, L.; Atri, M.; Bjarnason, G.A.; Burns, P.N. Dynamic Microbubble Contrast-enhanced US to Measure Tumor Response to Targeted Therapy: A Proposed Clinical Protocol with Results from Renal Cell Carcinoma Patients Receiving Antiangiogenic Therapy. *Radiology* **2011**, *260*, 581–590. [CrossRef]
31. Krajewski, K.M.; Shinagare, A.B. Novel imaging in renal cell carcinoma. *Curr. Opin. Urol.* **2016**, *26*, 388–395. [CrossRef]
32. Wang, H.Y.; Ding, H.J.; Chen, J.H.; Chao, C.H.; Lu, Y.Y.; Lin, W.Y.; Kao, C.H. Meta-analysis of the diagnostic performance of [18F]FDG-PET and PET/CT in renal cell carcinoma. *Cancer Imaging* **2012**, *12*, 464–474. [CrossRef]
33. Kayani, I.; Avril, N.; Bomanji, J.; Chowdhury, S.; Rockall, A.; Sahdev, A.; Nathan, P.; Wilson, P.; Shamash, J.; Sharpe, K.; et al. Sequential FDG-PET/CT as a biomarker of response to Sunitinib in metastatic clear cell renal cancer. *Clin. Cancer Res.* **2011**, *17*, 6021–6028. [CrossRef] [PubMed]
34. Nakaigawa, N.; Kondo, K.; Tateishi, U.; Minamimoto, R.; Kaneta, T.; Namura, K.; Ueno, D.; Kobayashi, K.; Kishida, T.; Ikeda, I.; et al. FDG PET/CT as a prognostic biomarker in the era of molecular-targeting therapies: Max SUVmax predicts survival of patients with advanced renal cell carcinoma. *BMC Cancer* **2016**, *16*, 67. [CrossRef] [PubMed]

35. Singh, H.; Arora, G.; Nayak, B.; Sharma, A.; Singh, G.; Kumari, K.; Jana, S.; Patel, C.; Pandey, A.K.; Seth, A.; et al. Semi-quantitative F-18-FDG PET/computed tomography parameters for prediction of grade in patients with renal cell carcinoma and the incremental value of diuretics. *Nucl. Med. Commun.* **2020**, *41*, 485–493. [CrossRef] [PubMed]
36. Zhu, H.; Zhao, S.; Zuo, C.; Ren, F. FDG PET/CT and CT Findings of Renal Cell Carcinoma With Sarcomatoid Differentiation. *AJR Am. J. Roentgenol.* **2020**, *215*, 645–651. [CrossRef] [PubMed]
37. Zhao, Y.; Wu, C.; Li, W.; Chen, X.; Li, Z.; Liao, X.; Cui, Y.; Zhao, G.; Liu, M.; Fu, Z. 2-[18F]FDG PET/CT parameters associated with WHO/ISUP grade in clear cell renal cell carcinoma. *Eur. J. Nucl. Med. Mol. Imaging* **2021**, *48*, 570–579. [CrossRef]
38. Nakajima, R.; Abe, K.; Kondo, T.; Tanabe, K.; Sakai, S. Clinical role of early dynamic FDG-PET/CT for the evaluation of renal cell carcinoma. *Eur. Radiol.* **2016**, *26*, 1852–1862. [CrossRef]
39. Ma, H.; Shen, G.; Liu, B.; Yang, Y.; Ren, P.; Kuang, A. Diagnostic performance of 18F-FDG PET or PET/CT in restaging renal cell carcinoma: A systematic review and meta-analysis. *Nucl. Med. Commun.* **2017**, *38*, 156–163. [CrossRef]
40. Hou, G.; Zhao, D.; Jiang, Y.; Zhu, Z.; Huo, L.; Li, F.; Cheng, W. Clinical utility of FDG PET/CT for primary and recurrent papillary renal cell carcinoma. *Cancer Imaging* **2021**, *21*, 25. [CrossRef]
41. Caldarella, C.; Muoio, B.; Isgrò, M.A.; Porfiri, E.; Treglia, G.; Giovanella, L. The role of fluorine-18-fluorodeoxyglucose positron emission tomography in evaluating the response to tyrosine-kinase inhibitors in patients with metastatic primary renal cell carcinoma. *Radiol. Oncol.* **2014**, *48*, 219–227. [CrossRef]
42. Weng, S.; DiNatale, R.G.; Silagy, A.; Mano, R.; Attalla, K.; Kashani, M.; Weiss, K.; Benfante, N.E.; Winer, A.G.; Coleman, J.A.; et al. The Clinicopathologic and Molecular Landscape of Clear Cell Papillary Renal Cell Carcinoma: Implications in Diagnosis and Management. *Eur. Urol.* **2021**, *79*, 468–477. [CrossRef]
43. Stillebroer, A.B.; Mulders, P.F.; Boerman, O.C.; Oyen, W.J.; Oosterwijk, E. Carbonic anhydrase IX in renal cell carcinoma: Implications for prognosis, diagnosis, and therapy. *Eur. Urol.* **2010**, *58*, 75–83. [CrossRef] [PubMed]
44. Divgi, C.R.; Uzzo, R.G.; Gatsonis, C.; Bartz, R.; Treutner, S.; Yu, J.Q.; Chen, D.; Carrasquillo, J.A.; Larson, S.; Bevan, P.; et al. Positron emission tomography/computed tomography identification of clear cell renal cell carcinoma: Results from the REDECT trial. *J. Clin. Oncol.* **2013**, *31*, 187–194. [CrossRef] [PubMed]
45. Turkbey, B.; Lindenberg, M.L.; Adler, S.; Kurdziel, K.A.; McKinney, Y.L.; Weaver, J.; Vocke, C.D.; Anver, M.; Bratslavsky, G.; Eclarinal, P.; et al. PET/CT imaging of renal cell carcinoma with (18)F-VM4-037: A phase II pilot study. *Abdom. Radiol.* **2016**, *41*, 109–118. [CrossRef] [PubMed]
46. Conroy, R. 89Zr-DFO-Girentuximab PET Agent Meets Specificity and Sensitivity End Points in Clear Cell RCC. Available online: https://www.cancernetwork.com/view/89zr-dfo-girentuximab-pet-agent-meets-specificity-and-sensitivity-end-points-in-clear-cell-rcc (accessed on 2 January 2023).
47. Merkx, R.I.J.; Lobeek, D.; Konijnenberg, M.; Jiménez-Franco, L.D.; Kluge, A.; Oosterwijk, E.; Mulders, P.F.A.; Rijpkema, M. Phase I study to assess safety, biodistribution and radiation dosimetry for (89)Zr-girentuximab in patients with renal cell carcinoma. *Eur. J. Nucl. Med. Mol. Imaging* **2021**, *48*, 3277–3285. [CrossRef]
48. Demirci, E.; Ocak, M.; Kabasakal, L.; Decristoforo, C.; Talat, Z.; Halaç, M.; Kanmaz, B. 68Ga-PSMA PET/CT imaging of metastatic clear cell renal cell carcinoma. *Eur. J. Nucl. Med. Mol. Imaging* **2014**, *41*, 1461–1462. [CrossRef] [PubMed]
49. Gorin, M.A.; Rowe, S.P.; Hooper, J.E.; Kates, M.; Hammers, H.J.; Szabo, Z.; Pomper, M.G.; Allaf, M.E. PSMA-Targeted 18F-DCFPyL PET/CT Imaging of Clear Cell Renal Cell Carcinoma: Results from a Rapid Autopsy. *Eur. Urol.* **2017**, *71*, 145–146. [CrossRef]
50. Rowe, S.P.; Gorin, M.A.; Hammers, H.J.; Som Javadi, M.; Hawasli, H.; Szabo, Z.; Cho, S.Y.; Pomper, M.G.; Allaf, M.E. Imaging of metastatic clear cell renal cell carcinoma with PSMA-targeted ^{18}F-DCFPyL PET/CT. *Ann. Nucl. Med.* **2015**, *29*, 877–882. [CrossRef]
51. Meyer, A.R.; Carducci, M.A.; Denmeade, S.R.; Markowski, M.C.; Pomper, M.G.; Pierorazio, P.M.; Allaf, M.E.; Rowe, S.P.; Gorin, M.A. Improved identification of patients with oligometastatic clear cell renal cell carcinoma with PSMA-targeted. *Ann. Nucl. Med.* **2019**, *33*, 617–623. [CrossRef]
52. Golan, S.; Aviv, T.; Groshar, D.; Yakimov, M.; Zohar, Y.; Prokocimer, Y.; Nadu, A.; Baniel, J.; Domachevsky, L.; Bernstine, H. Dynamic 68Ga-PSMA-11 PET/CT for the Primary Evaluation of Localized Renal Mass: A Prospective Study. *J. Nucl. Med.* **2021**, *62*, 773–778. [CrossRef]
53. Gao, J.; Xu, Q.; Fu, Y.; He, K.; Zhang, C.; Zhang, Q.; Shi, J.; Zhao, X.; Wang, F.; Guo, H. Comprehensive evaluation of 68Ga-PSMA-11 PET/CT parameters for discriminating pathological characteristics in primary clear-cell renal cell carcinoma. *Eur. J. Nucl. Med. Mol. Imaging* **2021**, *48*, 561–569. [CrossRef]
54. Muselaers, S.; Erdem, S.; Bertolo, R.; Ingels, A.; Kara, Ö.; Pavan, N.; Roussel, E.; Pecoraro, A.; Marchioni, M.; Carbonara, U.; et al. PSMA PET/CT in Renal Cell Carcinoma: An Overview of Current Literature. *J. Clin. Med.* **2022**, *11*, 1829. [CrossRef] [PubMed]
55. Sawicki, L.M.; Buchbender, C.; Boos, J.; Giessing, M.; Ermert, J.; Antke, C.; Antoch, G.; Hautzel, H. Diagnostic potential of PET/CT using a 68Ga-labelled prostate-specific membrane antigen ligand in whole-body staging of renal cell carcinoma: Initial experience. *Eur. J. Nucl. Med. Mol. Imaging* **2017**, *44*, 102–107. [CrossRef] [PubMed]
56. Urso, L.; Castello, A.; Rocca, G.C.; Lancia, F.; Panareo, S.; Cittanti, C.; Uccelli, L.; Florimonte, L.; Castellani, M.; Ippolito, C.; et al. Role of PSMA-ligands imaging in Renal Cell Carcinoma management: Current status and future perspectives. *J. Cancer Res. Clin. Oncol.* **2022**, *148*, 1299–1311. [CrossRef] [PubMed]
57. Toyama, Y.; Werner, R.A.; Ruiz-Bedoya, C.A.; Ordonez, A.A.; Takase, K.; Lapa, C.; Jain, S.K.; Pomper, M.G.; Rowe, S.P.; Higuchi, T. Current and future perspectives on functional molecular imaging in nephro-urology: Theranostics on the horizon. *Theranostics* **2021**, *11*, 6105–6119. [CrossRef] [PubMed]

58. Yin, Y.; Campbell, S.P.; Markowski, M.C.; Pierorazio, P.M.; Pomper, M.G.; Allaf, M.E.; Rowe, S.P.; Gorin, M.A. Inconsistent Detection of Sites of Metastatic Non-Clear Cell Renal Cell Carcinoma with PSMA-Targeted [18F]DCFPyL PET/CT. *Mol. Imaging Biol.* **2019**, *21*, 567–573. [CrossRef]
59. Deford-Watts, L.M.; Mintz, A.; Kridel, S.J. The potential of ^{11}C-acetate PET for monitoring the fatty acid synthesis pathway in Tumors. *Curr. Pharm. Biotechnol.* **2013**, *14*, 300–312. [CrossRef]
60. Oyama, N.; Okazawa, H.; Kusukawa, N.; Kaneda, T.; Miwa, Y.; Akino, H.; Fujibayashi, Y.; Yonekura, Y.; Welch, M.J.; Yokoyama, O. 11C-Acetate PET imaging for renal cell carcinoma. *Eur. J. Nucl. Med. Mol. Imaging* **2009**, *36*, 422–427. [CrossRef]
61. Oyama, N.; Ito, H.; Takahara, N.; Miwa, Y.; Akino, H.; Kudo, T.; Okazawa, H.; Fujibayashi, Y.; Komatsu, K.; Tsukahara, K.; et al. Diagnosis of complex renal cystic masses and solid renal lesions using PET imaging: Comparison of 11C-acetate and 18F-FDG PET imaging. *Clin. Nucl. Med.* **2014**, *39*, e208–e214. [CrossRef]
62. Ho, C.L.; Chen, S.; Ho, K.M.; Chan, W.K.; Leung, Y.L.; Cheng, K.C.; Wong, K.N.; Cheung, M.K.; Wong, K.K. Dual-tracer PET/CT in renal angiomyolipoma and subtypes of renal cell carcinoma. *Clin. Nucl. Med.* **2012**, *37*, 1075–1082. [CrossRef]
63. Nakajima, R.; Nozaki, S.; Kondo, T.; Nagashima, Y.; Abe, K.; Sakai, S. Evaluation of renal cell carcinoma histological subtype and fuhrman grade using 18F-fluorodeoxyglucose-positron emission tomography/computed tomography. *Eur. Radiol.* **2017**, *27*, 4866–4873. [CrossRef]
64. Marko, J.; Craig, R.; Nguyen, A.; Udager, A.M.; Wolfman, D.J. Chromophobe Renal Cell Carcinoma with Radiologic-Pathologic Correlation. *RadioGraphics* **2021**, *41*, 1408–1419. [CrossRef] [PubMed]
65. Rowe, S.P.; Gorin, M.A.; Solnes, L.B.; Ball, M.W.; Choudhary, A.; Pierorazio, P.M.; Epstein, J.I.; Javadi, M.S.; Allaf, M.E.; Baras, A.S. Correlation of 99mTc-sestamibi uptake in renal masses with mitochondrial content and multi-drug resistance pump expression. *EJNMMI Res.* **2017**, *7*, 80. [CrossRef] [PubMed]
66. Rowe, S.P.; Gorin, M.A.; Gordetsky, J.; Ball, M.W.; Pierorazio, P.M.; Higuchi, T.; Epstein, J.I.; Allaf, M.E.; Javadi, M.S. Initial experience using 99mTc-MIBI SPECT/CT for the differentiation of oncocytoma from renal cell carcinoma. *Clin. Nucl. Med.* **2015**, *40*, 309–313. [CrossRef] [PubMed]
67. Gorin, M.A.; Rowe, S.P.; Baras, A.S.; Solnes, L.B.; Ball, M.W.; Pierorazio, P.M.; Pavlovich, C.P.; Epstein, J.I.; Javadi, M.S.; Allaf, M.E. Prospective Evaluation of (99m)Tc-sestamibi SPECT/CT for the Diagnosis of Renal Oncocytomas and Hybrid Oncocytic/Chromophobe Tumors. *Eur. Urol.* **2016**, *69*, 413–416. [CrossRef]
68. Sistani, G.; Bjazevic, J.; Kassam, Z.; Romsa, J.; Pautler, S. The value of 99mTc-sestamibi single-photon emission computed tomography-computed tomography in the evaluation and risk stratification of renal masses. *Can. Urol. Assoc. J.* **2021**, *15*, 197–201. [CrossRef]
69. Asi, T.; Tuncali, M.; Tuncel, M.; Alkanat, N.E.İ.; Hazir, B.; Kösemehmetoğlu, K.; Baydar, D.E.; Akdoğan, B. The role of Tc-99m MIBI scintigraphy in clinical T1 renal mass assessment: Does it have a real benefit? *Urol. Oncol.* **2020**, *38*, 937.e911–937.e917. [CrossRef]
70. Zwanenburg, A.; Vallières, M.; Abdalah, M.A.; Aerts, H.J.W.L.; Andrearczyk, V.; Apte, A.; Ashrafinia, S.; Bakas, S.; Beukinga, R.J.; Boellaard, R.; et al. The Image Biomarker Standardization Initiative: Standardized Quantitative Radiomics for High-Throughput Image-based Phenotyping. *Radiology* **2020**, *295*, 328–338. [CrossRef]
71. Wang, W.; Cao, K.; Jin, S.; Zhu, X.; Ding, J.; Peng, W. Differentiation of renal cell carcinoma subtypes through MRI-based radiomics analysis. *Eur. Radiol.* **2020**, *30*, 5738–5747. [CrossRef]
72. Varghese, B.A.; Chen, F.; Hwang, D.H.; Cen, S.Y.; Desai, B.; Gill, I.S.; Duddalwar, V.A. Differentiation of Predominantly Solid Enhancing Lipid-Poor Renal Cell Masses by Use of Contrast-Enhanced CT: Evaluating the Role of Texture in Tumor Subtyping. *AJR Am. J. Roentgenol.* **2018**, *211*, W288–W296. [CrossRef]
73. Uhlig, J.; Biggemann, L.; Nietert, M.M.; Beißbarth, T.; Lotz, J.; Kim, H.S.; Trojan, L.; Uhlig, A. Discriminating malignant and benign clinical T1 renal masses on computed tomography: A pragmatic radiomics and machine learning approach. *Medicine* **2020**, *99*, e19725. [CrossRef]
74. Tanaka, T.; Huang, Y.; Marukawa, Y.; Tsuboi, Y.; Masaoka, Y.; Kojima, K.; Iguchi, T.; Hiraki, T.; Gobara, H.; Yanai, H.; et al. Differentiation of Small ($\leq$4 cm) Renal Masses on Multiphase Contrast-Enhanced CT by Deep Learning. *Am. J. Roentgenol.* **2020**, *214*, 605–612. [CrossRef] [PubMed]
75. Zabihollahy, F.; Schieda, N.; Krishna, S.; Ukwatta, E. Automated classification of solid renal masses on contrast-enhanced computed tomography images using convolutional neural network with decision fusion. *Eur. Radiol.* **2020**, *30*, 5183–5190. [CrossRef] [PubMed]
76. Oberai, A.; Varghese, B.; Cen, S.; Angelini, T.; Hwang, D.; Gill, I.; Aron, M.; Lau, C.; Duddalwar, V. Deep learning based classification of solid lipid-poor contrast enhancing renal masses using contrast enhanced CT. *Br. J. Radiol.* **2020**, *93*, 20200002. [CrossRef] [PubMed]
77. Xi, I.L.; Zhao, Y.; Wang, R.; Chang, M.; Purkayastha, S.; Chang, K.; Huang, R.Y.; Silva, A.C.; Vallières, M.; Habibollahi, P.; et al. Deep Learning to Distinguish Benign from Malignant Renal Lesions Based on Routine MR Imaging. *Clin. Cancer Res.* **2020**, *26*, 1944–1952. [CrossRef]
78. Cui, E.M.; Lin, F.; Li, Q.; Li, R.G.; Chen, X.M.; Liu, Z.S.; Long, W.S. Differentiation of renal angiomyolipoma without visible fat from renal cell carcinoma by machine learning based on whole-tumor computed tomography texture features. *Acta Radiol.* **2019**, *60*, 1543–1552. [CrossRef]

79. Lee, H.; Hong, H.; Kim, J.; Jung, D.C. Deep feature classification of angiomyolipoma without visible fat and renal cell carcinoma in abdominal contrast-enhanced CT images with texture image patches and hand-crafted feature concatenation. *Med. Phys.* **2018**, *45*, 1550–1561. [CrossRef]
80. Baghdadi, A.; Aldhaam, N.A.; Elsayed, A.S.; Hussein, A.A.; Cavuoto, L.A.; Kauffman, E.; Guru, K.A. Automated differentiation of benign renal oncocytoma and chromophobe renal cell carcinoma on computed tomography using deep learning. *BJU Int.* **2020**, *125*, 553–560. [CrossRef]
81. Duan, C.; Li, N.; Niu, L.; Wang, G.; Zhao, J.; Liu, F.; Liu, X.; Ren, Y.; Zhou, X. CT texture analysis for the differentiation of papillary renal cell carcinoma subtypes. *Abdom. Radiol.* **2020**, *45*, 3860–3868. [CrossRef]
82. Vendrami, C.L.; Velichko, Y.S.; Miller, F.H.; Chatterjee, A.; Villavicencio, C.P.; Yaghmai, V.; McCarthy, R.J. Differentiation of Papillary Renal Cell Carcinoma Subtypes on MRI: Qualitative and Texture Analysis. *Am. J. Roentgenol.* **2018**, *211*, 1234–1245. [CrossRef]
83. Cui, E.; Li, Z.; Ma, C.; Li, Q.; Lei, Y.; Lan, Y.; Yu, J.; Zhou, Z.; Li, R.; Long, W.; et al. Predicting the ISUP grade of clear cell renal cell carcinoma with multiparametric MR and multiphase CT radiomics. *Eur. Radiol.* **2020**, *30*, 2912–2921. [CrossRef]
84. Zhao, Y.; Chang, M.; Wang, R.; Xi, I.L.; Chang, K.; Huang, R.Y.; Vallières, M.; Habibollahi, P.; Dagli, M.S.; Palmer, M.; et al. Deep Learning Based on MRI for Differentiation of Low- and High-Grade in Low-Stage Renal Cell Carcinoma. *J. Magn. Reson. Imaging* **2020**, *52*, 1542–1549. [CrossRef]
85. Meng, X.; Shu, J.; Xia, Y.; Yang, R. A CT-Based Radiomics Approach for the Differential Diagnosis of Sarcomatoid and Clear Cell Renal Cell Carcinoma. *BioMed Res. Int.* **2020**, *2020*, 7103647. [CrossRef]
86. Goh, V.; Ganeshan, B.; Nathan, P.; Juttla, J.K.; Vinayan, A.; Miles, K.A. Assessment of response to tyrosine kinase inhibitors in metastatic renal cell cancer: CT texture as a predictive biomarker. *Radiology* **2011**, *261*, 165–171. [CrossRef]
87. Ursprung, S.; Beer, L.; Bruining, A.; Woitek, R.; Stewart, G.D.; Gallagher, F.A.; Sala, E. Radiomics of computed tomography and magnetic resonance imaging in renal cell carcinoma-a systematic review and meta-analysis. *Eur. Radiol.* **2020**, *30*, 3558–3566. [CrossRef]
88. Alessandrino, F.; Shinagare, A.B.; Bossé, D.; Choueiri, T.K.; Krajewski, K.M. Radiogenomics in renal cell carcinoma. *Abdom. Radiol.* **2019**, *44*, 1990–1998. [CrossRef] [PubMed]
89. Karlo, C.A.; Di Paolo, P.L.; Chaim, J.; Hakimi, A.A.; Ostrovnaya, I.; Russo, P.; Hricak, H.; Motzer, R.; Hsieh, J.J.; Akin, O. Radiogenomics of clear cell renal cell carcinoma: Associations between CT imaging features and mutations. *Radiology* **2014**, *270*, 464–471. [CrossRef] [PubMed]
90. Shinagare, A.B.; Vikram, R.; Jaffe, C.; Akin, O.; Kirby, J.; Huang, E.; Freymann, J.; Sainani, N.I.; Sadow, C.A.; Bathala, T.K.; et al. Radiogenomics of clear cell renal cell carcinoma: Preliminary findings of The Cancer Genome Atlas-Renal Cell Carcinoma (TCGA-RCC) Imaging Research Group. *Abdom. Imaging* **2015**, *40*, 1684–1692. [CrossRef] [PubMed]
91. Jamshidi, N.; Jonasch, E.; Zapala, M.; Korn, R.L.; Aganovic, L.; Zhao, H.; Tumkur Sitaram, R.; Tibshirani, R.J.; Banerjee, S.; Brooks, J.D.; et al. The Radiogenomic Risk Score: Construction of a Prognostic Quantitative, Noninvasive Image-based Molecular Assay for Renal Cell Carcinoma. *Radiology* **2015**, *277*, 114–123. [CrossRef] [PubMed]
92. Jamshidi, N.; Jonasch, E.; Zapala, M.; Korn, R.L.; Brooks, J.D.; Ljungberg, B.; Kuo, M.D. The radiogenomic risk score stratifies outcomes in a renal cell cancer phase 2 clinical trial. *Eur. Radiol.* **2016**, *26*, 2798–2807. [CrossRef]
93. Wang, Z.J.; Ohliger, M.A.; Larson, P.E.Z.; Gordon, J.W.; Bok, R.A.; Slater, J.; Villanueva-Meyer, J.E.; Hess, C.P.; Kurhanewicz, J.; Vigneron, D.B. Hyperpolarized 13C MRI: State of the Art and Future Directions. *Radiology* **2019**, *291*, 273–284. [CrossRef]
94. Mittlmeier, L.M.; Unterrainer, M.; Rodler, S.; Todica, A.; Albert, N.L.; Burgard, C.; Cyran, C.C.; Kunz, W.G.; Ricke, J.; Bartenstein, P.; et al. 18F-PSMA-1007 PET/CT for response assessment in patients with metastatic renal cell carcinoma undergoing tyrosine kinase or checkpoint inhibitor therapy: Preliminary results. *Eur. J. Nucl. Med. Mol. Imaging* **2021**, *48*, 2031–2037. [CrossRef] [PubMed]
95. Gorin, M.A.; Rowe, S.P. Kidney cancer: PSMA: A potential therapeutic target in RCC. *Nat. Rev. Urol.* **2017**, *14*, 646–647. [CrossRef] [PubMed]

cancers

MDPI

Article

The Association of Tumor Immune Microenvironment of the Primary Lesion with Time to Metastasis in Patients with Renal Cell Carcinoma: A Retrospective Analysis

Kazutoshi Fujita [1,*,†], Go Kimura [2,†], Toyonori Tsuzuki [3], Taigo Kato [4], Eri Banno [1], Akira Kazama [5], Ryo Yamashita [6], Yuto Matsushita [7], Daisuke Ishii [8], Tomoya Fukawa [9], Yuki Nakagawa [10], Tamaki Fukuyama [11], Fumikazu Sano [11], Yukihiro Kondo [2] and Hirotsugu Uemura [1]

1 Department of Urology, Kindai University Faculty of Medicine, Osaka 589-8511, Japan
2 Department of Urology, Nippon Medical School, Tokyo 113-8603, Japan
3 Department of Surgical Pathology, Aichi Medical University Hospital, Nagakute 480-1195, Japan
4 Department of Urology, Osaka University Graduate School of Medicine, Osaka 565-0871, Japan
5 Department of Urology, Division of Molecular Oncology, Niigata University Graduate School of Medical and Dental Sciences, Niigata 951-8510, Japan
6 Division of Urology, Shizuoka Cancer Center, Shizuoka 411-8777, Japan
7 Department of Urology, Hamamatsu University School of Medicine, Hamamatsu 431-3192, Japan
8 Department of Urology, Kitasato University of Medicine, Sagamihara 252-0374, Japan
9 Department of Urology, Tokushima University Graduate School of Biomedical Sciences, Tokushima 770-8503, Japan
10 Clinical Development Division, Chugai Pharmaceutical Co., Ltd., Tokyo 103-8324, Japan
11 Medical Affairs Division, Chugai Pharmaceutical Co., Ltd., Tokyo 103-8324, Japan
* Correspondence: kfujita@med.kindai.ac.jp; Tel.: +81-72-366-0221
† These authors contributed equally to this work.

Citation: Fujita, K.; Kimura, G.; Tsuzuki, T.; Kato, T.; Banno, E.; Kazama, A.; Yamashita, R.; Matsushita, Y.; Ishii, D.; Fukawa, T.; et al. The Association of Tumor Immune Microenvironment of the Primary Lesion with Time to Metastasis in Patients with Renal Cell Carcinoma: A Retrospective Analysis. *Cancers* **2022**, *14*, 5258. https://doi.org/10.3390/cancers14215258

Academic Editor: Evgeny Yakirevich

Received: 28 September 2022
Accepted: 24 October 2022
Published: 26 October 2022

Simple Summary: The association between the tumor immune microenvironment (TIME) of primary lesions and time to metastasis remains unknown. The aim of our retrospective study was to investigate the differences in the TIME of primary lesions based on time intervals to metastasis, mainly between the synchronous group (SG; metastasis within 3 months) and metachronous group (MG; metastasis after 3 months), and its association with clinicopathological parameters in patients with metastatic renal cell carcinoma (mRCC). SG showed more immunogenic feature of TIME (PD-L1 positivity, CD8+ TIL infiltration) and poor prognostic pathological features (WHO/ISUP grade 4, necrosis, lymphovascular invasion, infiltrative growth pattern, and sarcomatoid differentiation). In addition, we observed that the time to metastasis differed by TIME characteristics (PD-L1 status, immunophenotype), which were associated with the WHO/ISUP grade. The TIME of primary lesions could affect the time to metastasis.

Abstract: Biological or immunological differences in primary lesions between synchronous and metachronous metastatic renal cell carcinoma (mRCC) have been reported. However, the association between the tumor immune microenvironment (TIME) of primary lesions and time to metastasis remains unknown. We investigated the differences in the TIME of primary lesions based on time intervals to metastasis, mainly between the synchronous group (SG; metastasis within 3 months) and metachronous group (MG; metastasis after 3 months), and its association with clinicopathological parameters in patients with mRCC. Overall, 568 patients treated first-line with vascular endothelial growth factor receptor inhibitors comprised the analysis population (SG: N = 307 [54.0%]; MG: N = 261 [46.0%]). SG had a higher proportion of patients with poor prognostic pathological feature tumors: WHO/ISUP grade 4, necrosis, lymphovascular invasion, infiltrative growth pattern, and sarcomatoid differentiation. Regarding the TIME, more immunogenic features were seen in SG than MG, with a higher PD-L1 positivity and a lower proportion of the desert phenotype. This is the first study to examine the differences in the TIME of primary lesions in patients with mRCC based on the time intervals to metastasis. The TIME of primary lesions could affect the time to metastasis.

Keywords: TIME (tumor immune microenvironment); synchronous; metachronous; mRCC (metastatic renal cell carcinoma); PD-L1; immunophenotype

1. Introduction

Approximately two thirds of patients with renal cell carcinoma (RCC) have localized disease and undergo radical nephrectomy with curative intent. However, 20% of these patients develop distant metastasis, that is, metachronous metastasis. The remaining one third of the patients with RCC present with metastasis at diagnosis, that is, synchronous metastasis. The prognosis of synchronous and metachronous metastatic RCC (mRCC) is different [1,2], with the time from diagnosis to treatment of less than 1 year being one of the most important risk factors for poor survival in both International Metastatic Renal-Cell Carcinoma Database Consortium (IMDC) and Memorial Sloan-Kettering Cancer Center (MSKCC) prognostic models for mRCC [3,4].

Previous reports have documented biological or immunological differences in primary lesions between synchronous and metachronous mRCC [5,6]. However, the association between the tumor immune microenvironment (TIME) of the primary lesion, the source of metastasis, and the time to metastasis has not been studied.

Furthermore, the TIME also influences the efficacy of immune-oncology (IO) drugs, such as anti–programmed cell death protein 1 (PD-1)/programmed death ligand 1 (PD-L1) and anti–cytotoxic T-lymphocyte–associated protein 4 (CTLA-4) antibodies [7,8]. Currently, IO combination therapies, including IO plus IO (e.g., nivolumab plus ipilimumab) or IO plus vascular endothelial growth factor receptor inhibitor (VEGFRi; e.g., pembrolizumab plus axitinib, avelumab plus axitinib, nivolumab plus cabozantinib, and pembrolizumab plus lenvatinib), are the standard first-line (1L) therapy for patients with mRCC [9].

Development of IO therapy is expanding into the area of adjuvant RCC [10]. Pembrolizumab showed significant improvement in disease-free survival (DFS) in the high-risk RCC which included M1NED, defined as resection of the primary tumor and solid, isolated, soft-tissue metastases [11]. Thus, understanding the relationship between time to metastasis and the TIME of the primary lesion will help in selecting the optimal therapy.

The TIME can be described by the presence or absence of immune cells in the tumor area, the location of immune cells (infiltrated or excluded), the type of immune cells (anti-tumor immunity or immune suppressive cells, such as regulatory T cells and myeloid-derived suppressor cells), and the T cell status (activation, exhaustion, dysfunction) [12]. RCC is one of the immunogenic tumors, known as having a complex tumor microenvironment with immune suppressive cells [13]. We examined how PD-L1 expression and the immunophenotype of tumor-infiltrating immune cells in primary lesions differ by time to metastasis and its association with clinicopathological parameters in patients with mRCC.

2. Materials and Methods

2.1. Study Design and Outcomes

This report is based on an additional analysis of a dataset from a previous multicenter, retrospective study that compared overall survival (OS) by PD-L1 expression status in patients with recurrent or mRCC who had received systemic therapy (the ARCHERY study [14]). We investigated the differences in the TIME of the primary lesions between synchronous and metachronous mRCCs with different time intervals to metastasis.

2.2. Patients

A total of 770 patients with recurrent or mRCC who had started systemic therapy between January 2010 and December 2015 at 29 institutions in Japan were enrolled in the ARCHERY study. Of these, 381 patients underwent radical nephrectomy and 389 underwent cytoreductive nephrectomy. Only patients whose formalin-fixed paraffin-embedded nephrectomy specimens could be obtained were registered in this retrospective study.

Patients with other coexisting malignancies or those treated with checkpoint inhibitors as 1L systemic therapy were excluded [14].

An exploratory study was conducted using the ARCHERY database (the JEWEL study), in which patients with an unknown International Society of Urologic Pathologists (ISUP) grade in the ARCHERY study were excluded (N = 4). To improve the homogeneity of the population, we focused on patients from the ARCHERY study who had received 1L tyrosine kinase inhibitor (TKI; ie, VEGFRi) therapy (N = 569). The analysis population consisted of 568 patients, excluding one of the TKI-treated patients whose time to metastasis was not specified (Figure S1).

This study was registered in the UMIN Clinical Trials Registry (JEWEL study, UMIN000043415) and conducted with the approval of the institutional review board of the 29 study facilities. Furthermore, we obtained approval from the institutional review board of MINS (Reference number: 210204; approval date: 18 February 2021), a nonprofit organization. Informed consent was obtained from all participants, and the study was conducted in accordance with the Declaration of Helsinki.

2.3. *Assessment of Histology and Immune Status*

Hematoxylin-eosin–stained specimens were evaluated by two central pathologists. PD-L1 expression in tumor-infiltrating immune cells was analyzed using the VENTANA SP142 assay (Ventana Medical Systems, Inc., Tucson, AZ, USA, #740-4859). According to PD-L1 expression on immune cells (ICs), patients were classified as either PD-L1 negative (IC0 [PD-L1-expressing IC < 1%]) or PD-L1 positive (IC1 [IC: ≥1% but <5%], IC2 [IC: ≥5% but <10%], or IC3 [IC: ≥10%]) [15]. Three distinct immunophenotypes—inflamed (many CD8-positive T cells diffusely infiltrate into tumor cell nests), excluded (CD8-positive T cells infiltrate around tumor cell nests but not into them), and desert (no or few CD8-positive T cells infiltrate around and/or into tumor cell nests)—were identified using CD8 immunostaining [16]. The immunohistochemical assessments were performed during the central pathological review, and PD-L1 expression was evaluated independently by two central pathologists.

2.4. *Time to Metastasis*

Time to metastasis was defined as the time from the date of the "initial diagnosis" to the "date of metastasis." Information on whether the patient had distant metastasis at the time of initial diagnosis was collected from electronic data capture. For patients who did not have distant metastasis at the time of initial diagnosis, the "date of distant metastasis" was defined as the "date of recurrence." For patients who had distant metastasis at the time of initial diagnosis, the "date of distant metastasis" was defined as the "date of initial diagnosis."

Patients who had metastasis at the time of nephrectomy and patients who experienced metastasis within 3 months after nephrectomy were categorized as the "synchronous group" (SG). Patients who experienced metastasis 3 months after nephrectomy were categorized as the "metachronous group" (MG). There is no consensus on when the onset of metastasis should be considered metachronous—previous reports have referred to onset after the time of diagnosis (>0 months) [17], after 3 months [2], and after 6 months [5]; the 3-month cutoff was chosen for this study. Patients in the MG were further categorized into four subgroups based on the time interval from initial diagnosis to recurrence: 3–12 months, >12–24 months, >24 months–5 years, and >5 years after nephrectomy.

2.5. *Statistical Analysis*

Baseline characteristics were summarized by time-to-metastasis categories, and standardized differences (SDs) were calculated as a measure of the difference between the two groups. A logistic model was constructed with categorized time to metastasis (SG/MG) as the response variable and baseline characteristics as the candidate explanatory variables to explore baseline variables relevant to time-to-metastasis categorization. Stepwise selection began with no variables selected and was used for explanatory variable selection in the multivariable logistic model. Variables that met the entry or removal criteria were added or removed until a stable set of explanatory variables was obtained. The Wald test was used to determine explanatory variables in the multivariable logistic model, with two significance levels: $\alpha = 0.05$ for variable addition and $\alpha = 0.05$ for removal. Time-to-metastasis distributions were estimated using the Kaplan-Meier method for several baseline characteristics, with confidence intervals (CIs) of the median estimated using the Brookmeyer–Crowley method. Univariate Cox proportional hazards models were used to estimate the hazard ratios (HRs) and CIs. The mean, difference, and CIs of time to metastasis were also calculated as all patients were not censored. All *p*-values provided were interpreted in a descriptive manner.

3. Results

3.1. Baseline Characteristics of Patients with Synchronous and Metachronous mRCC

In total, 568 patients with mRCC were included in the analysis. Of these, 307 (54.0%) had synchronous metastasis and 261 (46.0%) developed metastasis 3 months after nephrectomy. The baseline characteristics at the time of initial diagnosis of SG and MG are shown in Table 1. The median age was 64.0 [range: 23, 87] years in SG and 64.0 [30, 85] years in MG. The distribution of World Health Organization (WHO)/ISUP grades was different between SG (grade 1/2: 29.6%, grade 3: 38.1%, and grade 4: 32.2%) and MG (grade 1/2: 47.5%, grade 3: 37.5%, and grade 4: 14.9%). The proportion of patients with grade 1/2 tumors was lower in SG than in MG (29.6% vs. 47.5%, SD: −0.4), while the proportion of patients with grade 4 tumors was higher in SG than in MG (32.2% vs. 14.9%, SD: 0.4). The proportions of patients with other poor prognostic pathological features (necrosis [50.8% vs. 34.9%, SD: 0.3], lymphovascular invasion (LVI) [30.6% vs. 19.2%, SD: 0.3], sarcomatoid differentiation [16.3% vs. 5.4%, SD: 0.4], and infiltrative growth pattern [28.3% vs. 21.1%, SD: 0.2]) were higher in SG than in MG.

Table 1. Baseline characteristics at the time of initial diagnosis of synchronous and metachronous metastatic renal cell carcinoma.

Characteristic, n (%)	Synchronous [a] (N = 307)	Metachronous [b] (N = 261)	Total (N = 568)	*p*-Value [c]	Standardized Difference
Sex [d]					
Male	238 (77.5)	196 (75.1)	434 (76.4)	0.497	0.1
Female	69 (22.5)	65 (24.9)	134 (23.6)		−0.1
Age					
Mean (standard deviation)	63.2 (10.9)	62.8 (10.3)	63.0 (10.6)	0.623	0.0
Median [range]	64.0 [23, 87]	64.0 [30, 85]	64.0 [23, 87]		
Age category [d]					
<65 y	164 (53.4)	140 (53.6)	304 (53.5)	0.473	0.0
≥65 and <75 y	94 (30.6)	88 (33.7)	182 (32.0)		−0.1
≥75 y	49 (16.0)	33 (12.6)	82 (14.4)		0.1
Histology [d]					
Clear cell	286 (93.2)	241 (92.3)	527 (92.8)	0.706	0.0
Non–clear cell	21 (6.8)	20 (7.7)	41 (7.2)		0.0
Sarcomatoid component [d]					
Absent	257 (83.7)	247 (94.6)	504 (88.7)	<0.0001	−0.4
Present	50 (16.3)	14 (5.4)	64 (11.3)		0.4
Growth pattern [d]					
Expansive	97 (31.6)	110 (42.1)	207 (36.4)	0.021	−0.2
Infiltrative	87 (28.3)	55 (21.1)	142 (25.0)		0.2
Indeterminable	123 (40.1)	96 (36.8)	219 (38.6)		0.1

Table 1. *Cont.*

Characteristic, n (%)	Synchronous [a] (N = 307)	Metachronous [b] (N = 261)	Total (N = 568)	*p*-Value [c]	Standardized Difference
Fuhrman grade [d]					
Grade 1/2	78 (25.4)	107 (41.0)	185 (32.6)	<0.0001	−0.3
Grade 3	154 (50.2)	126 (48.3)	280 (49.3)		0.0
Grade 4	75 (24.4)	28 (10.7)	103 (18.1)		0.4
WHO/ISUP grade [d]					
Grade 1/2	91 (29.6)	124 (47.5)	215 (37.9)	<0.0001	−0.4
Grade 3	117 (38.1)	98 (37.5)	215 (37.9)		0.0
Grade 4	99 (32.2)	39 (14.9)	138 (24.3)		0.4
Necrosis [d]					
Absent	150 (48.9)	169 (64.8)	319 (56.2)	0.0007	−0.3
Present	156 (50.8)	91 (34.9)	247 (43.5)		0.3
Indeterminable	1 (0.3)	1 (0.4)	2 (0.4)		0.0
Lymphovascular invasion [d]					
Absent	195 (63.5)	194 (74.3)	389 (68.5)	0.0074	−0.2
Present	94 (30.6)	50 (19.2)	144 (25.4)		0.3
Indeterminable	18 (5.9)	17 (6.5)	35 (6.2)		0.0
TIME					
PD-L1 expression					
IC0	153 (49.8)	181 (69.3)	334 (58.8)	<0.0001	−0.4
IC1	88 (28.7)	56 (21.5)	144 (25.4)		0.2
IC2	38 (12.4)	16 (6.1)	54 (9.5)		0.2
IC3	28 (9.1)	8 (3.1)	36 (6.3)		0.3
PD-L1 expression [d]					
Negative [e]	153 (49.8)	181 (69.3)	334 (58.8)	<0.0001	−0.4
Positive [f]	154 (50.2)	80 (30.7)	234 (41.2)		0.4
Immunophenotype [d]					
Desert	109 (35.5)	133 (51.0)	242 (42.6)	0.0001	−0.3
Excluded	169 (55.0)	119 (45.6)	288 (50.7)		0.2
Inflamed	29 (9.4)	9 (3.4)	38 (6.7)		0.2

[a] Defined as metastasis ≤ 3 months of initial diagnosis of renal cell carcinoma. [b] Defined as metastasis diagnosed > 3 months after initial diagnosis. [c] *p*-values were calculated using the chi-square test for categorical variables and Kruskal-Wallis test for continuous variables. [d] Candidate explanatory variables in the multivariable logistic regression model of metachronous/synchronous renal cell carcinoma; results are shown in Table 2. [e] Defined as IC0. [f] Defined as IC1/2/3. ISUP, International Society of Urologic Pathologists; PD-L1, programmed death ligand 1; TIME, tumor immune microenvironment; WHO, World Health Organization.

Table 2. Multivariable logistic regression analysis of synchronous/metachronous metastasis.

Selected Variable [a]	Definition of OR [b]	Adjusted OR [b] [95% CI]	*p*-Value [c]
PD-L1 expression	Positive/Negative	1.76 [1.22, 2.55]	0.0026
WHO/ISUP grade	Grade 3/Grades 1, 2	1.38 [0.93, 2.05]	0.110
	Grade 4/Grades 1, 2	2.58 [1.59, 4.20]	0.0001
Lymphovascular invasion	Present/Absent	1.60 [1.06, 2.40]	0.024

[a] All candidate explanatory variables are described in Table 1 (sex, age category at initial diagnosis, PD-L1 expression, immunophenotype, histology, sarcomatoid component, growth pattern, Fuhrman grade, WHO/ISUP grade, necrosis, and lymphovascular invasion). Explanatory variable addition and removal were based on stepwise selection with $\alpha = 0.05$. [b] Odds for synchronous/metachronous. [c] *p*-value was calculated using the Wald test for a parameter of the multivariable logistic regression model. CI, confidence interval; ISUP, International Society of Urologic Pathologists; OR, odds ratio; PD-L1, programmed death ligand 1; WHO, World Health Organization.

SG showed higher PD-L1 positivity (50.2% vs. 30.7%, SD: 0.4) than MG. Regarding immunophenotypes, the proportion of the desert phenotype was lower (35.5% vs. 51.0%, SD: −0.3), while that of the inflamed phenotype was higher (9.4% vs. 3.4%, SD: 0.2) (Table 1) in SG compared with MG.

3.2. Multivariable Logistic Regression Analysis of Metachronous/Synchronous Metastasis

The multivariable logistic regression model selected PD-L1 expression, WHO/ISUP grade, and LVI as explanatory variables (Table 2).

Although the distribution of immunophenotype was different between SG and MG (Table 1), the immunophenotype was not selected as an explanatory variable in the model, possibly because of the strong association between immunophenotype and PD-L1 expression and WHO/ISUP grade. PD-L1 positivity showed a distinct difference (desert vs. excluded vs. inflamed: 9.1% vs. 62.5% vs. 84.2%, respectively). The proportions of patients with WHO/ISUP grade 1/2 (52.5% vs. 29.5% vs. 7.9%) and grade 4 (14.5% vs. 27.4% vs. 63.2%) tumors were also considerably different (Table S1).

3.3. TIME and Time to Metastasis

The median time to metastasis was 6.5 months (95% CI: 2.5, 8.3) in PD-L1–negative patients and 0 months (95% CI: not applicable [NA], NA) in PD-L1–positive patients (HR: 1.57, 95% CI: 1.33, 1.86, Figure 1A).

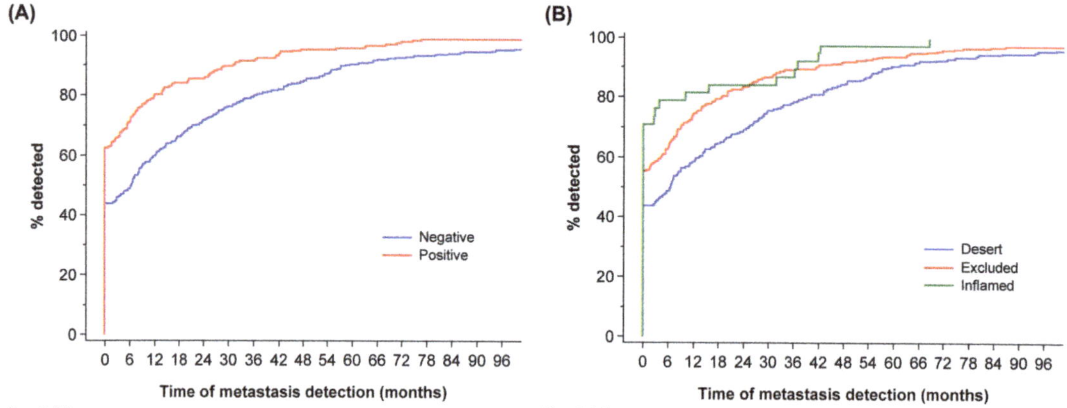

(A) No. at risk

Negative	334	171	134	113	94	79	69	60	52	43	32	28	25	23	20	18	16
Positive	234	68	46	37	34	24	20	16	11	11	10	8	5	3	3	3	3

PD-L1 Status	Mean [95% CI]	Difference [95% CI]	Median [95% CI]	Hazard Ratio [95% CI]	p-Value[a]
Negative	20.9 [17.7, 24.1]	Ref.	6.5 [2.5, 8.3]	Ref.	
Positive	9.5 [5.6, 13.3]	-11.4 [-16.4, -6.4]	0.0 [NA, NA]	1.57 [1.33, 1.86]	< 0.0001

(B) No. at risk

Desert	242	124	100	85	74	59	53	46	38	32	23	20	18	16	14	13	11
Excluded	288	107	73	59	48	38	31	28	24	21	18	15	12	10	9	8	8
Inflamed	38	8	7	6	6	6	5	2	1	1	1	1	0				

Immunophenotype	Mean [95% CI]	Difference [95% CI]	Median [95% CI]	Hazard Ratio [95% CI]	p-Value[a]
Desert	21.4 [17.6, 25.1]	Ref.	6.6 [0.0, 9.0]	Ref.	
Excluded	13.0 [9.5, 16.4]	-8.4 [-14.2, -2.5]	0.0 [0.0, 0.8]	1.35 [1.13, 1.60]	0.0007
Inflamed	7.7 [-1.9, 17.3]	-13.6 [-25.4, -1.9]	0.0 [NA, NA]	1.77 [1.26, 2.50]	0.0011

Figure 1. Kaplan-Meier curve of time to metastasis by (**A**) PD-L1 status or (**B**) immunophenotype. [a] *p*-values were calculated using the Wald test for a parameter of the univariate Cox proportional hazards model. CI, confidence interval; NA, not applicable; PD-L1, programmed death ligand 1; ref., reference.

The median time to metastasis was 6.6 months (95% CI: 0.0, 9.0) for desert, 0 months (95% CI: 0.0, 0.8) for excluded, and 0 months (95% CI: NA, NA) for inflamed immunophenotypes (HR desert vs. excluded: 1.35 [95% CI: 1.13, 1.60], HR desert vs. inflamed: 1.77 [95% CI: 1.26, 2.50], Figure 1B).

PD-L1 positivity and immunophenotype of the five subgroups categorized according to the time interval from initial diagnosis to metastasis (≤3 months, >3–12 months, >12–24 months, >24 months–5 years, and >5 years) are shown in Figure 2 and Table 3.

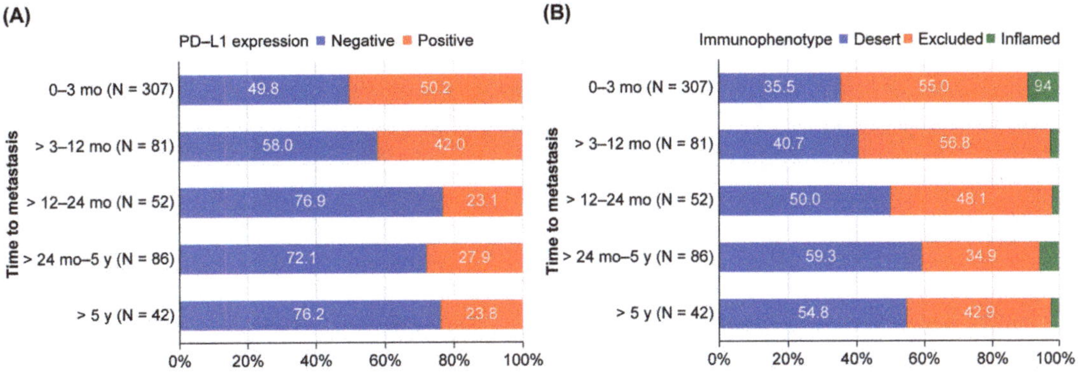

Figure 2. Distribution of immunological features by time to metastasis (100% stacked bar chart). (**A**) PD-L1 status by time from initial diagnosis to metastasis. (**B**) Immunophenotype by time from initial diagnosis to metastasis. PD-L1, programmed death ligand 1.

PD-L1 positivity was highest for time to metastasis ≤ 3 months (50.2%) and tended to decrease with increasing time of metastasis (>3–12 months: 42.0% and >12–24 months: 23.1%). However, it was similar in the groups with confirmed metastasis after 12 months (>12–24 months: 23.1%, >24 months–5 years: 27.9%, and >5 years: 23.8%; Figure 2A and Table 3). With a 12-month cutoff point for the time to metastasis, PD-L1 positivity was 48.5% in the ≤12 months group (N = 388) and 25.6% in the >12 months group (N = 180). The SD was 0.5 (Table S2).

The total number of patients with the inflamed phenotype was 38 (6.7%); therefore, the number of patients within each category was also quite small. The proportion of patients with the desert phenotype tended to increase with increasing time to metastasis (≤3 months: 35.5%, >3–12 months: 40.7%, >12–24 months: 50.0%, >24 months–5 years: 59.3%, and >5 years: 54.8%; Figure 2B and Table 3).

3.4. Pathological Features and Time from Initial Diagnosis to Metastasis

The pathological characteristics of the five subgroups categorized according to the time interval from initial diagnosis to metastasis (≤3 months, >3–12 months, >12–24 months, >24 months–5 years, and >5 years) are shown in Table 3.

The proportion of patients with different LVI and WHO ISUP grades, which were suggested to be related to metachronous/synchronous disease based on the multivariate analysis, showed a similar trend.

The proportion of patients with LVI differed between time to metastasis < 12 months and >12 months (≤3 months: 30.6% and >3–12 months: 28.4% vs. >12–24 months: 13.5%, >24 months–5 years: 15.1%, and >5 years: 16.7%). The proportion of patients with grade 1/2 tumors was also different when time to metastasis was <12 months vs. >12 months (≤3 months: 29.6% and >3–12 months: 25.9% vs. >12–24 months: 50.0%, >24 months–5 years: 59.3%, and >5 years: 61.9%; Table 3).

With a 12-month cutoff point for the time to metastasis, the proportion of patients with LVI was 30.2% in the ≤12 months group (N = 388) and 15.0% in the >12 months group (N = 180). The SD was 0.4. The proportions of patients with grade 1/2 (28.9% vs. 57.2%, SD: −0.6) and grade 4 (29.9% vs. 12.2%, SD: 0.4) tumors were also distinctly different (Table S2).

Table 3. Baseline characteristics by time from initial diagnosis to metastasis.

Characteristic, n (%)	Time from Initial Diagnosis to Metastasis					Total (N = 568)	*p*-Value [a]
	≤3 mo (N = 307)	>3–12 mo (N = 81)	>12–24 mo (N = 52)	>24 mo–5 y (N = 86)	>5 y (N = 42)		
Sex							
Male	238 (77.5)	60 (74.1)	39 (75.0)	63 (73.3)	34 (81.0)	434 (76.4)	0.832
Female	69 (22.5)	21 (25.9)	13 (25.0)	23 (26.7)	8 (19.0)	134 (23.6)	
Age							
Mean (standard deviation)	63.2 (10.9)	63.2 (11.4)	62.9 (10.0)	63.7 (9.8)	60.2 (9.5)	63.0 (10.6)	0.281
Median [range]	64.0 [23, 87]	65.0 [30, 85]	63.0 [32, 81]	66.0 [35, 81]	61.0 [38, 84]	64.0 [23, 87]	
Age category							
<65 y	164 (53.4)	39 (48.1)	31 (59.6)	41 (47.7)	29 (69.0)	304 (53.5)	0.179
≥65 and <75 y	94 (30.6)	27 (33.3)	15 (28.8)	34 (39.5)	12 (28.6)	182 (32.0)	
≥75 y	49 (16.0)	15 (18.5)	6 (11.5)	11 (12.8)	1 (2.4)	82 (14.4)	
Histology							
Clear cell	286 (93.2)	74 (91.4)	46 (88.5)	79 (91.9)	42 (100.0)	527 (92.8)	0.274
Non–clear cell	21 (6.8)	7 (8.6)	6 (11.5)	7 (8.1)	0 (0.0)	41 (7.2)	
Sarcomatoid component							
Absent	257 (83.7)	74 (91.4)	46 (88.5)	85 (98.8)	42 (100.0)	504 (88.7)	0.0002
Present	50 (16.3)	7 (8.6)	6 (11.5)	1 (1.2)	0 (0.0)	64 (11.3)	
Growth pattern							
Expansive	97 (31.6)	23 (28.4)	23 (44.2)	45 (52.3)	19 (45.2)	207 (36.4)	0.0044
Infiltrative	87 (28.3)	27 (33.3)	10 (19.2)	11 (12.8)	7 (16.7)	142 (25.0)	
Indeterminable	123 (40.1)	31 (38.3)	19 (36.5)	30 (34.9)	16 (38.1)	219 (38.6)	
Fuhrman grade							
Grade 1/2	78 (25.4)	21 (25.9)	20 (38.5)	42 (48.8)	24 (57.1)	185 (32.6)	<0.0001
Grade 3	154 (50.2)	46 (56.8)	24 (46.2)	40 (46.5)	16 (38.1)	280 (49.3)	
Grade 4	75 (24.4)	14 (17.3)	8 (15.4)	4 (4.7)	2 (4.8)	103 (18.1)	
WHO/ISUP grade							
Grade 1/2	91 (29.6)	21 (25.9)	26 (50.0)	51 (59.3)	26 (61.9)	215 (37.9)	<0.0001
Grade 3	117 (38.1)	43 (53.1)	13 (25.0)	28 (32.6)	14 (33.3)	215 (37.9)	
Grade 4	99 (32.2)	17 (21.0)	13 (25.0)	7 (8.1)	2 (4.8)	138 (24.3)	
Necrosis							
Absent	150 (48.9)	37 (45.7)	35 (67.3)	63 (73.3)	34 (81.0)	319 (56.2)	<0.0001
Present	156 (50.8)	43 (53.1)	17 (32.7)	23 (26.7)	8 (19.0)	247 (43.5)	
Indeterminable	1 (0.3)	1 (1.2)	0 (0.0)	0 (0.0)	0 (0.0)	2 (0.4)	
Lymphovascular invasion							
Absent	195 (63.5)	54 (66.7)	43 (82.7)	68 (79.1)	29 (69.0)	389 (68.5)	0.0081
Present	94 (30.6)	23 (28.4)	7 (13.5)	13 (15.1)	7 (16.7)	144 (25.4)	
Indeterminable	18 (5.9)	4 (4.9)	2 (3.8)	5 (5.8)	6 (14.3)	35 (6.2)	
TIME							
PD-L1 expression							
IC0	153 (49.8)	47 (58.0)	40 (76.9)	62 (72.1)	32 (76.2)	334 (58.8)	0.0005
IC1	88 (28.7)	22 (27.2)	9 (17.3)	18 (20.9)	7 (16.7)	144 (25.4)	
IC2	38 (12.4)	9 (11.1)	3 (5.8)	3 (3.5)	1 (2.4)	54 (9.5)	
IC3	28 (9.1)	3 (3.7)	0 (0.0)	3 (3.5)	2 (4.8)	36 (6.3)	
PD-L1 expression							
Negative [b]	153 (49.8)	47 (58.0)	40 (76.9)	62 (72.1)	32 (76.2)	334 (58.8)	<0.0001
Positive [c]	154 (50.2)	34 (42.0)	12 (23.1)	24 (27.9)	10 (23.8)	234 (41.2)	
Immunophenotype							
Desert	109 (35.5)	33 (40.7)	26 (50.0)	51 (59.3)	23 (54.8)	242 (42.6)	0.0008
Excluded	169 (55.0)	46 (56.8)	25 (48.1)	30 (34.9)	18 (42.9)	288 (50.7)	
Inflamed	29 (9.4)	2 (2.5)	1 (1.9)	5 (5.8)	1 (2.4)	38 (6.7)	

[a] *p*-values were calculated using the chi-square test for categorical variables and Kruskal-Wallis test for continuous variables. [b] Defined as IC0. [c] Defined as IC1/2/3. ISUP, International Society of Urologic Pathologists; PD-L1, programmed death ligand 1; TIME, tumor immune microenvironment; WHO, World Health Organization.

3.5. Clinical Characteristics at the Time of 1L Treatment

The clinical characteristics at the time of 1L treatment are shown in Table 4. The median age at the time of initial diagnosis was different between the groups: 64.0 [range: 23, 87] in SG and 68.0 [31, 89] in MG. The distribution of the IMDC risk groups was different between SG (favorable: 3.6%, intermediate: 64.8%, and poor: 31.6%) and MG (favorable: 39.5%, intermediate: 51.3%, and poor: 9.2%).

Table 4. Clinical characteristics at the time of first-line treatment of synchronous and metachronous metastatic renal cell carcinoma.

Characteristic, n (%)	Synchronous [a] (N = 307)	Metachronous [b] (N = 261)	Total (N = 568)	p-Value [c]	Standardized Difference
Sex					
Male	238 (77.5)	196 (75.1)	434 (76.4)	0.497	0.1
Female	69 (22.5)	65 (24.9)	134 (23.6)		−0.1
Age (year)					
Mean (standard deviation)	63.8 (10.9)	66.4 (10.3)	65.0 (10.7)	0.0032	−0.2
Median [range]	64.0 [23, 87]	68.0 [31, 89]	66.0 [23, 89]		
Age category					
<65 y	155 (50.5)	102 (39.1)	257 (45.2)	0.021	0.2
≥65 and <75 y	100 (32.6)	100 (38.3)	200 (35.2)		−0.1
≥75 y	52 (16.9)	59 (22.6)	111 (19.5)		−0.1
ECOG PS category					
0, 1	251 (81.8)	212 (81.2)	463 (81.5)	0.400	0.0
≥2	38 (12.4)	27 (10.3)	65 (11.4)		0.1
Unknown	18 (5.9)	22 (8.4)	40 (7.0)		−0.1
IMDC risk group					
Favorable	11 (3.6)	103 (39.5)	114 (20.1)	< 0.0001	−1.0
Intermediate	199 (64.8)	134 (51.3)	333 (58.6)		0.3
Poor	97 (31.6)	24 (9.2)	121 (21.3)		0.6
White blood cells					
≤ULN	253 (82.4)	231 (88.5)	484 (85.2)	0.107	−0.2
>ULN	49 (16.0)	26 (10.0)	75 (13.2)		0.2
Indeterminable	5 (1.6)	4 (1.5)	9 (1.6)		0.0
Neutrophils					
≤ULN	219 (71.3)	187 (71.6)	406 (71.5)	0.144	0.0
>ULN	63 (20.5)	42 (16.1)	105 (18.5)		0.1
Indeterminable	25 (8.1)	32 (12.3)	57 (10.0)		−0.1
Neutrophil–lymphocyte ratio					
<2.9	135 (44.0)	131 (50.2)	266 (46.8)	0.021	−0.1
≥2.9	147 (47.9)	97 (37.2)	244 (43.0)		0.2
Indeterminable	25 (8.1)	33 (12.6)	58 (10.2)		−0.1
CRP (mg/dL)					
<0.3	92 (30.0)	129 (49.4)	221 (38.9)	<0.0001	−0.4
≥0.3	199 (64.8)	114 (43.7)	313 (55.1)		0.4
Indeterminable	16 (5.2)	18 (6.9)	34 (6.0)		−0.1
Hemoglobin					
≥LLN	112 (36.5)	153 (58.6)	265 (46.7)	<0.0001	−0.5
<LLN	190 (61.9)	104 (39.8)	294 (51.8)		0.5
Indeterminable	5 (1.6)	4 (1.5)	9 (1.6)		0.0
Platelets					
≤ULN	258 (84.0)	243 (93.1)	501 (88.2)	0.0020	−0.3
>ULN	44 (14.3)	14 (5.4)	58 (10.2)		0.3
Indeterminable	5 (1.6)	4 (1.5)	9 (1.6)		0.0

Table 4. *Cont.*

Characteristic, n (%)	Synchronous [a] (N = 307)	Metachronous [b] (N = 261)	Total (N = 568)	*p*-Value [c]	Standardized Difference
Corrected serum calcium (mg/dL)					
≤10	250 (81.4)	223 (85.4)	473 (83.3)	0.058	−0.1
>10	39 (12.7)	18 (6.9)	57 (10.0)		0.2
Indeterminable	18 (5.9)	20 (7.7)	38 (6.7)		−0.1
Lactate dehydrogenase					
≤ULN × 1.5	273 (88.9)	241 (92.3)	514 (90.5)	0.370	−0.1
>ULN × 1.5	19 (6.2)	12 (4.6)	31 (5.5)		0.1
Indeterminable	15 (4.9)	8 (3.1)	23 (4.0)		0.1
Albumin					
≥LLN	95 (30.9)	128 (49.0)	223 (39.3)	<0.0001	−0.4
<LLN	199 (64.8)	123 (47.1)	322 (56.7)		0.4
Indeterminable	13 (4.2)	10 (3.8)	23 (4.0)		0.0

[a] Defined as metastasis ≤ 3 months of initial diagnosis of renal cell carcinoma. [b] Defined as metastasis diagnosed > 3 months after initial diagnosis of renal cell carcinoma. [c] *p*-values were calculated using the chi-square test for categorical variables and Kruskal-Wallis test for continuous variables. CRP, C-reactive protein; ECOG, Eastern Cooperative Oncology Group; IMDC, International Metastatic Renal Cell Carcinoma Database Consortium; LLN, lower limit of normal; PS, performance status; ULN, upper limit of normal.

Comparing the sites of metastasis in SG and MG, a higher proportion of lung (SG vs. MG, 69.1% vs. 55.6%, SD: 0.3) and lymph node (26.4% vs. 17.6%, SD: 0.2) involvement was observed in SG, while a higher proportion of pancreas (2.6% vs. 6.5%, SD: −0.2) involvement was observed in MG (Table S3).

The median OS after 1L VEGFRi treatment was 29.5 months (95% CI: 25.1, 32.5) in SG and 44.2 months (95% CI: 36.5, 51.1) in MG (HR, 0.74 [95% CI: 0.60, 0.91]) (Figure S2).

4. Discussion

Metastasis is seen in synchronous mRCC at the time of diagnosis, while metachronous metastasis develops after nephrectomy for localized RCC. In this study, we investigated the differences in PD-L1 expression and immunophenotype of tumor-infiltrating immune cells in primary lesions, mainly between SG and MG, based on time intervals to metastasis and its association with clinicopathological parameters in patients with mRCC.

4.1. The Difference in Baseline Characteristics between SG and MG

At the time of initial diagnosis, a higher proportion of poor prognostic pathological features were seen in SG compared with MG (WHO/ISUP grade 4, necrosis, LVI, infiltrative growth pattern, and sarcomatoid differentiation). Similarly, regarding the TIME, more immunogenic features were observed in SG than in MG: PD-L1 positivity (50.2% vs. 30.7%, SD: 0.4), proportion of desert phenotype (35.5% vs. 51.0%, SD: −0.3), and inflamed phenotype (9.4% vs. 3.4%, SD: 0.2). These pathological/immunological differences between SG and MG were consistent with previous reports and support poor prognosis in SG compared with MG [2,5,6].

4.2. Association between TIME and Time to Metastasis

Several previous reports have shown an association between TIME and pathological grade; T-cell infiltration with the expression of several checkpoint proteins (e.g., PD-1 and PD-L1/2) was observed to be associated with a more aggressive phenotype, higher pathological grade, and sarcomatoid differentiation [18–21]. In addition, T-cell infiltrating tumor types with checkpoint protein expression were considered to have a worse prognosis than those with the immune desert phenotype in both metastatic and localized RCC [20,22,23]. We observed that PD-L1 positivity was associated with a shorter time to metastasis. Among immunophenotypes, the desert phenotype was associated with a longer time to metastasis than the excluded phenotype. Moreover, an association between PD-L1 expression, immunophenotype, and ISUP grade was observed. Thus, our observations were consistent with those of previous reports [18–23].

In most other tumors, including melanoma, colorectal cancer, and lung cancer, the immunogenic features of the TIME (tumor infiltration by CD8+ T cells; inflamed phenotype) are considered to be markers of good prognosis [24,25], while they are associated with poor prognosis in RCC [20]. Although this is thought to be due to the strong association between the TIME and pathological grade in RCC, a more detailed study on the TIME is needed.

We examined the association between the TIME and synchronous/metachronous disease, including other pathological features. PD-L1 expression, WHO/ISUP grade, and LVI were associated with synchronous/metachronous disease. To our knowledge, this is the first report to show this association, and we consider the results to be a recommendation to evaluate each of the TIME and pathological features for better understanding of synchronous/metachronous RCC.

4.3. Clinical Importance of Time to Metastasis and TIME

We observed that PD-L1 positivity differed between metastases within 12 months and metastases > 12 months. However, this observation was not very obvious in the distribution of the immunophenotypes.

A trend of differential prevalence within and after 12 months of time to metastasis was also observed for pathological features other than PD-L1 expression. LVI and WHO/ISUP grade 1/2, which were selected in the synchronous/metachronous logistic model, showed similar trends as PD-L1 expression. All SDs with a 12-month cutoff point for the time to metastasis supported this observation.

These results suggest that even if a metastasis is categorized as metachronous, if the time to metastasis is <12 months, the tumor environment is close to that of a synchronous metastasis. Specifically, the tumor environment of a metastasis that develops within 12 months is immunologically more tumor-infiltrating lymphocyte (TIL)-infiltrated and PD-L1-expressing and pathologically of a higher grade than the tumor environment of a metastasis that develops after 12 months.

The MSKCC risk classification in the cytokine era [3] and the IMDC risk classification in the targeted therapy era [4] are most widely used in the prognostic factor models of mRCC. Both classifications include "time from diagnosis to treatment of less than 12 months" as a risk factor. The "time from diagnosis to treatment" is "the time from initial diagnosis to confirmation of metastasis." Therefore, the observation that both PD-L1 positivity and pathological features tended to differ at about 12 months in our study might support the setting of 12 months as the cutoff point in IMDC and MSKCC. The IMDC/MSKCC criteria do not include pathological grade or other pathological features that are considered as prognostic factors, but this may be because of employing the 12-month cutoff as a factor to explain pathological features.

4.4. Clinical Importance of TIME in the Current IO Era

The rapid evolution from TKI to immunotherapy with checkpoint inhibitors has dramatically changed treatment outcomes in mRCC [9]. The results from different studies also suggest a trend toward different subgroups of patients benefiting from checkpoint inhibitors and VEGFRi. In the post hoc analysis of the CheckMate 214 trial (nivolumab plus ipilimumab vs. sunitinib), although the overall response rate (ORR) of sunitinib decreased as the number of risk factors increased (50–16%), the ORR of nivolumab plus ipilimumab was consistent across any number of IMDC risk factors (ranging from 39% to 44%) [26]. Furthermore, while longer OS and a higher ORR were observed with nivolumab plus ipilimumab than with sunitinib across tumor PD-L1 expression levels, the magnitude of benefit was higher in the population with 1% or greater PD-L1 expression [7]. In addition, in RCC with sarcomatoid histology, which is known to be associated with a poor response to VEGFRi, both avelumab plus axitinib and nivolumab plus ipilimumab showed better efficacy outcomes compared with sunitinib in the post hoc analysis [8,27].

The post hoc biomarker analysis of iMmotion150 (atezolizumab plus bevacizumab, atezolizumab vs. sunitinib) and iMmotion151 (atezolizumab plus bevacizumab vs. sunitinib) based on gene expression associated with "tumor angiogenesis," "pre-existing immunity," and "immunosuppressive myeloid inflammation" showed that these signatures were differentially associated with progression-free survival (PFS)/OS across treatments [21,28]. Favorable trends of PFS/OS for atezolizumab plus bevacizumab were observed in tumors characterized by high T-effector gene expression, but not in tumors characterized by high angiogenic gene expression [21,28]. The immunophenotype evaluated in our study assessed the localization of TILs relative to the tumor area [20] and may represent the status of the pre-existing immunity.

In summary, IMDC risk, pathological features (sarcomatoid histology and tumor angiogenesis), and the TIME (status of pre-existing immunity, PD-L1 expression, and T-effector gene expression) are considered key factors in selecting 1L treatment for mRCC. With the advent of IO combination therapies, it is important to select treatment according to tumor type [29], but the TIME has not yet been evaluated in clinical practice.

Our study showed that the TIME of synchronous metastasis or tumors with shorter time to recurrence was immunologically more TIL-infiltrated and PD-L1-expressing and pathologically of a higher grade than that of metachronous metastases. We also observed that the IMDC risk distribution at the time of the 1L therapy was worse in SG than in MG (favorable/intermediate/poor: 3.6%/64.8%/31.6% vs. 39.5%/51%/9.2%), and the OS under VEGFRi therapy was shorter in SG than in MG (median OS: 29.5 months vs. 44.2 months, HR: 0.74 [95% CI: 0.60, 0.91]). Thus, our study showed that the timing of metastasis is related to the TIME of the primary tumor and OS under VEGFRi treatment. As immune markers have not been evaluated in the current clinical practice of RCC, it is possible that the timing of metastasis reflects the TIME and may be useful in selecting 1L treatment.

4.5. Limitations

This study had several limitations. First, the patients included in this study underwent nephrectomy and received VEGFRi as 1L systemic therapy. However, since we evaluated the time to metastasis before systemic therapy and not OS, the results were not affected by systemic therapy. The response to IO drugs is unknown in this cohort. In this study, surgical specimens were used in all patients, and about half of the specimens were from cytoreductive nephrectomy. However, considering the results of the CARMENA study [30] and the fact that there are a certain number of patients who cannot undergo surgery, we thought it necessary to examine the use of needle biopsy specimens. Second, we analyzed the immune status of the primary tumors. The TIME of metastatic lesions, which is the target of systemic therapy, especially in the MG, might be different from the TIME of the primary lesions at the time of nephrectomy. Only patients who underwent radical nephrectomy were included in the SG group. The TIME of SG patients who cannot undergo radical nephrectomy might be different from that of patients who can undergo cytoreductive nephrectomy. Third, owing to the retrospective nature of the study, unmeasured confounding factors and/or selection bias could have affected the study results. Fourth, PD-L1 expression status was analyzed in tumor-infiltrating immune cells but not in tumor cells. Factors that were not considered in this study, such as PD-L1 expression status in tumor cells and gene expression [31,32] or mutation analysis, should be investigated in the future.

5. Conclusions

In conclusion, this is the first study to examine the differences in the TIME characteristics of primary lesions in patients with mRCC based on the time to metastasis. We demonstrated that PD-L1 expression in tumor-infiltrating immune cells and immunophenotypes of the primary lesions differed according to the time to recurrence. A synchronous or shorter time to recurrence was associated with increased PD-L1 expression in tumor-infiltrating immune cells, more inflamed and excluded phenotypes, and fewer desert phenotypes. In the IO era, an understanding of the TIME of primary lesions might provide useful insights regarding the choice of 1L systemic therapy in patients with mRCC.

Supplementary Materials: The following supporting information can be downloaded at: https://www.mdpi.com/article/10.3390/cancers14215258/s1, Table S1: Association between PD-L1 status, immunophenotype, and WHO/ISUP grade; Table S2: Comparison of baseline characteristics at the time of initial diagnosis by time to metastasis (≤12 months vs. >12 months); Table S3: Metastatic sites of synchronous and metachronous metastatic renal cell carcinoma; Figure S1: Study population; Figure S2: Overall survival after first-line tyrosine kinase inhibitor treatment by time from diagnosis to metastasis (synchronous/metachronous).

Author Contributions: Conceptualization, K.F. and G.K.; data curation, T.T.; formal analysis, Y.N.; funding acquisition, T.F. (Tamaki Fukuyama); investigation, K.F., G.K., T.T., T.K., E.B., A.K., R.Y., Y.M., D.I., T.F. (Tomoya Fukawaand) and Y.K.; methodology, Y.N.; project administration, F.S.; supervision, H.U.; writing—original draft, K.F., G.K., Y.N. and T.F. (Tamaki Fukuyama); writing—review & editing, all authors. All authors have read and agreed to the published version of the manuscript.

Funding: This study was funded by Chugai Pharmaceuticals Co., Ltd.

Institutional Review Board Statement: This study was registered in the UMIN Clinical Trials Registry (UMIN000034131) and conducted with the approval of the institutional review board of the 29 study facilities. Furthermore, we obtained approval from the institutional review board of MINS (Reference number: 210204; approval date: 18 February 2021), a nonprofit organization. This study was conducted in accordance with the principles of the Declaration of Helsinki.

Informed Consent Statement: Informed consent was obtained from all the participants involved in the study.

Data Availability Statement: Qualified researchers may request access to individual patient-level data through the clinical study data request platform (www.clinicalstudydatarequest.com). For further details on Chugai's Data Sharing Policy and how to request access to related clinical study documents, see here (www.chugai-pharm.co.jp/english/profile/rd/ctds_request.html).

Acknowledgments: The authors thank the patients who participated in the trial and their families, as well as the investigators and staff at all participating centers. We also thank Tetsutaro Hamano (P4 Statistics Co., Ltd.) for performing all the statistical analysis. Ohe Chisato, (Department of Pathology and Laboratory Medicine, Kansai Medical University, Japan) contributed to the ARCHERY study by assessing PD-L1 expression of samples. Editorial support for this manuscript was provided by Deepa Sundararajan of Cactus Life Sciences (part of Cactus Communications, Mumbai, India).

Conflicts of Interest: K.F. reports grants or contracts from Pfizer Japan Inc. and Yakult Bio-Science and payment or honoraria for lectures, presentations, speakers' bureaus, manuscript writing, or educational events from Astellas Pharma Inc., AstraZeneca K.K., Bayer Pharmaceutical Ltd., Bristol-Myers Squibb K.K., Janssen Pharmaceuticals, Kissei Pharmaceutical, MSD K.K., Novartis Pharma K.K., Ono Pharmaceutical Co., Ltd., Pfizer Japan Inc., and Takeda Pharmaceutical Company Limited. G.K. reports payment or honoraria for lectures, presentations, speakers' bureaus, manuscript writing, or educational events from Bayer Yakuhin Ltd., Bristol-Myers Squibb K.K., Chugai Pharmaceutical Co., Ltd., Eisai Co., Ltd., Merck Biopharma, MSD K.K., Ono Pharmaceutical Co., Ltd., Sanofi K.K., and Takeda Pharmaceutical Company Limited. T.T. reports funding from Chugai Pharmaceutical Co., Ltd. Y.N. holds stock or stock options and is a full-time employee of Chugai Pharmaceutical Co., Ltd. T.F. (Tamaki Fukuyama) is a full-time employee of Chugai Pharmaceutical Co., Ltd. F.S. holds stock or stock options and is a full-time employee of Chugai Pharmaceutical Co., Ltd. H.U. reports grants from Astellas Pharma Inc., Daiichi-Sankyo Co., Ltd., Janssen Pharmaceuticals, Kissei Pharmaceutical, Nippon Shinyaku Co., Ono Pharmaceutical Co., Ltd., Sanofi K.K., Taiho Pharmaceutical Co., Ltd., and Takeda Pharmaceutical Company Limited; contracts from AstraZeneca K.K.; consulting fees from AstraZeneca K.K., Bayer Pharmaceutical Ltd., Chugai Pharmaceutical Co., Ltd., Janssen Pharmaceuticals, MSD K.K., Ono Pharmaceutical Co., Ltd., Pfizer Japan Inc., Sanofi K.K., and Takeda Pharmaceutical Company Limited; and speakers' bureau fees from Astellas Pharma Inc., AstraZeneca K.K., Bayer Pharmaceutical Ltd., Bristol-Myers Squibb K.K., Janssen Pharmaceuticals,

MSD K.K., Ono Pharmaceutical Co., Ltd., Pfizer Japan Inc., and Sanofi K.K., T.K., E.B., A.K., R.Y., Y.M., D.I., T.F. (Tomoya Fukawaand) and Y.K. declare no conflicts of interest.

References

1. Dabestani, S.; Thorstenson, A.; Lindblad, P.; Harmenberg, U.; Ljungberg, B.; Lundstam, S. Renal cell carcinoma recurrences and metastasis in primary non-metastatic patients: A population-based study. *World J. Urol.* **2016**, *34*, 1081–1086. [CrossRef] [PubMed]
2. Donskov, F.; Xie, W.; Overby, A.; Wells, J.C.; Fraccon, A.P.; Sacco, C.S.; Porta, C.; Stukalin, I.; Lee, J.L.; Koutsoukos, K.; et al. Synchronous versus metachronous metastatic disease: Impact of time to metastasis on patient outcome-Results from the International Metastatic Renal Cell Carcinoma Database Consortium. *Eur. Urol. Oncol.* **2020**, *3*, 530–539. [CrossRef] [PubMed]
3. Motzer, R.J.; Mazumdar, M.; Bacik, J.; Berg, W.; Amsterdam, A.; Ferrara, J. Survival and prognostic stratification of 670 patients with advanced renal cell carcinoma. *J. Clin. Oncol.* **1999**, *17*, 2530–2540. [CrossRef] [PubMed]
4. Heng, D.Y.; Xie, W.; Regan, M.M.; Harshman, L.C.; Bjarnason, G.A.; Vaishampayan, U.N.; Mackenzie, M.; Wood, L.; Donskov, F.; Tan, M.H.; et al. External validation and comparison with other models of the International Metastatic Renal-Cell Carcinoma Database Consortium prognostic model: A population-based study. *Lancet Oncol.* **2013**, *14*, 141–148. [CrossRef]
5. Kammerer-Jacquet, S.F.; Brunot, A.; Pladys, A.; Bouzille, G.; Dagher, J.; Medane, S.; Peyronnet, B.; Mathieu, R.; Verhoest, G.; Bensalah, K.; et al. Synchronous metastatic clear-cell renal cell carcinoma: A distinct morphologic, immunohistochemical, and molecular phenotype. *Clin. Genitourin. Cancer* **2017**, *15*, e1–e7. [CrossRef]
6. Turajlic, S.; Xu, H.; Litchfield, K.; Rowan, A.; Chambers, T.; Lopez, J.I.; Nicol, D.; O'Brien, T.; Larkin, J.; Horswell, S.; et al. Tracking cancer evolution reveals constrained routes to metastasis: TRACERx Renal. *Cell* **2018**, *173*, 581–594.e12. [CrossRef] [PubMed]
7. Motzer, R.J.; Tannir, N.M.; McDermott, D.F.; Arén Frontera, O.; Melichar, B.; Choueiri, T.K.; Plimack, E.R.; Barthélémy, P.; Porta, C.; George, S.; et al. Nivolumab plus ipilimumab versus sunitinib in advanced renal-cell carcinoma. *N. Engl. J. Med.* **2018**, *378*, 1277–1290. [CrossRef]
8. Choueiri, T.K.; Larkin, J.; Pal, S.; Motzer, R.J.; Rini, B.I.; Venugopal, B.; Alekseev, B.; Miyake, H.; Gravis, G.; Bilen, M.A.; et al. Efficacy and correlative analyses of avelumab plus axitinib versus sunitinib in sarcomatoid renal cell carcinoma: Post hoc analysis of a randomized clinical trial. *ESMO Open* **2021**, *6*, 100101. [CrossRef]
9. Tran, J.; Ornstein, M.C. Clinical Review on the management of metastatic renal cell carcinoma. *JCO Oncol. Pract.* **2022**, *18*, 187–196. [CrossRef]
10. Yu, E.M.; Linville, L.; Rosenthal, M.; Aragon-Ching, J.B. A Contemporary Review of Immune Checkpoint Inhibitors in Advanced Clear Cell Renal Cell Carcinoma. *Vaccines* **2021**, *9*, 919. [CrossRef]
11. Choueiri, T.K.; Tomczak, P.; Park, S.H.; Venugopal, B.; Ferguson, T.; Chang, Y.H.; Hajek, J.; Symeonides, S.N.; Lee, J.L.; Powles, T. KEYNOTE-564 Investigators. Adjuvant Pembrolizumab after Nephrectomy in Renal-Cell Carcinoma. *N. Engl. J. Med.* **2021**, *385*, 683–694. [CrossRef]
12. Baghban, R.; Roshangar, L.; Jahanban-Esfahlan, R.; Seidi, K.; Ebrahimi-Kalan, A.; Jaymand, M.; Kolahian, S.; Javaheri, T.; Zare, P. Tumor microenvironment complexity and therapeutic implications at a glance. *Cell Commun. Signal.* **2020**, *18*, 59. [CrossRef] [PubMed]
13. Díaz-Montero, C.M.; Rini, B.I.; Finke, J.H. The immunology of renal cell carcinoma. *Nat. Rev. Nephrol.* **2020**, *16*, 721–735. [CrossRef] [PubMed]
14. Uemura, M.; Nakaigawa, N.; Sassa, N.; Tatsugami, K.; Harada, K.; Yamasaki, T.; Matsubara, N.; Yoshimoto, T.; Nakagawa, Y.; Fukuyama, T.; et al. Prognostic value of programmed death-ligand 1 status in Japanese patients with renal cell carcinoma. *Int J. Clin. Oncol.* **2021**, *26*, 2073–2084. [CrossRef] [PubMed]
15. Herbst, R.S.; Soria, J.C.; Kowanetz, M.; Fine, G.D.; Hamid, O.; Gordon, M.S.; Sosman, J.A.; McDermott, D.F.; Powderly, J.D.; Gettinger, S.N.; et al. Predictive correlates of response to the anti-PD-L1 antibody MPDL3280A in cancer patients. *Nature* **2014**, *515*, 563–567. [CrossRef] [PubMed]
16. Mariathasan, S.; Turley, S.J.; Nickles, D.; Castiglioni, A.; Yuen, K.; Wang, Y.; Kadel, E.E., III; Koeppen, H.; Astarita, J.L.; Cubas, R.; et al. TGFβ attenuates tumour response to PD-L1 blockade by contributing to exclusion of T cells. *Nature* **2018**, *554*, 544–548. [CrossRef] [PubMed]
17. Choi, S.Y.; Yoo, S.; You, D.; Jeong, I.G.; Song, C.; Hong, B.; Hong, J.H.; Ahn, H.; Kim, C.S. Prognostic factors for survival of patients with synchronous or metachronous brain metastasis of renal cell carcinoma. *Clin. Genitourin. Cancer* **2017**, *15*, 717–723. [CrossRef]
18. Kawashima, A.; Kanazawa, T.; Kidani, Y.; Yoshida, T.; Hirata, M.; Nishida, K.; Nojima, S.; Yamamoto, Y.; Kato, T.; Hatano, K.; et al. Tumour grade significantly correlates with total dysfunction of tumour tissue-infiltrating lymphocytes in renal cell carcinoma. *Sci. Rep.* **2020**, *10*, 6220. [CrossRef]
19. Kawakami, F.; Sircar, K.; Rodriguez-Canales, J.; Fellman, B.M.; Urbauer, D.L.; Tamboli, P.; Tannir, N.M.; Jonasch, E.; Wistuba, I.I.; Wood, C.G.; et al. Programmed cell death ligand 1 and tumor-infiltrating lymphocyte status in patients with renal cell carcinoma and sarcomatoid dedifferentiation. *Cancer* **2017**, *123*, 4823–4831. [CrossRef]
20. Giraldo, N.A.; Becht, E.; Pagès, F.; Skliris, G.; Verkarre, V.; Vano, Y.; Mejean, A.; Saint-Aubert, N.; Lacroix, L.; Natario, I.; et al. Orchestration and prognostic significance of immune checkpoints in the microenvironment of primary and metastatic renal cell cancer. *Clin. Cancer Res.* **2015**, *21*, 3031–3040. [CrossRef]

21. McDermott, D.F.; Huseni, M.A.; Atkins, M.B.; Motzer, R.J.; Rini, B.I.; Escudier, B.; Fong, L.; Joseph, R.W.; Pal, S.K.; Reeves, J.A.; et al. Clinical activity and molecular correlates of response to atezolizumab alone or in combination with bevacizumab versus sunitinib in renal cell carcinoma. *Nat. Med.* **2018**, *24*, 749–757. [CrossRef] [PubMed]
22. Shen, M.; Chen, G.; Xie, Q.; Li, X.; Xu, H.; Wang, H.; Zhao, S. Association between PD-L1 expression and the prognosis and clinicopathologic features of renal cell carcinoma: A systematic review and meta-analysis. *Urol. Int.* **2020**, *104*, 533–541. [CrossRef]
23. Giraldo, N.A.; Becht, E.; Vano, Y.; Petitprez, F.; Lacroix, L.; Validire, P.; Sanchez-Salas, R.; Ingels, A.; Oudard, S.; Moatti, A.; et al. Tumor-infiltrating and peripheral blood T-cell immunophenotypes predict early relapse in localized clear cell renal cell carcinoma. *Clin. Cancer Res.* **2017**, *23*, 4416–4428. [CrossRef] [PubMed]
24. Fridman, W.H.; Pagès, F.; Sautès-Fridman, C.; Galon, J. The immune contexture in human tumours: Impact on clinical outcome. *Nat. Rev. Cancer* **2012**, *12*, 298–306. [CrossRef] [PubMed]
25. Barnes, T.A.; Amir, E. HYPE or HOPE: The prognostic value of infiltrating immune cells in cancer. *Br. J. Cancer* **2017**, *117*, 451–460. [CrossRef] [PubMed]
26. Escudier, B.; Motzer, R.J.; Tannir, N.M.; Porta, C.; Tomita, Y.; Maurer, M.A.; McHenry, M.B.; Rini, B.I. Efficacy of nivolumab plus ipilimumab according to number of IMDC risk factors in CheckMate 214. *Eur. Urol.* **2020**, *77*, 449–453. [CrossRef] [PubMed]
27. Tannir, N.M.; Signoretti, S.; Choueiri, T.K.; McDermott, D.F.; Motzer, R.J.; Flaifel, A.; Pignon, J.C.; Ficial, M.; Frontera, O.A.; George, S.; et al. Efficacy and safety of nivolumab plus ipilimumab versus sunitinib in first-line treatment of patients with advanced sarcomatoid renal cell carcinoma. *Clin. Cancer Res.* **2021**, *27*, 78–86. [CrossRef] [PubMed]
28. Motzer, R.J.; Powles, T.; Atkins, M.B.; Escudier, B.; McDermott, D.F.; Alekseev, B.Y.; Lee, J.L.; Suarez, C.; Stroyakovskiy, D.; de Giorgi, U.; et al. Final overall survival and molecular analysis in IMmotion151, a phase 3 trial comparing atezolizumab plus bevacizumab vs. sunitinib in patients with previously untreated metastatic renal cell carcinoma. *JAMA Oncol.* **2022**, *8*, 275–280. [CrossRef]
29. Motzer, R.J.; Choueiri, T.K.; McDermott, D.F.; Powles, T.; Vano, Y.A.; Gupta, S.; Yao, J.; Han, C.; Ammar, R.; Papillon-Cavanagh, S.; et al. Biomarker analysis from CheckMate 214: Nivolumab plus ipilimumab versus sunitinib in renal cell carcinoma. *J. Immunother. Cancer* **2022**, *10*, e004316. [CrossRef]
30. Méjean, A.; Ravaud, A.; Thezenas, S.; Colas, S.; Beauval, J.B.; Bensalah, K.; Geoffrois, L.; Thiery-Vuillemin, A.; Cormier, L.; Escudier, B. Sunitinib Alone or after Nephrectomy in Metastatic Renal-Cell Carcinoma. *N. Engl. J. Med.* **2018**, *379*, 417–427. [CrossRef]
31. Roldán, F.L.; Izquierdo, L.; Ingelmo-Torres, M.; Lozano, J.J.; Carrasco, R.; Cuñado, A.; Reig, O.; Mengual, L.; Alcaraz, A. Prognostic Gene Expression-Based Signature in Clear-Cell Renal Cell Carcinoma. *Cancers* **2022**, *14*, 3754. [CrossRef] [PubMed]
32. Shi, Z.; Zheng, J.; Liang, Q.; Liu, Y.; Yang, Y.; Wang, R.; Wang, M.; Zhang, Q.; Xuan, Z.; Sun, H.; et al. Identification and Validation of a Novel Ferroptotic Prognostic Genes-Based Signature of Clear Cell Renal Cell Carcinoma. *Cancers* **2022**, *14*, 4690. [CrossRef] [PubMed]

cancers

MDPI

Review

Understanding the Tumor Immune Microenvironment in Renal Cell Carcinoma

Daniel D. Shapiro [1,2,*,†], Brendan Dolan [1,†], Israa A. Laklouk [3], Sahar Rassi [1], Taja Lozar [4], Hamid Emamekhoo [5], Andrew L. Wentland [6], Meghan G. Lubner [6] and Edwin Jason Abel [1]

1 Department of Urology, The University of Wisconsin School of Medicine and Public Health, Madison, WI 53726, USA
2 Division of Urology, William S. Middleton Memorial Veterans Hospital, Madison, WI 53705, USA
3 Department of Pathology, The University of Wisconsin School of Medicine and Public Health, Madison, WI 53705, USA
4 McArdle Laboratory for Cancer Research, Department of Oncology, University of Wisconsin School of Medicine and Public Health, Madison, WI 53726, USA
5 Department of Medical Oncology, The University of Wisconsin School of Medicine and Public Health, Madison, WI 53726, USA
6 Department of Radiology, The University of Wisconsin School of Medicine and Public Health, Madison, WI 53726, USA
* Correspondence: ddshapiro@wisc.edu; Tel.: +1-920-740-3911
† These authors contributed equally to this work.

Simple Summary: Over the last decade there has been a significant increase in the number of therapies that activate the host's immune system to target and eliminate renal cell carcinoma (RCC) tumors. The superior efficacy of these agents demonstrates that host control of RCC requires a robust and effective immune response. Given the increasing incorporation of immune activating agents into routine clinical practice for the management of RCC, it is important for researchers and clinicians to understand the characteristics of the immune microenvironment in RCC tumors. The purpose of this review is to describe the concepts of the anti-tumor immune response to RCC and to provide a detailed summary of the current understanding of the immune response to RCC tumor development and progression. Additionally, this article explores how components of the immune microenvironment are being used to predict response to therapy and patient survival.

Abstract: Scientific understanding of how the immune microenvironment interacts with renal cell carcinoma (RCC) has substantially increased over the last decade as a result of research investigations and applying immunotherapies, which modulate how the immune system targets and eliminates RCC tumor cells. Clinically, immune checkpoint inhibitor therapy (ICI) has revolutionized the treatment of advanced clear cell RCC because of improved outcomes compared to targeted molecular therapies. From an immunologic perspective, RCC is particularly interesting because tumors are known to be highly inflamed, but the mechanisms underlying the inflammation of the tumor immune microenvironment are atypical and not well described. While technological advances in gene sequencing and cellular imaging have enabled precise characterization of RCC immune cell phenotypes, multiple theories have been suggested regarding the functional significance of immune infiltration in RCC progression. The purpose of this review is to describe the general concepts of the anti-tumor immune response and to provide a detailed summary of the current understanding of the immune response to RCC tumor development and progression. This article describes immune cell phenotypes that have been reported in the RCC microenvironment and discusses the application of RCC immunophenotyping to predict response to ICI therapy and patient survival.

Keywords: renal cell carcinoma; immune microenvironment; tumor immunology; immune checkpoint inhibitor; immunotherapy

Citation: Shapiro, D.D.; Dolan, B.; Laklouk, I.A.; Rassi, S.; Lozar, T.; Emamekhoo, H.; Wentland, A.L.; Lubner, M.G.; Abel, E.J. Understanding the Tumor Immune Microenvironment in Renal Cell Carcinoma. *Cancers* **2023**, *15*, 2500. https://doi.org/10.3390/cancers15092500

Academic Editor: José I. López

Received: 31 March 2023
Revised: 24 April 2023
Accepted: 24 April 2023
Published: 27 April 2023

1. Introduction

In 2022, renal cell carcinoma (RCC) accounted for most of the estimated 79,000 new diagnoses of kidney cancer and 13,920 deaths, making it one of the most common malignancies in the United States [1,2]. For clinically localized or locally advanced disease, surgery is the primary treatment, but tumors with aggressive features, such as high tumor grade, presence of tumor thrombus, tumor necrosis and peritumoral fat invasion, may have higher rates of metastatic progression [3,4]. Consequently, RCC has been the focus of tumor microenvironment (TME) research to understand mechanisms of pathogenesis and to inform treatment approaches. Previously, angiogenic components of the TME in RCC emerged as promising targets, with improved survival demonstrated in metastatic RCC (mRCC) using tyrosine kinase inhibitors [5]. More recently, immune checkpoint inhibitors (ICI) have shown superior outcomes compared to tyrosine kinase inhibitor and mTOR inhibitor therapies in advanced RCC clinical trials [6–9]. Understanding the immunologic aspects of the RCC TME will be important to developing new therapeutic targets and prognostic signatures. Although ICIs have shown success for RCC treatment [10,11], our understanding of the mechanisms of anti-tumor immune response and immune dysregulation in RCC remains limited. This review aims to summarize our understanding of the basic concepts of RCC tumor-immune biology and discusses in detail the attempts to use components of the RCC immune microenvironment as prognostic and predictive biomarkers.

2. The Immune Response to Cancer

The tumor immune microenvironment is shaped by the response of the immune system to tumor cell development, which is distinct from the response to external pathogens. For example, recognition of "non-self" antigens on bacteria allows targeting and removal while avoiding autoimmunity. However, tumor cells are unique because non-synonymous somatic mutations of host genes produce "neoantigens" that may be recognized by immune cells as foreign, triggering a host immune response, which has been described as the "cancer-immunity cycle" (Figure 1) [12]. Tumor cell recognition is initiated primarily by dendritic cells (DCs), which act as the main antigen presenting cells (APCs) within the body. Neoantigens are captured and processed by DCs in the tumor microenvironment, which has migrated to the tumor under the influence of pro-inflammatory signals [13]. Certain types of mutations, such as frameshift mutations, may result in significantly altered peptide sequences leading to more robust immune responses [14]. Measures of tumor mutational burden (TMB) have been used as an indirect measure of neoantigen production and have been found to be associated with survival and improved response to ICI therapy in some tumors, such as melanoma and NSCLC; however, prior studies have suggested that TMB is less predictive of ICI success in RCC [14,15].

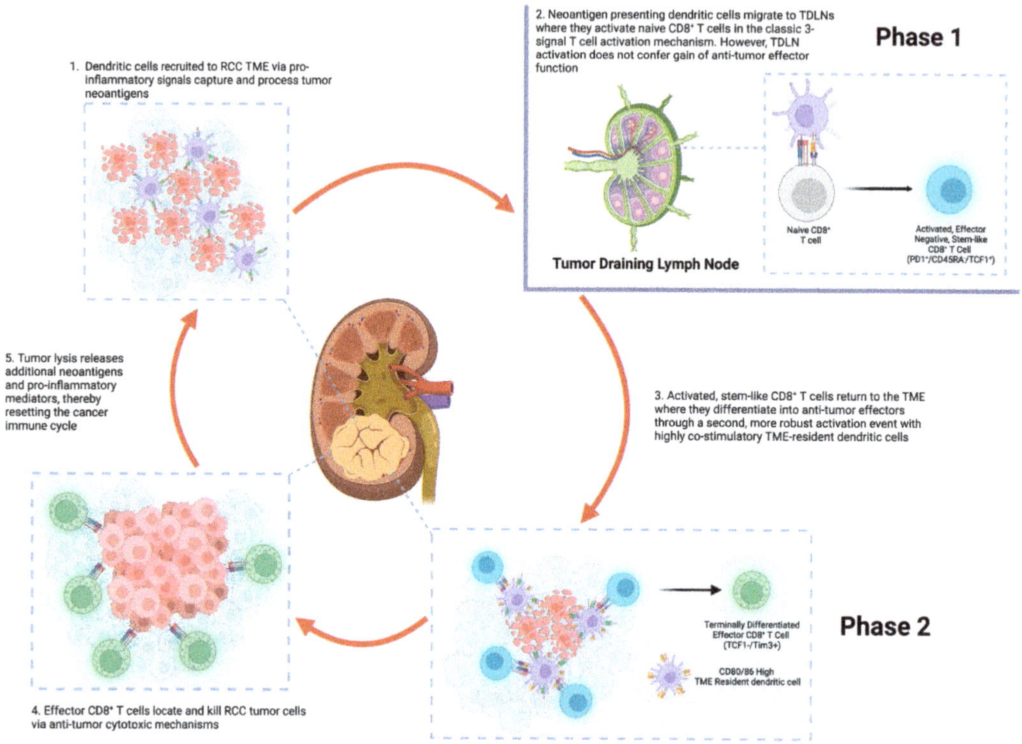

Figure 1. The renal cell carcinoma cancer immune cycle. Depiction of the immune response cycle to renal cell carcinoma. Dendritic cells (DCs) and other antigen presenting cells (APCs) are recruited to the tumor microenvironment where renal cell carcinoma (RCC) tumor neoantigens are captured, processed and then presented by DCs and APCs on MHC I and MHC II molecules (step 1). DCs and other APCs migrate to tumor draining lymph nodes (TDLNs), where they present tumor neoantigens to T cells (step 2). Naïve CD8$^+$ T cells undergo a two-phase process to gain effector functions (as proposed by Prokhnevska et al. [16]). In phase 1, naïve CD8$^+$ T cells become activated stem-like CD8$^+$ T cells, which migrate to the tumor microenvironment and have the ability to replicate but do not yet possess effector functions. In phase 2, the stem-like CD8$^+$ T cells differentiate into effector cells by a second co-stimulatory signal from tumor resident DCs. These terminally differentiated effector CD8$^+$ T cells target and destroy tumor cells (step 4). This releases more RCC neoantigens, further potentiating the cancer immune cycle (step 5).

After neoantigen capture and processing by DCs and other APCs, cells migrate to tumor draining lymph nodes and present neoantigens to T cells, which become primed and activated leading to effector T cell responses. One critical step of the cancer-immunity cycle is the adaptive immune system recognizing that neoantigens are foreign and bypassing central tolerance. Activation of T cells triggers both effector and regulatory responses, and the balance of these two responses determines the overall tumor destruction. After activation, effector cells infiltrate into the tumor through specialized vascular endothelial cells with the aid of pro-inflammatory cytokines and chemokines. Accordingly, T cells recognize cancer cells expressing neoantigens loaded on MHC class I molecules with the T cell receptor (TCR) which ultimately results in killing the cancer cell [12,13]. CD8$^+$ T cells release death-inducing granules containing granzymes and perforin which can create cancer cell membrane pores allowing enzyme entry and cell death [17]. Additionally, CD8$^+$ can induce apoptosis by activating the FasL/Fas pathway, which activates caspases and endonucleases, leading to fragmentation of cancer cell DNA [17]. Tumor cell death releases more tumor neoantigens, which stimulates and reinforces the cancer-immune cycle. Acting as a negative balance or checkpoint, immune regulatory signals can dampen the host adaptive immune response and produce a pro-tumorigenic microenvironment. For example, expression of PD-L1 on tumor cells binds to PD-1 on CD8+ T cells resulting in reduced cytotoxicity, proliferation and response to TCR stimulation [13]. Inhibition of negative immune regulatory responses (by blocking PD-1 or CTLA-4 signaling) is the central tenant of ICI therapy and has been successful in improving immune destruction and clearance of multiple tumor types including RCC.

The recognition of tumor neoantigens and creation of an effector response within the tumor draining lymph nodes has been dogma and extrapolated from research investigating response to viral infections. More recent studies suggest that the adaptive immune response to tumors may be more complex. Prokhnevska et al. demonstrated a two-phase process in a mouse model of RCC [16]. First, $CD8^+$ T cells are primed within the tumor draining lymph node. These stem-like $CD8^+$ T cells (expressing TCF1 and PD-1) proliferate but do not express effector molecules associated with tumor cell killing. The stem-like $CD8^+$ T cells then migrate to the tumor via chemokine signaling and express co-stimulatory receptors, maintaining a stem-like phenotype until a second phase of co-stimulation by APCs generates an effector transcriptional program necessary for tumor cell killing (Figure 1) [16]. This two phase effector T cell activation is thought to have implications for ICI therapy. Blocking PD-1 may promote activation and proliferation of stem-like $CD8^+$ T cells, but a second stimulatory signal is required from APCs within the TME, which leads to an effector $CD8^+$ T cell response and tumor killing. The necessity of a second stimulatory signal could help explain why there is significant variation in responses to anti-PD-1/PD-L1 therapies among different RCC tumors and between patients [16].

Heterogeneity and distribution of different immune cells within the TME is vast and varies significantly among different tumors [18]. In addition, the immune microenvironment composition within tumors changes as tumors evolve over time. More advanced tumors often develop a more immunosuppressive phenotype with decreased involvement of effector T cells [19,20]. For example, as colorectal tumors evolve from stage T1 to T4, T cell tumor infiltration significantly decreases [19]. In RCC, more advanced stages are associated with an increase of exhausted CD8+ T cell phenotypes and expression of an immunosuppressive M2 macrophage phenotype [20]. Although the distribution of immune cells within the TME has been frequently shown to be prognostic of clinical outcomes [18,21], it is critical to understand that tumor-immune interactions are dynamic and vary over time.

3. Cell Types of the Tumor Immune Microenvironment

The cancer microenvironment, including the immune microenvironment is a dynamic pathological ecosystem [22,23]. Several organs and tissues including the bone marrow, blood, spleen and tumor-draining lymph nodes form an interconnected immunological network to help create and deliver immune cells to the TME. The distribution of immune cell phenotypes within the TME influence tumor growth and patient outcomes. There are two main categories of immune cells: cells that comprise the innate and adaptive immune system. Broadly, components of the innate immune system include macrophages, dendritic cells, neutrophils, myeloid derived suppressor cells, natural killer (NK) cells, eosinophils, basophils, mast cells and innate lymphoid cells [24]. Components of the adaptive immune system include T cells (e.g., $CD4^+$ and $CD8^+$ T cells) and B cells. Many of these cell types have significant plasticity and can gain different phenotypes depending on the surrounding microenvironment signals and cellular interactions.

Macrophages are commonly observed immune cells in the RCC TME. These tumor associated macrophages (TAMs) have significant plasticity and vary between the M1 and M2 phenotypes. M1 cells have a tumor suppressive phenotype and express proinflammatory cytokines, which potentiate T_H1 responses [24,25]. M2 macrophages tend to have a pro-tumorigenic, immunosuppressive phenotype [24]. Tumor cells can influence which phenotype a macrophage adopts by secreting IL-4 promoting the M2 phenotype, which has demonstrated the ability to remodel the surrounding tumor stroma, promoting tumor invasion [24,25]. Additionally, TAMs may repress cytotoxic T cells through a number of proposed mechanisms. These include depletion of L-arginine and reactive oxygen species generation, or they may modulate T cell function through expression of regulatory cytokines, such as IL-10 and TGF-β [26]. In general, RCC tumors with high infiltration of TAMs may limit anti-tumor T cell responses and are associated with a poor prognosis [27,28].

Dendritic cells are the primary innate immune cells that activate T cell mediated immune responses [29,30]. Dendritic cells are considered "professional antigen presenting cells" and uptake, process and present antigens through MHC class I and MHC class II molecules. Dendritic cells stimulate T cell activity through secretion of cytokines that induces T cell differentiation [30]. Other immune cells, such as B cells and macrophages, can also present antigens to T cells, but DCs are the most active T cell stimulating cells. Dendritic cells are broadly classified into classical DCs (cDCs), which can be further categorized as cDC1 and cDC2 subtypes, plasmacytoid DCs (pDCs) and monocyte-derived inflammatory DCs (moDCs) [31]. Similar to macrophages, the DC phenotype exists on a spectrum and DCs infiltrating tumors can be immunogenic or tumor tolerogenic. Tumors can also secrete cytokines to induce particular DC transcriptional and metabolic pathways that promote a tolerant environment such that DC will drive T_H2 and T_{reg} responses, such as pathways that involve IDO, Arg1, iNOS, and STAT3 [24]. Given their phenotypic plasticity, the role of DCs in tumor control is variable. As with other cell types, the DC phenotype may be tumor stage dependent, and as tumors become more advanced, DCs may shift from a tumor suppressive to tumor promoting phenotype [24].

Another important member of the innate immune system is the natural killer (NK) cell. These innate lymphoid cells have cytotoxic effector functions [24,25]. Tumor cells are typically destroyed by $CD8^+$ T cells, which rely on the expression of MHC class I molecule expression on the tumor cell surface for T cell recognition. Tumor cells can evade immune detection through the mutation or down regulation of MHC class I molecules. NK cells recognize and target cells that lack MHC class I expression in the TME as a mechanism to prevent immune evasion; however, NK cells are less efficient at destroying tumor cells compared to $CD8^+$ T cells [13]. Gene expression data from The Cancer Genome Atlas (TCGA) revealed that ccRCC tumors had the highest probability of immune infiltration for several immune cell types, including NK cells compared to other tumor types [32]. Other studies using RCC patient-derived xenografts also demonstrate RNA signatures for inflamed RCC tumors, including tumors with heavy NK cell infiltration [33]. NK cells express a number of inhibitory immune checkpoint proteins similar to T cells including TIM-3, LAG-3 and TIGIT (T cell immunoglobulin and ITIM domain) as well as stimulatory checkpoints, such as 4-1BB, which may be potential targets for future therapeutic exploitation [34].

T cells are a major component of the adaptive immune response to RCC, which may determine how a tumor progresses over time [21]. Two major classes of T cells exist expressing either CD4 or CD8 cell surface proteins. $CD4^+$ T "helper" cells can be divided into various phenotypic subtypes. T_H1 cells are associated with an anti-tumor phenotype and help to stimulate a cytotoxic $CD8^+$ T cell response via the production of IL-2 and IFNγ. Conversely, T_H2 and T_{reg} $CD4^+$ cells have more pro-tumorigenic phenotypes and lead to a reduced antitumor response through the production of inhibitory cytokines (e.g., IL-10, TGFβ and IL-35). T follicular helper (T_{FH}) cells are another $CD4^+$ subtype that express BCL-6, which have been found in tertiary lymphoid structures and interact with B cells to stimulate anti-tumor antibody production. T regulatory (T_{reg}) cells express CD4, CD25 and FOXP3 [35], which primarily function to promote peripheral tolerance and prevent autoimmunity secondary to chronic T cell activation as seen in the TME [36]. These cells are hypothesized to help promote tumor growth by suppressing cytotoxic effector cells [35]. While these cells appear to have a pro-tumorigenic function, their presence has been associated with both a poor and favorable prognosis, as their presence may be indicative of an ongoing robust cytotoxic T cell response to the tumor [36]. T_{reg} cells within the TME have been demonstrated to promote loss of effector $CD8^+$ T cell function through secretion of IL-35 leading to T cell exhaustion and upregulation of multiple inhibitory receptors (PD-1, TIM-3, LAG-3) on $CD8^+$ T cells [37].

The primary anti-tumor effector cells are $CD8^+$ T cells [38], which can be divided into different phenotypes based on differential gene expression. These phenotypes include naïve, effector, memory and exhausted $CD8^+$ T cell phenotypes. Effector T cells are critical to anti-tumor immunity and express genes associated with cytotoxic activity, including *GZMA*, *GZMB* and *NKG7* [38]. Historically, it was believed that $CD8^+$ T cells gain effector functions within tumor draining lymph nodes (TDLN), as is seen during viral infections, but recent work in RCC and prostate tumors demonstrates that some $CD8^+$ T cells may undergo a two-phase activation process starting in the TDLN and ending with a second stimulatory signal in the TME [16]. $CD8^+$ T cells recognize tumor cells through the T cell receptor (TCR) interacting with neoantigens loaded on MHC class I molecules present on the tumor cell surface.

Higher neoantigen load produced by homogenous tumor cells has been demonstrated to promote more effective immune surveillance in lung cancer. The investigators speculate that higher neoantigen heterogeneity may yield a lower antigen "dose", which may not effectively stimulate an immune response [15]. This concept is relevant to RCC as these tumors tend to be highly heterogenous. Intratumoral heterogeneity may potentially result in a spatially heterogenous neoantigen load and a less effective $CD8^+$ T cell response. Over time and as the tumor progresses, $CD8^+$ T cells will begin to exhibit an "exhausted" phenotype within the TME. This appears to be a gradual change and not necessarily a distinct group of $CD8^+$ cells. $CD8^+$ T cells will begin to express various levels and combinations of inhibitory receptors, such as TIM-3, CTLA-4, LAG-3 and PD-1, and stimulation of these receptors leads to contraction of the immune response [25,38]. Modulation of the exhausted $CD8^+$ T cell phenotype through inhibition of these receptors (such as PD-1 and CTLA-4 blockade) is the basis for ICI therapy and is the current first line systemic therapy for metastatic and adjuvant high risk ccRCC.

Another critical member of the adaptive immune response to tumors are B cells, which can be subdivided into naïve, IgM memory, switched memory, germinal center, plasmablast, plasma cells and B regulatory (B_{reg}) cells [39]. B cells localize in tumor draining lymph nodes and in ectopic lymphoid structures called tertiary lymphoid structures (TLS) in the TME [40]. Within TLS, B cells can undergo maturation from naïve B cells to memory B cells and plasma cells (Figure 2), which then propagate into the tumor bed. B cell phenotypes within the TME are highly heterogenous and B cells can exhibit both pro- and anti-tumorigenic functions. Anti-tumorigenic functions of B cells include the production of tumor specific antibodies and T cell priming and activation as well as direct cytotoxic activity. Conversely, B_{reg} cells can produce cytokines, such as IL-10, which inhibit cytotoxic $CD8^+$ T cell activity [41]. B cells can also lead to immune tolerance and secrete cytokines that promote cancer cell growth [40]. In general, B cell infiltration has been correlated with improved clinical outcomes among multiple tumor types [41]; however, in ccRCC, higher expression of B cell associated genes was associated with worse prognosis based on analysis of TCGA gene expression data [42]. Additionally, increased B_{reg} infiltration has been associated with a poor prognosis in various tumor types [41,43].

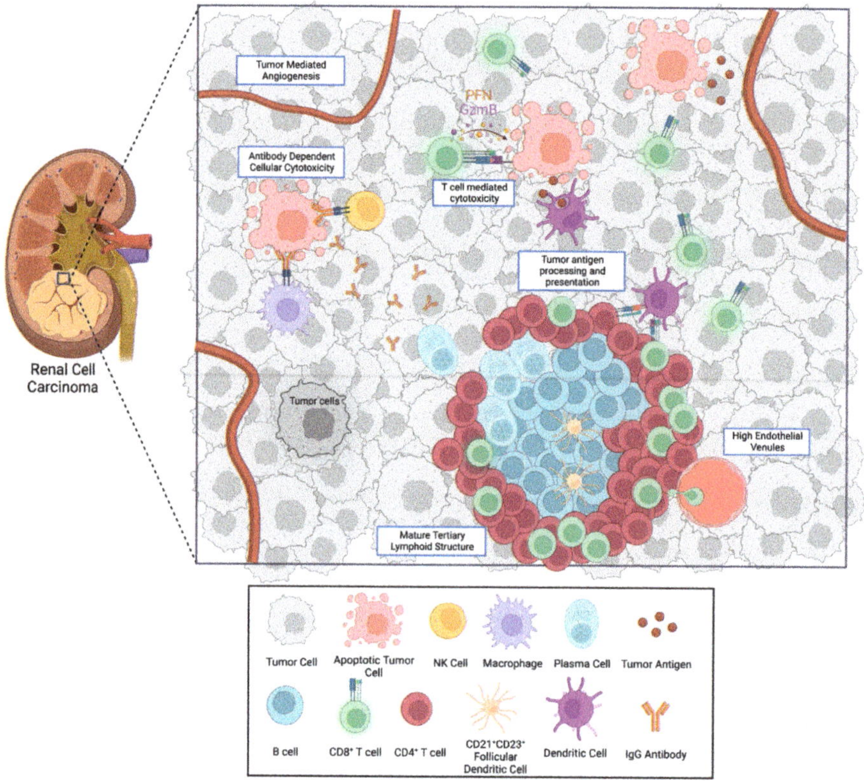

Figure 2. Components of the renal cell carcinoma tumor immune microenvironment. This figure demonstrates some of the many immune components of the tumor microenvironment. Mature tertiary lymphoid structures are composed of a germinal center containing follicular dendritic cells, B cells which differentiate into plasma cells secreting antibodies and a surrounding T cell zone composed of CD4$^+$ and CD8$^+$ T cells, which can be activated by antigen presenting cells. High endothelial venules (expressing PNAd) are also present within TLS, which are specialized capillaries that aid in lymphocyte migration. Plasma cells migrate from the TLS along fibroblast tracks via chemokine signaling [44] and continue to produce antibodies that bind to RCC tumor cells. This allows for antibody dependent cellular cytotoxicity (ADCC) when cells such as NK cells and macrophages recognize and destroy antibody bound tumor cells. Additionally, cytotoxic CD8$^+$ T cells target tumor cells presenting neoantigens loaded on MHC class I molecules (i.e., T cell mediated cytotoxicity). Binding of these MHC class I molecules to the T cell receptor leads to the release of granzymes and perforin from the T cell, which creates pores in the tumor cell membrane and subsequent tumor cell death leading to further tumor neoantigen release.

4. Characteristics of the Renal Cell Carcinoma Tumor Immune Microenvironment

RCC is known to be one of the most highly immune infiltrated solid tumors containing a heterogeneous population of infiltrating immune cells [45,46]. Using the TCGA, a study by Rooney et al. used an RNA-based metric of immune cytolytic activity by measuring expression of granzyme A (*GZMA*) and perforin (*PRF1*), which are upregulated in activated CD8$^+$ T cells. Among 18 different untreated primary tumor subtypes, ccRCC demonstrated the highest cytolytic activity of all tumors [47]. These findings were confirmed in a similar study that evaluated 19 different cancer types using the TCGA and demonstrated ccRCC and lung adenocarcinoma to have the highest T cell and overall immune infiltration gene signature score [48].

Different histologic subtypes of RCC have significantly different immune microenvironments, highlighting the tumor cell's ability to generate or alter the immune response differently among the various RCC subtypes. Clear cell RCC has the highest degree of immune infiltration compared to either papillary or chromophobe RCC based on TCGA immune gene signature analyses [49]. Chromophobe RCC is associated with an increased T_H17 gene signature; while IL-8 and CD56 NK cell gene signatures are increased in papillary RCC [49]. In an effort to identify a prognostic biomarker for RCC, Ricketts et al. identified that the T_H2 gene signature, among various other immune cell gene signatures, was the only one correlating with poor survival among all RCC subtypes, including clear cell, papillary and chromophobe [49]. Conversely, the T_H17 gene signature was associated with prolonged survival in ccRCC only [49].

Chevrier et al. conducted a detailed analysis of the RCC immune microenvironment by performing mass cytometry on 73 ccRCC tumors and 5 matched normal kidney samples, which provided a comprehensive "immune atlas" of RCC tumors [50]. Tumors within this study consisted of all grades and also included four patients with metastatic disease. The analysis demonstrated T cells to be the most common immune cell within the RCC immune microenvironment (composing 51% of immune cells), and myeloid cells (31%), NK cells (9%) and B cells (4%) were the next most common. Among T cells, various phenotypes were detected, including 11 different $CD8^+$ T cell phenotypes and 8 $CD4^+$ T cell phenotypes. Tumors demonstrated broad expression of PD-1 across T cell phenotypes; however, other inhibitory receptors, such as CTLA-4, 4-1BB and TIM-3, had more variable expression, which has implications related to therapeutic targeting of PD-1 versus other inhibitory molecules that may be involved with T cell exhaustion [50]. This study also characterized 17 different tumor associated macrophage (TAM) phenotypes, and demonstrated the transition between immature circulating monocytes to mature tissue macrophages. TAMs with high CD38 expression were associated with higher T_{reg} cells or exhausted T cells, which may indicate that $CD38^+$ TAMs modulate T cell activity in the RCC TME. Importantly, different TAM populations demonstrated prognostic ability with clinical metrics [50].

Using paired single-cell RNA and T cell receptor sequencing in four ICI-naïve and two ICI-treated metastatic ccRCC patients, Krishna et al. dissected the RCC immune microenvironment, finding 5 $CD8^+$ T cell phenotypic clusters, 5 $CD4^+$ T cell clusters, 2 monocyte clusters, 3 dendritic cell clusters and 4 TAM clusters. Multiregional tumor sampling also demonstrated significant intratumoral immune diversity, with certain tumor regions with extensive T cell infiltration, while other portions were T cell excluded or TAM infiltrated. Interestingly, an ICI-resistant tumor was characterized by T cell exclusion and had the highest prevalence of TAM populations compared to all other tumors evaluated, raising the possibility that an immunosuppressive TAM and T cell exclusion phenotype may result in ICI resistance [27]. This study also demonstrated that tissue-resident $CD8^+$ T cells (expressing CD69, ZNF683, CD103, PCD1 and LAG3) may promote anti-tumor immunity and appear to expand after ICI administration. The investigators speculate that tissue-resident T cells may swiftly recognize neoantigens and produce cytotoxic effector molecules, similar to their ability to rapidly respond to pathogenic infections [27].

Renal cell carcinoma tumors have significant TME spatial heterogeneity which impacts the density and phenotypes of immune cells across the tumor. Investigations of spatial heterogeneity have demonstrated that the degree of T cell exhaustion strongly correlates with the cell's location within the tumor. The physical location of the immune cells may drive T cell receptor clonotype enrichment to a greater extent than the tumor's somatic mutation heterogeneity [51]. A study by Li et al. demonstrated that T cell clonotypes were enriched in only certain regions across the tumor, and these regions had negligible differences in somatic mutations [51]. Additionally, particular TAM phenotypes appear to enrich in either the tumor core or invasive front. The study also demonstrated that T cell clones appear to infiltrate tumors and undergo a gradual phenotypic transformation from activation to dysfunction. Taken together, these data suggest that sampling a single tumor region will unlikely reflect the extent of T cell (i.e., T cell receptor) clonal expansion across the entire tumor. The study also demonstrated that the location of the tumor and immune cell interaction impacts the ability of the tumor to progress. Renal cell carcinoma cells expressing an epithelial-mesenchymal transition (EMT) gene program, which is critical for metastatic competence, were more abundant at the tumor/normal tissue interface. These EMT-high RCC cells colocalized with IL-1β expressing macrophages, which can break down the surrounding collagen-rich stroma. This suggests that macrophages may help to promote tumor invasion [51].

How cells interact and organize in the TME is important for an effective immune response. Tertiary lymphoid structures (TLSs) have been increasingly recognized as possible sites of tumor antigen recognition and generation of adaptive immune responses locally within the TME [44,52–64]. Tertiary lymphoid structures are organized aggregates of immune cells that arise, unlike lymph nodes, postnatally in non-lymphoid tissues [62]. The organization of TLS consists of an inner zone of $CD20^+$ B cells surrounded by T cells, mimicking lymph follicles of secondary lymphoid organs (Figure 2) [55,62]. T cells within TLSs consist of $CD4^+$ T_{FH}, T_H1, T_{regs} and $CD8^+$ cytotoxic T cells. $CD21^+CD23^+$ follicular dendritic cells are also present. The TLS is surrounded by specialized blood vessels called high endothelial venules (HEVs) that promote lymphocyte extravasation into the TME [54]. Tertiary lymphoid structures appear to have the ability to generate adaptive immune responses to tumor cells independent of more distant tumor draining lymph nodes. Mice lacking secondary lymphoid organs can still generate antitumor immune responses including effective T cell priming within intratumor TLS [55]. In RCC specifically, spatial transcriptomic analyses of primary tumor samples have shown TLSs to be sites of plasma cell maturation, which then migrate from TLS into the nearby TME guided by local fibroblast chemokines [44]. The plasma cells secrete IgG antibody which then binds and induces tumor cell apoptosis, possibly through a macrophage induced mechanism as evidenced by the higher infiltration of CD68+ macrophages surrounding IgG bound tumor cells [44]. Tertiary lymphoid structures and B cells localized within TLSs also appear to promote ICI response. Higher expression of B cell related genes was shown to be present in metastatic RCC tumors responsive to ICI. These intratumoral B cells were shown to localize within TLSs, and patients with higher TLS density were more likely to respond to ICI therapy [64].

One common research strategy is to categorize the intensity and type of immune responses within individual RCC tumors to describe the spectrum of immune responses. The phenotypes typically range from tumors that have a severely limited or absent immune infiltration to tumors that are heavily infiltrated with immune cells. Table 1 lists tumor immune microenvironment classification systems and the relative prognosis associated with each class. The classifications often not only incorporate the quantity of immune cell infiltration but also the degree of activation of individual immune cells. For example, Clark et al. characterized the immune infiltration using transcriptomic profiles into four categories: (1) $CD8^+$ inflamed tumors which had high $CD8^+$ infiltration with an effector gene signature characterized by increased IFNγ signaling; (2) $CD8^-$ inflamed tumors characterized by an innate immune signature with increased dendritic and macrophage cells in the TME; (3) VEGF immune desert tumors with a high stromal gene signature, increased endothelial cells and enrichment of an angiogenesis signature; and (4) metabolic immune desert tumors, which had low immune and stromal scores as well as increased mTOR signaling and a unique metabolic profile. The lack of immune cells in the metabolic immune desert tumors suggests that a hypoxic, nutrient poor microenvironment may be immunosuppressive [65]. This demonstrates just one of many classification systems, which in general demonstrate a spectrum of immune infiltration, and no consensus has been reached as to which classification system is most prognostic or predictive of systemic therapy response. These classification systems also suggest an evolution of the immune response to tumors over time. Further studies have investigated how the immune response evolves as tumors progress.

Table 1. Tumor microenvironment classification systems.

Study	RCC Subtype	Stages Studied	Method of Categorization	Tumor Microenvironment Categorization	Category Characteristics	Category Prognosis
Senbabaoglu et al., 2016 [48]	Clear cell	All Stages	TCGA based gene expression signatures	Non-infiltrated	Low T cell infiltration, low stromal score, increased metabolism and mitochondrial related genes	Best
				Heterogeneously Infiltrated	Increased angiogenesis-related gene expression, c-KIT and Smad1	Intermediate
				T cell Enriched	High T cell infiltration, high granzyme B and IFNγ expression	Worst
Giraldo et al., 2017 [66]	Clear cell	Localized RCC	Multiparametric flow cytometric immunophenotypic analysis	Immune activated	Increased $CD8^+$ clonality, increased cytotoxic genes	Best
				Immune silent	Low levels tumor infiltrating lymphocytes	Intermediate
				Immune regulated	M2-rich, poorly cytotoxic, increased T_{reg}	Worst
Clark et al., 2019 [65]	Clear cell	All Stages	Transcriptomic and proteomic microenvironment signatures	VEGF immune desert	Elevated stromal score, endothelial enrichment	Best
				$CD8^-$ Inflamed	Innate immune signature, fibroblast signature	Intermediate
				Metabolic immune desert	Low immune and stromal scores, elevated mitochondrial, OXPHOS, glycolysis protein expression	Intermediate
				$CD8^+$ inflamed	High $CD8^+$ infiltration, *BAP1* mutations, CD38 expression, IFNγ signaling	Worst
Hakimi et al., 2019 [67]	Clear cell	Metastatic	Somatic mutation analysis, RNA and protein expression	Cluster 3	High *PBRM1* mutations, high angiogenesis score, moderate immune infiltration	Best
				Cluster 2	Moderate angiogenesis score, moderate immune infiltration	Intermediate
				Cluster 1	Lowest angiogenesis score, lowest immune infiltration	Intermediate
				Cluster 4	High *BAP1* mutation, moderate angiogenesis score, high immune infiltration, higher PD-L1 expression	Worst
Braun et al., 2020 [68]	Clear cell	Metastatic	Integrated genetic, transcriptomic and immunopathologic analysis	Immune Excluded	5-fold more $CD8^+$ T cells at the tumor margin than in tumor center	No difference in prognosis
				Immune Desert	Not excluded and below 25th percentile for $CD8^+$ T cells, 50 cells/mm^2 in the tumor center	
				Immune Infiltrated	Not excluded and ≥25th percentile for $CD8^+$ T cells in the tumor center. Enriched for M1 macrophages, $CD4^+$ memory T cells, NK cells	

4.1. Immune Microenvironment Changes during Tumor Development and Treatment

The immune microenvironment is dynamic, and the phenotype of immune cells within the tumor changes over the course of tumor development. Recent investigations have focused on how the immune cell phenotypes alter over the course of tumor development [34,51,69]. Braun et al. evaluated how the immune cell phenotypes change from early to advanced (i.e., metastatic) disease stages. The investigators profiled immune cells from ccRCC tumors using single-cell transcriptomics from 13 patients with advancing disease stage [20]. The study demonstrated within T cell populations, there was marked transcriptional heterogeneity consisting of 19 individual populations of T cells including T_{reg} cells, $CD4^+$ with activated or central memory phenotypes, $CD8^+$ tissue resident memory cells, and a large population of exhausted $CD8^+$ T cells. Six of the $CD8^+$ populations expressed markers of exhaustion such as PD-1, *TOX* and TIM-3 at high or moderate levels, indicating a terminally exhausted phenotype. Another group of $CD8^+$ T cells expressed cytotoxic molecules consistent with an effector population. A group of three $CD8^+$ clusters expressed *ZNF683* (HOBIT), *PRDM1* (BLIMP-1) and *ITGAE* (CD103), indicative of a tissue-resident memory phenotype. Among the $CD4^+$ T cell population, 3 broad phenotypes were identified including an activated, central memory and T_{reg} population. Among early stage ccRCC tumors, effector $CD8^+$ T cells were enriched while metastatic ccRCC tumors were enriched for terminally exhausted $CD8^+$ T cells. With advancing disease stage, there was an increase in expression of inhibitory checkpoints such as PD-1, TIM-3 and LAG-3. Markers of progenitor or stem-like T cells, including TCF1 and T-bet, were increased early but decreased with disease progression. Overall, these data demonstrate a progressive T cell exhaustion with advancing disease stage. Tumors begin with an initial infiltration of cytotoxic $CD8^+$ T cells followed by progressive immune dysfunction leading to terminally exhausted $CD8^+$ T cells in advanced disease stages [20]. Interestingly, terminally exhausted $CD8^+$ T cells were highly clonal (based on TCR sequencing) and represented a high proportion of the total number of T cells within the TME, suggesting a possible limited number of neoantigens are being targeted by T cells [20]. The findings by Braun et al. were confirmed by another study demonstrating that T cell clones entering ccRCC tumors undergo a phenotypic transition over time from an effector state to progressive dysfunction and ultimately terminal exhaustion with advancing disease stage [51].

Tumor associated macrophages (TAMs) also shift phenotypically with advancing disease stage. Early-stage tumors contain TAMs expressing more proinflammatory genes; however, TAMs shift to an anti-inflammatory gene signature with advancing disease. Additionally, the expression of M2 genes, which are associated with a pro-tumorigenic microenvironment, increases with metastatic disease [20]. M2 TAMs interact with $CD8^+$ T cells through expression of multiple T cell immune checkpoint ligands, including PD-L1, CD80 and CD86 (which bind CTLA-4), CD155 (binds TIGIT) and Galectin-9 (binds TIM-3), and these interactions are associated with a worse overall survival [20]. Modulation of terminally exhausted T cells by inhibiting molecules, such as PD-1, PD-L1 and CTLA-4, is the basis for ICI therapy, which has led to significant improvements in the survival of patients with advanced RCC.

The immune microenvironment additionally shifts in response to ICI therapy. Bi et al. investigated immune microenvironment changes before and after ICI therapy, using single-cell transcriptomic dissection of mRCC tumor biopsies [70]. TAMs in responders to ICI therapy shift to a pro-inflammatory state, possibly secondary to IFNγ, produced by $CD8^+$ T cells. Also, $CD8^+$ T cells express higher levels of co-inhibitory receptors and effector molecules such as GZMB, PRF1 and IFNG [70]. Despite this significantly more inflammatory immune microenvironment in responders to ICI therapy, these tumors also had a profound increase in $CD8^+$ T cell checkpoint genes and TAM anti-inflammatory signaling genes, suggesting a possible route to eventual ICI resistance [70].

4.2. Renal Cell Carcinoma Tumor Antigens and Genomic Correlations with Immune Response

As discussed, tumor neoantigens are critical components for immune cell recognition. Tumor mutational burden (TMB) is often used as a surrogate marker for neoantigen production and has been predictive of ICI responsiveness and degree of immune infiltration in multiple tumor types. However, RCC tumors only demonstrate a relatively moderate TMB, which has not been reliably predictive of RCC response to ICI therapy [34,71,72]. Compared to melanoma and non-small cell lung cancer, which harbor 10–400 mutations/megabase, ccRCC only harbors about 1.1 mutation/megabase. Although measured TMB is lower, RCC still demonstrates one of the highest levels of immune infiltration, particularly T cell infiltration, in the TME compared to other tumor types [71]. One possible explanation for this may be how TMB has been previously quantified, with the most common method being measurement of single nucleotide variants (SNVs). However, SNVs may not always produce inflammatory neoantigens which might confound TMB data for RCC. Mutations caused by insertions and deletions ("indels") are much more likely to cause frameshifts leading to altered protein structures and formation of neoantigens that are recognized by immune cells. While RCC may have lower TMB, indels are highly abundant in RCC, which may lead to a higher neoantigen load and immune infiltration [73,74], despite relatively low measured SNV.

Beyond TMB, investigations have been performed linking particular RCC genotypes to the degree of tumor immune infiltration. The central pathway of tumor development in ccRCC is caused by the loss or inactivation of the tumor suppressor VHL located on chromosome 3p. Loss of VHL is often associated with loss of other tumor suppressor genes including PBRM1, BAP1 and SETD2, which are also located on chromosome 3p [75]. VHL inactivation causes stabilization of hypoxia inducible factors, HIF-1α and HIF-2α [75,76]. HIF activation causes expression of genes regulating angiogenesis, glycolysis and apoptosis [75]. HIF proteins also impact the tumor immune microenvironment [77]. In an immune competent murine autochthonous model of ccRCC, *Hif2a* deletion led to increased expression of immune infiltration mRNA signatures, antigen presentation and interferon activity. Additionally, greater effector CD8$^+$ T cell infiltration (expressing markers of CD69 and perforin) was detected. In general, mice with *HIF2a* deletions developed less tumors compared to *HIF1a* mutants, which may be related to greater anti-tumor immune response secondary to *HIF2a* deletion [77]. In summary, activation of HIF-2α in ccRCC may act as a suppressor of T cell inflammation [77,78].

Mutations in *PBRM1* and *BAP1* are also critical driver events of ccRCC tumor development and are known to be prognostic for outcomes in RCC [75]. Studies have investigated their impact on the immune microenvironment. A cohort of metastatic ccRCC patients had whole exome sequencing (WES) of tumor tissue. Tumors harboring *PBRM1* loss of function (LOF) mutations were more likely to respond to ICI therapy [71]. Additionally, *PBRM1* LOF mutations were associated with lower expression of immune-inhibitory ligands within the tumor [71]. A separate study demonstrated that mRCC primary tumors containing truncating *PBRM1* mutations were less immune infiltrated overall and had a lower total CD8$^+$ T cell infiltration compared to tumors with intact *PBRM1* [68]. Loss of PBRM1 function causes a reduction in IFNγ-STAT1 signaling in both murine and human RCC cell lines; thus, PBRM1 inactivation leads to a less immunogenic TME [79]. Conversely, tumors harboring *BAP1* mutations have been associated with heavy CD8$^+$ T cell inflammation [33,65].

4.3. The Predictive and Prognostic Capability of the Tumor Immune Microenvironment in RCC

One goal of immune microenvironment research has been to identify immune based biomarkers that have the ability to prognosticate patient survival as well as predict response to ICI or anti-angiogenic therapies. Multiple investigations have identified potential biomarkers with occasionally conflicting results [32,34,48,49,65–69,73,80–83]. Table 2 lists some immune based biomarkers evaluated in multiple studies [84]. Studies have been heterogeneous in terms of the stage of tumors evaluated (early versus advanced stage), number of samples taken from individual tumors, whether the primary or metastatic tumor was sampled, and differences in how the immune microenvironment was evaluated (based on bulk or single cell sequencing signatures, immunofluorescence, biopsies versus whole slides, etc.). These differences make comparisons of potential biomarkers among studies difficult but there are some consistent themes within the available data.

Table 2. Renal cell carcinoma immune based biomarkers.

Biomarker	Description	Investigational Use	Limitations
PD-L1 [6,10,84]	Cell surface protein	• Prognostic biomarker • Predictive biomarker for ICI response	• Studies failed to show utility as either prognostic or predictive • Tumor spatial heterogeneity, lack multi-region sampling • Cellular expression heterogeneity • Lack of standardized assay
Tumor Mutational Burden [80]	Number of non-synonymous somatic mutations in tumor DNA	• Predictive biomarker for ICI response	• Studies failed to show predictive capacity • RCC has low TMB • Lack of standardized assays
Tumor Neoantigen Burden [80]	Number of immunogenic tumor-specific proteins (neoantigens) predicted by quantity of somatic mutations	• Predictive biomarker for ICI response	• Studies failed to show predictive capacity • Lack of standardized predictive algorithms to determine neoantigen load
Renal 101 Immuno signature [85]	26-gene immune based signature	• Prognostic of progression free survival in setting of ICI therapy	• Has not been used outside of clinical trial setting
T-effector score [80]	Five gene signature (CD8A, EOMES, PRF1, IFNG, CD274)	• Prognostic of progression free survival in setting of ICI + anti-VEGF therapy	• Has not been used outside of clinical trial setting
Myeloid inflammation [80]	Six gene signature (IL-6, CXCL1, CXCL2, CXCL3, CXCL8, PTGS2)	• Prognostic of progression free survival in setting of ICI + anti-VEGF therapy	• Has not been used outside of clinical trial setting

Multiple studies have evaluated the immune microenvironment's relation to patient prognosis. Renal cell carcinoma tumors that are heavily infiltrated with T cells have been associated with a worse prognosis compared to many other solid tumors in which T cell infiltration is often associated with a favorable prognosis [32,34,48]. Clear cell RCC tumors that are heavily infiltrated with T cells as measured by gene-based signatures are correlated with higher stage and grade as well as worse cancer specific survival [48]. Clark et al. demonstrated across a range of ccRCC tumor stages, from localized to metastatic, tumors classified as $CD8^+$ inflamed, were associated with worse overall survival and other poor prognostic features, including higher *BAP1* mutations, higher grade tumors and increased *PD-1/PD-L1* expression [65]. Similarly, Giraldo et al. evaluated the immune microenvironment of metastatic ccRCC tumors and demonstrated patients with high $CD8^+$ infiltration had about a twofold increase in risk of mortality compared to tumors with low $CD8^+$ infiltration [82]. Another study of localized ccRCC tumors demonstrated that the ratio between effector T cell to T_{reg} cells was positively correlated with a lower rate of recurrence in localized ccRCC tumors [83]. An exhaustive phenotypic characterization of immune cells within localized ccRCC tumors using flow cytometric analysis showed that tumors at high risk of early progression contained a population of $CD8^+$ $PD-1^+$ cells that co-expressed TIM-3 and LAG-3 and exhibited an exhausted phenotype with decreased cytotoxic potential, polyclonality and enrichment of M2 TAMs [66]. These tumors, termed "immune regulated", may be ideal for post-nephrectomy adjuvant ICI therapy [66,86]. Thus, not only the density of particular immune cells, for example $CD8^+$ T cells, but also their phenotype (i.e., activated, exhausted) is critical in determining their prognostic capability. Currently, no immune-based gene signatures or biomarkers have been identified for routine clinical use to improve clinical decision making.

In addition to prognosis, studies have evaluated the ability of the immune microenvironment to predict response to systemic therapies for RCC, most often in the metastatic setting. A phase II trial, IMmotion150, evaluated anti-PD-L1 therapy (atezolizumab) with or without an anti-VEGF antibody (bevacizumab) compared to the tyrosine kinase inhibitor, sunitinib. One theory is that VEGF, which is upregulated in ccRCC, may enhance cancer immune evasion and blocking VEGF with bevacizumab may improve the immune activation triggered by atezolizumab [86]. As part of this trial, investigators evaluated gene signatures of the TME and how they correlated with clinical outcomes. The trial demonstrated the addition of bevacizumab to atezolizumab improved progression free survival compared to either atezolizumab or sunitinib alone among tumors with ≥1% PD-L1 positivity [80]. Among tumors with a high $T_{effector}$ gene signature (defined by increased expression of *CD8A, EOMES, PRF1, IFNG, CD274*), the combination of atezolizumab + bevacizumab had significantly improved progression free survival compared to sunitinib monotherapy. This was further confirmed by the phase III IMmotion151 trial, which again demonstrated that tumors with a $T_{effector}$ signature had an improved objective response rate and progression free survival when treated with atezolizumab + bevacizumab compared to sunitinib [87]. A study by Hakimi et al. evaluated 409 tumors by unsupervised clustering of microarray data from patients with metastatic ccRCC treated with either pazopanib or sunitinib as part of the COMPARZ trial [67]. The study identified 4 biologically distinct molecular "clusters" with different therapeutic responses. Cluster 4 had the worst overall and progression free survival probability, and this cluster was associated with less frequent *PBRM1* mutations as well as enrichment for *TP53* and *BAP1* mutations. In regard to the immune microenvironment, cluster 4 had significant enrichment for inflammatory signatures including IFNγ gene signatures as well as the highest total immune infiltrate of the 4 clusters. Immune deconvolution analysis demonstrated that macrophages were the dominant immune population within this cluster. Macrophage infiltration, particularly the M2 phenotype, was associated with worse overall survival among all patients. When determining response to TKI therapy, patients with a high angiogenesis gene signature and low macrophage infiltration had the best response to TKI therapy, thus serving as a predictive biomarker among patients within this trial but having not been further validated [67].

Carlisle et al. evaluated how T cell response in RCC was affected by ICI therapy and how a patient's preexisting immune response to RCC may correlate with response to ICI therapy [81]. Investigators collected peripheral blood before and after ICI therapy administration in 36 mRCC patients (27 clear cell and 9 non-clear cell patients). Patients with increased expansion of $HLA\text{-}DR^+CD38^+CD8^+$ T cells had the largest reductions in tumor size and longest progression free survival. This T cell phenotype expressed high levels of cytotoxic effector genes, including perforin-1, GZMB and IFNγ. Additional immunofluorescence studies of the primary tumors demonstrated that tumors harboring greater infiltration of $CD8^+$ T cells, more $TCF\text{-}1^+CD8^+$ T cells and more $MHCII^+$ cells were more likely to benefit from ICI therapy [81]. In regard to $TCF\text{-}1^+CD8^+$ T cells, a prior study had demonstrated that kidney tumors harbor these cells in areas of APCs. $TCF\text{-}1^+CD8^+$ T cells have stem-like properties, allowing them to produce more effector $CD8^+$ T cells [88]. The presence of these intra-tumoral niches predict the magnitude of T cell infiltration and patient survival. Tumors that progress appear to lose these intra-tumoral niches, suggesting a possible mechanism of tumor immune escape [88]. Overall, similar to biomarkers for prognostication, a definitive, clinically useful predictive immune based biomarker has yet to be developed.

5. Future Directions

As further knowledge is gained exploring the tumor immune microenvironment, this may lead to successful new therapies exploiting the tumor-immune interaction, including treatments targeting tumor specific antigens, multi-antigen vaccines, personalized peptide vaccines and engineered T cell therapies [89]. Future therapies to overcome the inhibitory immune checkpoints beyond CTLA-4 and PD-1/PD-L1, such as therapies targeting LAG-3, TIM-3 and TIGIT, may be clinically beneficial since immune exhaustion and subsequent tumor escape remains critical for RCC progression. Additionally, investigations are underway to target immune metabolic pathways and stimulatory checkpoints (4-1BB and OX40), which may enhance immunotherapy responses [34].

Another potentially impactful factor shaping the immune response to renal cell carcinoma is the host microbiome. Numerous studies have shown that the gut microbiome influences the immune response to tumors, and therapeutic responses may be improved via its modulation [90–93]. Specific bacteria can stimulate the production of pro-inflammatory cytokines or anti-inflammatory cytokines. The byproducts of microbial metabolism can act as carcinogens. Also, the microbiome can educate immune cells, thus determining an individual's overall immune response [94]. A recent phase 1 randomized trial by Dizman et al. randomized 30 treatment-naïve patients with mRCC to nivolumab and ipilimumab, with or without daily oral CBM588. CBM588 is a bifidogenic live bacterial product. Progression free survival was significantly longer in patients receiving ICI therapy with CBM588 (12.7 vs. 2.5 months, HR 0.15 95% CI 0.05–0.47, $p = 0.001$) with no significant difference in toxicity between study arms [95]. These data reinforce similar findings from patients with non-small cell lung cancer receiving CBM588. Possible mechanisms include increased propionate, a short chain fatty acid that has anti-tumor properties. The microbiota may upregulate chemokines, such as those involved in dendritic and cytotoxic T cell recruitment. These findings warrant further validation. Continued investigation of the host microbiome will likely help develop therapies to augment current and future immune activating agents.

It is necessary to explore a variety of clinical and behavioral variables that may impact tumorigenesis, immune response and treatment outcomes. For example, one interesting area of research includes the role of the circadian rhythm in the immune response to RCC. Clock genes, such as *CLOCK*, *BMAL1*, *PER* and *CRY*, regulate many processes, including cell division and metabolism [96]. Some studies have shown that alteration of these genes is linked to poor prognosis in RCC. Additionally, the circadian rhythm can influence the RCC immune microenvironment, which can impact tumor surveillance and response to therapy [96].

A possible future experimental roadmap to understand the immune microenvironment in RCC will likely include the following components: comprehensive molecular profiling including single-cell RNA sequencing and spatial transcriptomics/proteomics to identify different immune cell populations and activation status, validation of these populations through flow cytometry and immunohistochemistry, functional evaluation of immune cell populations both in vitro and in vivo, immune checkpoint analysis to determine the expression of checkpoint molecules in RCC tumor samples and their association with patient outcomes and tumor infiltrating lymphocyte isolation and characterization with T cell receptor sequencing to assess clonality and antigen specificity. Additionally, investigators should perform preclinical testing of novel immunotherapies developed from the insights gained from the prior steps. This should be followed by clinical translation of these novel immunotherapies to patients to generate new or synergistic therapies that improve patient survival and quality of life. Finally, studying long term outcomes of current and new therapies will help develop biomarkers for treatment response.

6. Conclusions

Renal cell carcinoma is one of the most highly immune infiltrated tumors and often responds to ICI therapy. Increasing our understanding of the RCC tumor immune microenvironment is critical for future progress [89]. Immune based biomarkers that are prognostic and predictive are continually being investigated. Tissue heterogeneity significantly limits the development of broadly applicable biomarkers, and further efforts to understand the intra- and inter-tumoral immune microenvironment heterogeneity are underway. Future studies should focus on other RCC histologic subtypes as the majority of studies to date have been conducted in the clear cell subtype. Finally, research should continue to investigate the impact of localized therapy, such as surgery, ablation and radiation on the tumor-immune interaction, and how these therapies can be effectively combined with systemic therapies to understand the optimal multidisciplinary approach to RCC treatment that improves patient survival and quality of life.

Author Contributions: Conceptualization, D.D.S., B.D. and E.J.A.; writing—original draft preparation, D.D.S., B.D. and E.J.A.; writing—review and editing, D.D.S., B.D., I.A.L., S.R., T.L., H.E., A.L.W., M.G.L. and E.J.A.; visualization, D.D.S. and B.D. All authors have read and agreed to the published version of the manuscript.

Funding: This research received no external funding.

Acknowledgments: Figures were constructed using BioRender.com (accessed on 4 January 2023).

Conflicts of Interest: The authors declare no conflict of interest.

References

1. Seer Cancer Stat Facts: Kidney and Renal Pelvis Cancer. National Cancer Institute: Bethesda, MD, USA, 2022. Available online: https://seer.Cancer.Gov/statfacts/html/kidrp.Html (accessed on 5 April 2023).
2. Siegel, R.L.; Miller, K.D.; Fuchs, H.E.; Jemal, A. Cancer statistics, 2022. *CA A Cancer J. Clin.* **2022**, *72*, 7–33. [CrossRef] [PubMed]
3. Mattila, K.E.; Laajala, T.D.; Tornberg, S.V.; Kilpeläinen, T.P.; Vainio, P. A three-feature prediction model for metastasis-free survival after surgery of localized clear cell renal cell carcinoma. *Sci. Rep.* **2021**, *11*, 8650. [CrossRef] [PubMed]
4. Leibovich, B.C.; Blute, M.L.; Cheville, J.C.; Lohse, C.M.; Frank, I.; Kwon, E.D.; Weaver, A.L.; Parker, A.S.; Zincke, H. Prediction of progression after radical nephrectomy for patients with clear cell renal cell carcinoma: A stratification tool for prospective clinical trials. *Cancer* **2003**, *97*, 1663–1671. [CrossRef] [PubMed]
5. Motzer, R.J.; Hutson, T.E.; Tomczak, P.; Michaelson, M.D.; Bukowski, R.M.; Oudard, S.; Negrier, S.; Szczylik, C.; Pili, R.; Bjarnason, G.A.; et al. Overall Survival and Updated Results for Sunitinib Compared with Interferon Alfa in Patients With Metastatic Renal Cell Carcinoma. *J. Clin. Oncol.* **2009**, *27*, 3584–3590. [CrossRef]
6. Motzer, R.J.; Tannir, N.M.; McDermott, D.F.; Aren Frontera, O.; Melichar, B.; Choueiri, T.K.; Plimack, E.R.; Barthélémy, P.; Porta, C.; George, S.; et al. Nivolumab plus Ipilimumab versus Sunitinib in Advanced Renal-Cell Carcinoma. *N. Engl. J. Med.* **2018**, *378*, 1277–1290. [CrossRef]
7. Rini, B.I.; Plimack, E.R.; Stus, V.; Gafanov, R.; Hawkins, R.; Nosov, D.; Pouliot, F.; Alekseev, B.; Soulières, D.; Melichar, B.; et al. Pembrolizumab plus Axitinib versus Sunitinib for Advanced Renal-Cell Carcinoma. *N. Engl. J. Med.* **2019**, *380*, 1116–1127. [CrossRef]
8. Motzer, R.J.; Rini, B.I.; McDermott, D.F.; Aren Frontera, O.; Hammers, H.J.; Carducci, M.A.; Salman, P.; Escudier, B.; Beuselinck, B.; Amin, A.; et al. Nivolumab plus ipilimumab versus sunitinib in first-line treatment for advanced renal cell carcinoma: Extended follow-up of efficacy and safety results from a randomised, controlled, phase 3 trial. *Lancet Oncol.* **2019**, *20*, 1370–1385. [CrossRef]
9. Motzer, R.J.; Escudier, B.; George, S.; Hammers, H.J.; Srinivas, S.; Tykodi, S.S.; Sosman, J.A.; Plimack, E.R.; Procopio, G.; McDermott, D.F.; et al. Nivolumab versus everolimus in patients with advanced renal cell carcinoma: Updated results with long-term follow-up of the randomized, open-label, phase 3 checkmate 025 trial. *Cancer* **2020**, *126*, 4156–4167. [CrossRef]
10. Choueiri, T.K.; Tomczak, P.; Park, S.H.; Venugopal, B.; Ferguson, T.; Chang, Y.-H.; Hajek, J.; Symeonides, S.N.; Lee, J.L.; Sarwar, N.; et al. Adjuvant Pembrolizumab after Nephrectomy in Renal-Cell Carcinoma. *N. Engl. J. Med.* **2021**, *385*, 683–694. [CrossRef]
11. Powles, T.; Tomczak, P.; Park, S.H.; Venugopal, B.; Ferguson, T.; Symeonides, S.N.; Hajek, J.; Gurney, H.; Chang, Y.-H.; Lee, J.L.; et al. Pembrolizumab versus placebo as post-nephrectomy adjuvant therapy for clear cell renal cell carcinoma (KEYNOTE-564): 30-month follow-up analysis of a multicentre, randomised, double-blind, placebo-controlled, phase 3 trial. *Lancet Oncol.* **2022**, *23*, 1133–1144. [CrossRef]
12. Chen, D.S.; Mellman, I. Oncology Meets Immunology: The Cancer-Immunity Cycle. *Immunity* **2013**, *39*, 1–10. [CrossRef] [PubMed]

13. Becht, E.; Giraldo, N.A.; Germain, C.; de Reyniès, A.; Laurent-Puig, P.; Zucman-Rossi, J.; Dieu-Nosjean, M.-C.; Sautès-Fridman, C.; Fridman, W.H. Immune Contexture, Immunoscore, and Malignant Cell Molecular Subgroups for Prognostic and Theranostic Classifications of Cancers. *Adv. Immunol.* **2016**, *130*, 95–190. [CrossRef] [PubMed]
14. DiNatale, R.G.; Hakimi, A.A.; Chan, T.A. Genomics-based immuno-oncology: Bridging the gap between immunology and tumor biology. *Hum. Mol. Genet.* **2020**, *29*, R214–R225. [CrossRef] [PubMed]
15. McGranahan, N.; Furness, A.J.S.; Rosenthal, R.; Ramskov, S.; Lyngaa, R.; Saini, S.K.; Jamal-Hanjani, M.; Wilson, G.A.; Birkbak, N.J.; Hiley, C.T.; et al. Clonal neoantigens elicit T cell immunoreactivity and sensitivity to immune checkpoint blockade. *Science* **2016**, *351*, 1463–1469. [CrossRef]
16. Prokhnevska, N.; Cardenas, M.A.; Valanparambil, R.M.; Sobierajska, E.; Barwick, B.G.; Jansen, C.; Moon, A.R.; Gregorova, P.; Delbalzo, L.; Greenwald, R.; et al. CD8+ T cell activation in cancer comprises an initial activation phase in lymph nodes followed by effector differentiation within the tumor. *Immunity* **2022**, *56*, 107–124. [CrossRef]
17. Raskov, H.; Orhan, A.; Christensen, J.P.; Gögenur, I. Cytotoxic CD8+ T cells in cancer and cancer immunotherapy. *Br. J. Cancer* **2021**, *124*, 359–367. [CrossRef]
18. Fridman, W.H.; Pagès, F.; Sautès-Fridman, C.; Galon, J. The immune contexture in human tumours: Impact on clinical outcome. *Nat. Rev. Cancer* **2012**, *12*, 298–306. [CrossRef]
19. Bindea, G.; Mlecnik, B.; Tosolini, M.; Kirilovsky, A.; Waldner, M.; Obenauf, A.C.; Angell, H.; Fredriksen, T.; Lafontaine, L.; Berger, A.; et al. Spatiotemporal Dynamics of Intratumoral Immune Cells Reveal the Immune Landscape in Human Cancer. *Immunity* **2013**, *39*, 782–795. [CrossRef]
20. Braun, D.A.; Street, K.; Burke, K.P.; Cookmeyer, D.L.; Denize, T.; Pedersen, C.B.; Gohil, S.H.; Schindler, N.; Pomerance, L.; Hirsch, L.; et al. Progressive immune dysfunction with advancing disease stage in renal cell carcinoma. *Cancer Cell* **2021**, *39*, 632–648.e8. [CrossRef]
21. Binnewies, M.; Roberts, E.W.; Kersten, K.; Chan, V.; Fearon, D.F.; Merad, M.; Coussens, L.M.; Gabrilovich, D.I.; Ostrand-Rosenberg, S.; Hedrick, C.C.; et al. Understanding the tumor immune microenvironment (TIME) for effective therapy. *Nat. Med.* **2018**, *24*, 541–550. [CrossRef]
22. Hanahan, D. Hallmarks of Cancer: New Dimensions. *Cancer Discov.* **2022**, *12*, 31–46. [CrossRef] [PubMed]
23. Luo, W. Nasopharyngeal carcinoma ecology theory: Cancer as multidimensional spatiotemporal "unity of ecology and evolution" pathological ecosystem. *Theranostics* **2023**, *13*, 1607–1631. [CrossRef] [PubMed]
24. Hinshaw, D.C.; Shevde, L.A. The tumor microenvironment innately modulates cancer progression. *Cancer Res.* **2019**, *79*, 4557–4566. [CrossRef] [PubMed]
25. Vano, Y.; Giraldo, N.A.; Fridman, W.H.; Sautès-Fridman, C. *Oncoimmunology, A Practical Guide for Cancer Immunotherapy*; Springer: Berlin/Heidelberg, Germany, 2017; pp. 5–21.
26. Nixon, B.G.; Kuo, F.; Ji, L.; Liu, M.; Capistrano, K.; Do, M.; Franklin, R.A.; Wu, X.; Kansler, E.R.; Srivastava, R.M.; et al. Tumor-associated macrophages expressing the transcription factor IRF8 promote T cell exhaustion in cancer. *Immunity* **2022**, *55*, 2044–2058.e5. [CrossRef]
27. Krishna, C.; DiNatale, R.G.; Kuo, F.; Srivastava, R.M.; Vuong, L.; Chowell, D.; Gupta, S.; Vanderbilt, C.; Purohit, T.A.; Liu, M.; et al. Single-cell sequencing links multiregional immune landscapes and tissue-resident T cells in ccRCC to tumor topology and therapy efficacy. *Cancer Cell* **2021**, *39*, 662–677.e6. [CrossRef]
28. Shen, H.; Liu, J.; Chen, S.; Ma, X.; Ying, Y.; Li, J.; Wang, W.; Wang, X.; Xie, L. Prognostic Value of Tumor-Associated Macrophages in Clear Cell Renal Cell Carcinoma: A Systematic Review and Meta-Analysis. *Front. Oncol.* **2021**, *11*, 657318. [CrossRef]
29. Conejo-Garcia, J.R.; Rutkowski, M.R.; Cubillos-Ruiz, J.R. State-of-the-art of regulatory dendritic cells in cancer. *Pharmacol. Ther.* **2016**, *164*, 97–104. [CrossRef]
30. Verneau, J.; Sautés-Fridman, C.; Sun, C.-M. Dendritic cells in the tumor microenvironment: Prognostic and theranostic impact. *Semin. Immunol.* **2020**, *48*, 101410. [CrossRef]
31. Collin, M.; Bigley, V. Human dendritic cell subsets: An update. *Immunology* **2018**, *154*, 3–20. [CrossRef]
32. Varn, F.S.; Wang, Y.; Mullins, D.W.; Fiering, S.; Cheng, C. Systematic Pan-Cancer Analysis Reveals Immune Cell Interactions in the Tumor Microenvironment. *Cancer Res.* **2017**, *77*, 1271–1282. [CrossRef]
33. Wang, T.; Lu, R.; Kapur, P.; Jaiswal, B.S.; Hannan, R.; Zhang, Z.; Pedrosa, I.; Luke, J.J.; Zhang, H.; Goldstein, L.D.; et al. An Empirical Approach Leveraging Tumorgrafts to Dissect the Tumor Microenvironment in Renal Cell Carcinoma Identifies Missing Link to Prognostic Inflammatory Factors. *Cancer Discov.* **2018**, *8*, 1142–1155. [CrossRef] [PubMed]
34. Braun, D.A.; Bakouny, Z.; Hirsch, L.; Flippot, R.; Van Allen, E.M.; Wu, C.J.; Choueiri, T.K. Beyond conventional immune-checkpoint inhibition—Novel immunotherapies for renal cell carcinoma. *Nat. Rev. Clin. Oncol.* **2021**, *18*, 199–214. [CrossRef] [PubMed]
35. McRitchie, B.R.; Akkaya, B. Exhaust the exhausters: Targeting regulatory t cells in the tumor microenvironment. *Front. Immunol.* **2022**, *13*, 940052. [CrossRef] [PubMed]
36. Ahrends, T.; Borst, J. The opposing roles of cd4+ t cells in anti-tumour immunity. *Immunology* **2018**, *154*, 582–592. [CrossRef]
37. Turnis, M.E.; Sawant, D.V.; Szymczak-Workman, A.L.; Andrews, L.P.; Delgoffe, G.M.; Yano, H.; Beres, A.J.; Vogel, P.; Workman, C.J.; Vignali, D.A. Interleukin-35 limits anti-tumor immunity. *Immunity* **2016**, *44*, 316–329. [CrossRef]
38. Van der Leun, A.M.; Thommen, D.S.; Schumacher, T.N. Cd8+ t cell states in human cancer: Insights from single-cell analysis. Nature reviews. *Cancer* **2020**, *20*, 218–232.

39. Fridman, W.H.; Meylan, M.; Petitprez, F.; Sun, C.-M.; Italiano, A.; Sautès-Fridman, C. B cells and tertiary lymphoid structures as determinants of tumour immune contexture and clinical outcome. *Nat. Rev. Clin. Oncol.* **2022**, *19*, 441–457. [CrossRef]
40. Downs-Canner, S.M.; Meier, J.; Vincent, B.G.; Serody, J.S. B Cell Function in the Tumor Microenvironment. *Annu. Rev. Immunol.* **2022**, *40*, 169–193. [CrossRef]
41. Lauss, M.; Donia, M.; Svane, I.M.; Jönsson, G. B cells and tertiary lymphoid structures: Friends or foes in cancer immunotherapy? *Clin. Cancer Res.* **2022**, *28*, 1751–1758. [CrossRef]
42. Iglesia, M.D.; Parker, J.S.; Hoadley, K.; Serody, J.S.; Perou, C.; Vincent, B.G. Genomic Analysis of Immune Cell Infiltrates Across 11 Tumor Types. *J. Natl. Cancer Inst.* **2016**, *108*, djw144. [CrossRef]
43. Murakami, Y.; Saito, H.; Shimizu, S.; Kono, Y.; Shishido, Y.; Miyatani, K.; Matsunaga, T.; Fukumoto, Y.; Ashida, K.; Sakabe, T.; et al. Increased regulatory B cells are involved in immune evasion in patients with gastric cancer. *Sci. Rep.* **2019**, *9*, 13083. [CrossRef]
44. Meylan, M.; Petitprez, F.; Becht, E.; Bougoüin, A.; Pupier, G.; Calvez, A.; Giglioli, I.; Verkarre, V.; Lacroix, G.; Verneau, J.; et al. Tertiary lymphoid structures generate and propagate anti-tumor antibody-producing plasma cells in renal cell cancer. *Immunity* **2022**, *55*, 527–541.e5. [CrossRef] [PubMed]
45. Jonasch, E.; Walker, C.L.; Rathmell, W.K. Clear cell renal cell carcinoma ontogeny and mechanisms of lethality. *Nat. Rev. Nephrol.* **2021**, *17*, 245–261. [CrossRef] [PubMed]
46. Becht, E.; Giraldo, N.A.; Lacroix, L.; Buttard, B.; Elarouci, N.; Petitprez, F.; Selves, J.; Laurent-Puig, P.; Sautes-Fridman, C.; Fridman, W.H.; et al. Estimating the population abundance of tissue-infiltrating immune and stromal cell populations using gene expression. *Genome Biol.* **2016**, *17*, 218. [CrossRef]
47. Rooney, M.S.; Shukla, S.A.; Wu, C.J.; Getz, G.; Hacohen, N. Molecular and Genetic Properties of Tumors Associated with Local Immune Cytolytic Activity. *Cell* **2015**, *160*, 48–61. [CrossRef] [PubMed]
48. Şenbabaoğlu, Y.; Gejman, R.S.; Winer, A.G.; Liu, M.; Van Allen, E.M.; de Velasco, G.; Miao, D.; Ostrovnaya, I.; Drill, E.; Luna, A.; et al. Tumor immune microenvironment characterization in clear cell renal cell carcinoma identifies prognostic and immunotherapeutically relevant messenger RNA signatures. *Genome Biol.* **2016**, *17*, 231. [CrossRef] [PubMed]
49. Ricketts, C.J.; Cubas, A.A.D.; Fan, H.; Smith, C.C.; Lang, M.; Reznik, E.; Bowlby, R.; Gibb, E.A.; Akbani, R.; Beroukhim, R.; et al. The cancer genome atlas comprehensive molecular characterization of renal cell carcinoma. *Cell Rep.* **2018**, *23*, 313–326.e5. [CrossRef]
50. Chevrier, S.; Levine, J.H.; Zanotelli, V.R.T.; Silina, K.; Schulz, D.; Bacac, M.; Ries, C.H.; Ailles, L.; Jewett, M.A.S.; Moch, H.; et al. An Immune Atlas of Clear Cell Renal Cell Carcinoma. *Cell* **2017**, *169*, 736–749.e18. [CrossRef]
51. Li, R.; Ferdinand, J.R.; Loudon, K.W.; Bowyer, G.S.; Laidlaw, S.; Muyas, F.; Mamanova, L.; Neves, J.B.; Bolt, L.; Fasouli, E.S.; et al. Mapping single-cell transcriptomes in the intra-tumoral and associated territories of kidney cancer. *Cancer Cell* **2022**, *40*, 1583–1599.e10. [CrossRef]
52. Calderaro, J.; Petitprez, F.; Becht, E.; Laurent, A.; Hirsch, T.Z.; Rousseau, B.; Luciani, A.; Amaddeo, G.; Derman, J.; Charpy, C.; et al. Intra-tumoral tertiary lymphoid structures are associated with a low risk of early recurrence of hepatocellular carcinoma. *J. Hepatol.* **2019**, *70*, 58–65. [CrossRef]
53. Moussion, C.; Girard, J.-P. Dendritic cells control lymphocyte entry to lymph nodes through high endothelial venules. *Nature* **2011**, *479*, 542–546. [CrossRef] [PubMed]
54. Dieu-Nosjean, M.-C.; Goc, J.; Giraldo, N.A.; Sautès-Fridman, C.; Fridman, W.H. Tertiary lymphoid structures in cancer and beyond. *Trends Immunol.* **2014**, *35*, 571–580. [CrossRef] [PubMed]
55. Goc, J.; Fridman, W.-H.; Sautès-Fridman, C.; Dieu-Nosjean, M.-C. Characteristics of tertiary lymphoid structures in primary cancers. *Oncoimmunology* **2013**, *2*, e26836. [CrossRef] [PubMed]
56. Johansson-Percival, A.; He, B.; Li, Z.-J.; Kjellén, A.; Russell, K.; Li, J.; Larma, I.; Ganss, R. De novo induction of intratumoral lymphoid structures and vessel normalization enhances immunotherapy in resistant tumors. *Nat. Immunol.* **2017**, *18*, 1207–1217. [CrossRef] [PubMed]
57. Johansson-Percival, A.; Ganss, R. Therapeutic Induction of Tertiary Lymphoid Structures in Cancer Through Stromal Remodeling. *Front. Immunol.* **2021**, *12*, 674375. [CrossRef] [PubMed]
58. Kazanietz, M.G.; Durando, M.; Cooke, M. CXCL13 and Its Receptor CXCR5 in Cancer: Inflammation, Immune Response, and Beyond. *Front. Endocrinol.* **2019**, *10*, 471. [CrossRef] [PubMed]
59. Kroeger, D.R.; Milne, K.; Nelson, B.H. Tumor-Infiltrating Plasma Cells Are Associated with Tertiary Lymphoid Structures, Cytolytic T-Cell Responses, and Superior Prognosis in Ovarian Cancer. *Clin. Cancer Res.* **2016**, *22*, 3005–3015. [CrossRef]
60. Sautès-Fridman, C.; Dimberg, A.; Verma, V. Editorial: Tertiary Lymphoid Structures: From Basic Biology to Translational Impact in Cancer. *Front. Immunol.* **2022**, *13*, 870862. [CrossRef]
61. Sautès-Fridman, C.; Petitprez, F.; Calderaro, J.; Fridman, W.H. Tertiary lymphoid structures in the era of cancer immunotherapy. *Nat. Rev. Cancer* **2019**, *19*, 307–325. [CrossRef]
62. Schumacher, T.N.; Thommen, D.S. Tertiary lymphoid structures in cancer. *Science* **2022**, *375*, eabf9419. [CrossRef]
63. Vaghjiani, R.G.; Skitzki, J.J. Tertiary Lymphoid Structures as Mediators of Immunotherapy Response. *Cancers* **2022**, *14*, 3748. [CrossRef] [PubMed]
64. Helmink, B.A.; Reddy, S.M.; Gao, J.; Zhang, S.; Basar, R.; Thakur, R.; Yizhak, K.; Sade-Feldman, M.; Blando, J.; Han, G.; et al. B cells and tertiary lymphoid structures promote immunotherapy response. *Nature* **2020**, *577*, 549–555. [CrossRef] [PubMed]

65. Clark, D.J.; Dhanasekaran, S.M.; Petralia, F.; Pan, J.; Song, X.; Hu, Y.; da Veiga Leprevost, F.; Reva, B.; Lih, T.-S.M.; Chang, H.-Y.; et al. Integrated Proteogenomic Characterization of Clear Cell Renal Cell Carcinoma. *Cell* **2019**, *179*, 964–983.e31. [CrossRef] [PubMed]
66. Giraldo, N.A.; Becht, E.; Vano, Y.; Petitprez, F.; Lacroix, L.; Validire, P.; Sanchez-Salas, R.; Ingels, A.; Oudard, S.; Moatti, A.; et al. Tumor-Infiltrating and Peripheral Blood T-cell Immunophenotypes Predict Early Relapse in Localized Clear Cell Renal Cell Carcinoma. *Clin. Cancer Res.* **2017**, *23*, 4416–4428. [CrossRef] [PubMed]
67. Hakimi, A.A.; Voss, M.H.; Kuo, F.; Sanchez, A.; Liu, M.; Nixon, B.G.; Vuong, L.; Ostrovnaya, I.; Chen, Y.-B.; Reuter, V.; et al. Transcriptomic Profiling of the Tumor Microenvironment Reveals Distinct Subgroups of Clear Cell Renal Cell Cancer: Data from a Randomized Phase III Trial. *Cancer Discov.* **2019**, *9*, 510–525. [CrossRef]
68. Braun, D.A.; Hou, Y.; Bakouny, Z.; Ficial, M.; Angelo, M.S.; Forman, J.; Ross-Macdonald, P.; Berger, A.C.; Jegede, O.A.; Elagina, L.; et al. Interplay of somatic alterations and immune infiltration modulates response to PD-1 blockade in advanced clear cell renal cell carcinoma. *Nat. Med.* **2020**, *26*, 909–918. [CrossRef]
69. Remark, R.; Alifano, M.; Cremer, I.; Lupo, A.; Dieu-Nosjean, M.-C.; Riquet, M.; Crozet, L.; Ouakrim, H.; Goc, J.; Cazes, A.; et al. Characteristics and Clinical Impacts of the Immune Environments in Colorectal and Renal Cell Carcinoma Lung Metastases: Influence of Tumor Origin. *Clin. Cancer Res.* **2013**, *19*, 4079–4091. [CrossRef]
70. Bi, K.; He, M.X.; Bakouny, Z.; Kanodia, A.; Napolitano, S.; Wu, J.; Grimaldi, G.; Braun, D.A.; Cuoco, M.S.; Mayorga, A.; et al. Tumor and immune reprogramming during immunotherapy in advanced renal cell carcinoma. *Cancer Cell* **2021**, *39*, 649–661.e5. [CrossRef]
71. Miao, D.; Margolis, C.A.; Gao, W.; Voss, M.H.; Li, W.; Martini, D.J.; Norton, C.; Bossé, D.; Wankowicz, S.M.; Cullen, D.; et al. Genomic correlates of response to immune checkpoint therapies in clear cell renal cell carcinoma. *Science* **2018**, *359*, 801–806. [CrossRef]
72. Samstein, R.M.; Lee, C.-H.; Shoushtari, A.N.; Hellmann, M.D.; Shen, R.; Janjigian, Y.Y.; Barron, D.A.; Zehir, A.; Jordan, E.J.; Omuro, A.; et al. Tumor mutational load predicts survival after immunotherapy across multiple cancer types. *Nat. Genet.* **2019**, *51*, 202–206. [CrossRef]
73. Vuong, L.; Kotecha, R.R.; Voss, M.H.; Hakimi, A.A. Tumor Microenvironment Dynamics in Clear-Cell Renal Cell Carcinoma. *Cancer Discov.* **2019**, *9*, 1349–1357. [CrossRef] [PubMed]
74. Turajlic, S.; Litchfield, K.; Xu, H.; Rosenthal, R.; McGranahan, N.; Reading, J.L.; Wong, Y.N.S.; Rowan, A.; Kanu, N.; Al Bakir, M.; et al. Insertion-and-deletion-derived tumour-specific neoantigens and the immunogenic phenotype: A pan-cancer analysis. *Lancet Oncol.* **2017**, *18*, 1009–1021. [CrossRef] [PubMed]
75. Shapiro, D.D.; Virumbrales-Muñoz, M.; Beebe, D.J.; Abel, E.J. Models of Renal Cell Carcinoma Used to Investigate Molecular Mechanisms and Develop New Therapeutics. *Front. Oncol.* **2022**, *12*, 871252. [CrossRef] [PubMed]
76. Hsieh, J.J.; Purdue, M.P.; Signoretti, S.; Swanton, C.; Albiges, L.; Schmidinger, M.; Heng, D.Y.; Larkin, J.; Ficarra, V. Renal cell carcinoma. *Nat. Rev. Dis. Prim.* **2017**, *3*, 17009. [CrossRef]
77. Hoefflin, R.; Harlander, S.; Schäfer, S.; Metzger, P.; Kuo, F.; Schönenberger, D.; Adlesic, M.; Peighambari, A.; Seidel, P.; Chen, C.-Y.; et al. HIF-1α and HIF-2α differently regulate tumour development and inflammation of clear cell renal cell carcinoma in mice. *Nat. Commun.* **2020**, *11*, 4111. [CrossRef]
78. Xiong, Y.; Liu, L.; Xia, Y.; Qi, Y.; Chen, Y.; Chen, L.; Zhang, P.; Kong, Y.; Qu, Y.; Wang, Z.; et al. Tumor infiltrating mast cells determine oncogenic HIF-2α-conferred immune evasion in clear cell renal cell carcinoma. *Cancer Immunol. Immunother.* **2019**, *68*, 731–741. [CrossRef]
79. Liu, X.-D.; Kong, W.; Peterson, C.B.; McGrail, D.J.; Hoang, A.; Zhang, X.; Lam, T.; Pilie, P.G.; Zhu, H.; Beckermann, K.E.; et al. PBRM1 loss defines a nonimmunogenic tumor phenotype associated with checkpoint inhibitor resistance in renal carcinoma. *Nat. Commun.* **2020**, *11*, 2135. [CrossRef]
80. McDermott, D.F.; Huseni, M.A.; Atkins, M.B.; Motzer, R.J.; Rini, B.I.; Escudier, B.; Fong, L.; Joseph, R.W.; Pal, S.K.; Reeves, J.A.; et al. Clinical activity and molecular correlates of response to atezolizumab alone or in combination with bevacizumab versus sunitinib in renal cell carcinoma. *Nat. Med.* **2018**, *24*, 749–757. [CrossRef]
81. Carlisle, J.W.; Jansen, C.S.; Cardenas, M.A.; Sobierajska, E.; Reyes, A.M.; Greenwald, R.; Del Balzo, L.; Prokhnevska, N.; Kucuk, O.; Carthon, B.C.; et al. Clinical outcome following checkpoint therapy in renal cell carcinoma is associated with a burst of activated CD8 T cells in blood. *J. Immunother. Cancer* **2022**, *10*, e004803. [CrossRef]
82. Giraldo, N.A.; Becht, E.; Pagès, F.; Skliris, G.P.; Verkarre, V.; Vano, Y.; Mejean, A.; Saint-Aubert, N.; Lacroix, L.; Natario, I.; et al. Orchestration and Prognostic Significance of Immune Checkpoints in the Microenvironment of Primary and Metastatic Renal Cell Cancer. *Clin. Cancer Res.* **2015**, *21*, 3031–3040. [CrossRef]
83. Ghatalia, P.; Gordetsky, J.; Kuo, F.; Dulaimi, E.; Cai, K.Q.; Devarajan, K.; Bae, S.; Naik, G.; Chan, T.A.; Uzzo, R.; et al. Prognostic impact of immune gene expression signature and tumor infiltrating immune cells in localized clear cell renal cell carcinoma. *J. Immunother. Cancer* **2019**, *7*, 139. [CrossRef]
84. Cotta, B.H.; Choueiri, T.K.; Cieslik, M.; Ghatalia, P.; Mehra, R.; Morgan, T.M.; Palapattu, G.S.; Shuch, B.; Vaishampayan, U.; Van Allen, E.; et al. Current Landscape of Genomic Biomarkers in Clear Cell Renal Cell Carcinoma. *Eur. Urol.* 2023, in press. [CrossRef]

85. Motzer, R.J.; Robbins, P.B.; Powles, T.; Albiges, L.; Haanen, J.B.; Larkin, J.; Mu, X.J.; Ching, K.A.; Uemura, M.; Pal, S.K.; et al. Avelumab plus axitinib versus sunitinib in advanced renal cell carcinoma: Biomarker analysis of the phase 3 JAVELIN Renal 101 trial. *Nat. Med.* **2020**, *26*, 1733–1741. [CrossRef] [PubMed]
86. Wallin, J.J.; Bendell, J.C.; Funke, R.; Sznol, M.; Korski, K.; Jones, S.; Hernandez, G.; Mier, J.; He, X.; Hodi, F.S.; et al. Atezolizumab in combination with bevacizumab enhances antigen-specific T-cell migration in metastatic renal cell carcinoma. *Nat. Commun.* **2016**, *7*, 12624. [CrossRef] [PubMed]
87. Motzer, R.J.; Banchereau, R.; Hamidi, H.; Powles, T.; McDermott, D.; Atkins, M.B.; Escudier, B.; Liu, L.-F.; Leng, N.; Abbas, A.R.; et al. Molecular Subsets in Renal Cancer Determine Outcome to Checkpoint and Angiogenesis Blockade. *Cancer Cell* **2020**, *38*, 803–817.e4. [CrossRef] [PubMed]
88. Jansen, C.S.; Prokhnevska, N.; Master, V.A.; Sanda, M.G.; Carlisle, J.W.; Bilen, M.A.; Cardenas, M.; Wilkinson, S.; Lake, R.; Sowalsky, A.G.; et al. An intra-tumoral niche maintains and differentiates stem-like CD8 T cells. *Nature* **2019**, *576*, 465–470. [CrossRef]
89. Xu, Y.; Miller, C.P.; Warren, E.H.; Tykodi, S.S. Current status of antigen-specific T-cell immunotherapy for advanced renal-cell carcinoma. *Hum. Vaccines Immunother.* **2021**, *17*, 1882–1896. [CrossRef]
90. Gopalakrishnan, V.; Spencer, C.N.; Nezi, L.; Reuben, A.; Andrews, M.C.; Karpinets, T.V.; Prieto, P.A.; Vicente, D.; Hoffman, K.; Wei, S.C.; et al. Gut microbiome modulates response to anti–PD-1 immunotherapy in melanoma patients. *Science* **2018**, *359*, 97–103. [CrossRef]
91. Routy, B.; le Chatelier, E.; DeRosa, L.; Duong, C.P.M.; Alou, M.T.; Daillère, R.; Fluckiger, A.; Messaoudene, M.; Rauber, C.; Roberti, M.P.; et al. Gut microbiome influences efficacy of PD-1–based immunotherapy against epithelial tumors. *Science* **2018**, *359*, 91–97. [CrossRef]
92. Derosa, L.; Hellmann, M.D.; Spaziano, M.; Halpenny, D.; Fidelle, M.; Rizvi, H.; Long, N.; Plodkowski, A.J.; Arbour, K.C.; Chaft, J.E.; et al. Negative association of antibiotics on clinical activity of immune checkpoint inhibitors in patients with advanced renal cell and non-small-cell lung cancer. *Ann. Oncol.* **2018**, *29*, 1437–1444. [CrossRef]
93. Sivan, A.; Corrales, L.; Hubert, N.; Williams, J.B.; Aquino-Michaels, K.; Earley, Z.M.; Benyamin, F.W.; Lei, Y.M.; Jabri, B.; Alegre, M.-L.; et al. Commensal Bifidobacterium promotes antitumor immunity and facilitates anti-PD-L1 efficacy. *Science* **2015**, *350*, 1084–1089. [CrossRef]
94. Park, E.M.; Chelvanambi, M.; Bhutiani, N.; Kroemer, G.; Zitvogel, L.; Wargo, J.A. Targeting the gut and tumor microbiota in cancer. *Nat. Med.* **2022**, *28*, 690–703. [CrossRef] [PubMed]
95. Dizman, N.; Meza, L.; Bergerot, P.; Alcantara, M.; Dorff, T.; Lyou, Y.; Frankel, P.; Cui, Y.; Mira, V.; Llamas, M.; et al. Nivolumab plus ipilimumab with or without live bacterial supplementation in metastatic renal cell carcinoma: A randomized phase 1 trial. *Nat. Med.* **2022**, *28*, 704–712. [CrossRef] [PubMed]
96. Santoni, M.; Molina-Cerrillo, J.; Santoni, G.; Lam, E.T.; Massari, F.; Mollica, V.; Mazzaschi, G.; Rapoport, B.L.; Grande, E.; Buti, S. Role of Clock Genes and Circadian Rhythm in Renal Cell Carcinoma: Recent Evidence and Therapeutic Consequences. *Cancers* **2023**, *15*, 408. [CrossRef] [PubMed]

cancers

MDPI

Commentary

Role of Clock Genes and Circadian Rhythm in Renal Cell Carcinoma: Recent Evidence and Therapeutic Consequences

Matteo Santoni [1], Javier Molina-Cerrillo [2], Giorgio Santoni [3], Elaine T. Lam [4], Francesco Massari [5], Veronica Mollica [5], Giulia Mazzaschi [6], Bernardo L. Rapoport [7,8], Enrique Grande [9] and Sebastiano Buti [6,*]

1 Oncology Unit, Macerata Hospital, Via Santa Lucia 2, 62100 Macerata, Italy
2 Department of Medical Oncology, Hospital Ramón y Cajal, 28029 Madrid, Spain
3 Scuola di Scienze del Farmaco e dei Prodotti della Salute, Università di Camerino, 62032 Camerino, Italy
4 University of Colorado Anschutz Medical Campus, Aurora, CO 80045, USA
5 Medical Oncology, IRCCS Azienda Ospedaliero-Universitaria di Bologna, Via Albertoni-15, 40138 Bologna, Italy
6 Department of Medicine and Surgery, University of Parma, 43121 Parma, Italy
7 The Medical Oncology Centre of Rosebank, 129 Oxford Road, Saxonwold, Johannesburg 2196, South Africa
8 Department of Immunology, Faculty of Health Sciences, University of Pretoria, Corner Doctor Savage Road and Bophelo Road, Pretoria 0002, South Africa
9 Department of Medical Oncology, MD Anderson Cancer Center Madrid, 28033 Madrid, Spain
* Correspondence: sebabuti@libero.it or sebastiano.buti@unipr.it; Tel.: +39-0521-702314; Fax: +39-0521-995448

Simple Summary: Circadian rhythms are physical, mental, and behavioral changes that follow a 24-h cycle. These natural processes primarily respond to light and dark, and affect most living things, including animals, plants, and microbes. Circadian rhythm is also involved in the regulation of cellular differentiation and physiology as well as in the modulation of the immune system. Some genes controlling circadian rhythm may be implicated in the occurrence of common malignant cancers, including renal cell carcinoma. Recent studies showed that time-of-day infusion directly conditions the efficacy of immunotherapy in patients with cancer. Drugs targeting the circadian clock have been identified and their role in the era of immunotherapy should be investigated. In this review, we illustrate the role of clock genes in kidney cancer onset, progression and prognosis, and the potential therapeutic consequences of this emerging evidence.

Abstract: Circadian rhythm regulates cellular differentiation and physiology and shapes the immune response. Altered expression of clock genes might lead to the onset of common malignant cancers, including Renal Cell Carcinoma (RCC). Data from Cancer Genome Atlas (TCGA) indicate that clock genes *PER1-3*, *CRY2*, *CLOCK*, *NR1D2* and *RORα* are overexpressed in RCC tissues and correlate with patients' prognosis. The expression of clock genes could finely tune transcription factor activity in RCC and is associated with the extent of immune cell infiltration. The clock system interacts with hypoxia-induced factor-1α (HIF-1α) and regulates the circadian oscillation of mammalian target of rapamycin (mTOR) activity thereby conditioning the antitumor effect of mTOR inhibitors. The stimulation of natural killer (NK) cell activity exerted by the administration of interferon-α, a cornerstone of the first era of immunotherapy for RCC, relevantly varies according to circadian dosing time. Recent evidence demonstrated that time-of-day infusion directly affects the efficacy of immune checkpoint inhibitors in cancer patients. Compounds targeting the circadian clock have been identified and their role in the era of immunotherapy deserves to be further investigated. In this review, we aimed at addressing the impact of clock genes on the natural history of kidney cancer and their potential therapeutic implications.

Keywords: circadian rhythm; clock genes; renal cell carcinoma; immunotherapy; HIF

Citation: Santoni, M.; Molina-Cerrillo, J.; Santoni, G.; Lam, E.T.; Massari, F.; Mollica, V.; Mazzaschi, G.; Rapoport, B.L.; Grande, E.; Buti, S. Role of Clock Genes and Circadian Rhythm in Renal Cell Carcinoma: Recent Evidence and Therapeutic Consequences. *Cancers* **2023**, *15*, 408. https://doi.org/10.3390/cancers15020408

Academic Editors: José I. López and Claudia Manini

Received: 20 November 2022
Revised: 3 January 2023
Accepted: 5 January 2023
Published: 7 January 2023

1. Introduction

The circadian rhythm is a hierarchically cyclic system that regulates the daily oscillations of physiological processes and can respond to external environmental changes to maintain internal homeostasis [1]. Starting in the 1980s, studies culminating in the characterization of the first clock gene (CG) in *Drosophila melanogaster* paved the way for the characterization of additional genes and proteins, leading to what is presently known as the circadian clock (CC) [2].

CC genes control and sustain circadian rhythms in many pathophysiologic processes [3]. Disruptions in circadian rhythms are implicated in several pathologies such as diabetes, cardiometabolic and neurodegenerative disorders, and cancer. Circadian disturbance by shiftwork, jet lag, late night light exposure, and late-night food binging has been long linked to increased cancer risk. Furthermore, loss of circadian rhythmicity in patients has been associated with poor response to anti-cancer therapies and increased early mortality rates amongst cancer patients [4].

Renal Cell Carcinoma (RCC) is one of the most common urinary cancers worldwide, with a predicted increase in incidence in the coming years [5,6]. Partial or total nephrectomy is the gold standard curative treatment approach for patients with localized disease. Unfortunately, up to 30% of patients present with local or distant recurrence after surgery, thus requiring palliative systemic therapies [7,8]. From 2005–2015, tyrosine kinase inhibitors (TKIs) targeting vascular endothelial growth factor receptor (VEGFR) have represented the mainstay of metastatic renal cell carcinoma (mRCC) treatment [9]. Immunotherapy, as a single agent, in doublets, or in combination with anti-VEGFR TKIs, has rapidly become a cornerstone of the RCC therapeutic armamentarium since the approval of nivolumab in 2015, leading to a marked improvement in patients' quality of life (QoL) and survival [10,11]. Although crucial advancements have been made to cure this disease, most mRCC patients have primary or acquired resistance to targeted therapy and immunotherapy, thus underlining the necessity of developing novel, personalized therapeutic approaches.

In this review, we explored the role of clock genes in kidney cancer occurrence, progression, and prognosis and their potential therapeutic implications.

2. Role of Clock Genes in Cancer

Mounting evidence supports the existence of molecular interconnections between CC and cancer. Many recognized cancer hallmarks such as copious metabolic demands, a favorable inflammatory microenvironment, immune suppression, and resistance to cell death, have a well-established circadian component. Hence, it might be conceivable that oncogenic transformation may lead to malfunctioning of the CC, which in turn creates a homeostatic imbalance, facilitating cancer growth and progression; on the other hand, it could be also speculated that CC malfunction could predispose to oncogenic transformation.

The CC genes can be divided into two operation levels: systemic and cellular [11]. The central clock at a systemic scale, known as the "master" clock, is regulated by the central nervous system in the anterior hypothalamus [12] and is responsible for coordinating the cell-autonomous clocks in peripheral tissues and the brain in response to environmental stimuli [13]. At the cellular level, the CC genes are regulated by positive and negative transcription-translation loops which control the rhythmicity of cellular, metabolic, and physiologic events [14]. At the transcriptional level, the clock is driven by positive loop regulators: basic helix–loop–helix heterodimeric transcription factors regulate the expression of key circadian genes, which are the negative regulators of the circadian loop. Modifications of some circadian genes at the translational level regulate protein stability, control nuclear entry of repressor protein complexes, and impact the clock's autoregulatory feedback loops [14].

There are eight core CC genes involved with the circadian rhythm: Period1 (*PER1*), period2 (*PER2*), period3 (*PER3*), cryptochrome1 (*CRY1*), cryptochrome2 (*CRY2*), aryl hydrocarbon receptor nuclear translocator-like protein 1 (*BMAL1*), neuronal PAS domain protein 2 (*NPAS2*), and circadian locomotor output cycles protein kaput (*CLOCK*). The *PER1*, *PER2*, and *PER3* genes regulate cell growth, proliferation, and apoptosis [15,16], *CRY1* and *CRY2* regulate the transcription G1/S and G2/M cell cycle checkpoints [17], while *BMAL1* and *NPAS2* inhibit the proliferation and invasion of cancer cells by suppressing the c-Myc transcription factor [18]. The *CLOCK* gene enhances VEGF-mediated angiogenesis in cancer cells and metastatic invasion by interacting with HIF-1α/BMAL1 [19]. Interestingly, down-regulation of *PER1-3, CRY2, NPAS2, BMAL1,* as well as *CLOCK* gene expression, correlates with high histological grade and poor prognosis [20] and short survival in different cancers [21,22].

The epithelial-to-mesenchymal transition (EMT) is a crucial step in cancer progression and enables cancer cell metastasis. Low expression of *PER2* led to the activation of EMT genes *TWIST1* and *SLUG* and promoted cancer metastasis [23]. In addition, some components of the CC have a significant antitumor effect through cell cycle arrest, the DNA damage response, and correlation with essential pathways, including the p53 [24].

Investigations are currently ongoing to evaluate other CC genes, such as casein kinase 1ε (*CK1ε*), receptor subfamily 1 group D member 1/2 (*NR1D1/2*), RAR-related orphan receptor α and β (*RORα/β*), timeless (*TIMELESS*) and timeless-interacting protein (*TIPIN*). To date, although far more aware of their essential role in feedback loop regulation, data on the real contribution of these CC genes in cancer development and progression are still not sufficient or conclusive [25].

3. Role of Circadian Clock Genes in Renal Cell Carcinoma Tumorigenesis and Prognosis

Animal models of genetic disruption of CC genes have been strongly associated with different cancers, including RCC, prostate, breast, colon, liver, pancreas, ovary, and lung malignancies [26]. In Wilms tumors, rare kidney cancers that primarily affects children, the expression of CLOCK protein is dramatically reduced in tumor cells, suggesting that the CC molecular axis may be disrupted in dedifferentiation-mediated embryonal tumors [27].

In RCC, the CC circuitry is deregulated, and the altered expression of CC genes might contribute to tumor onset and progression (Figure 1).

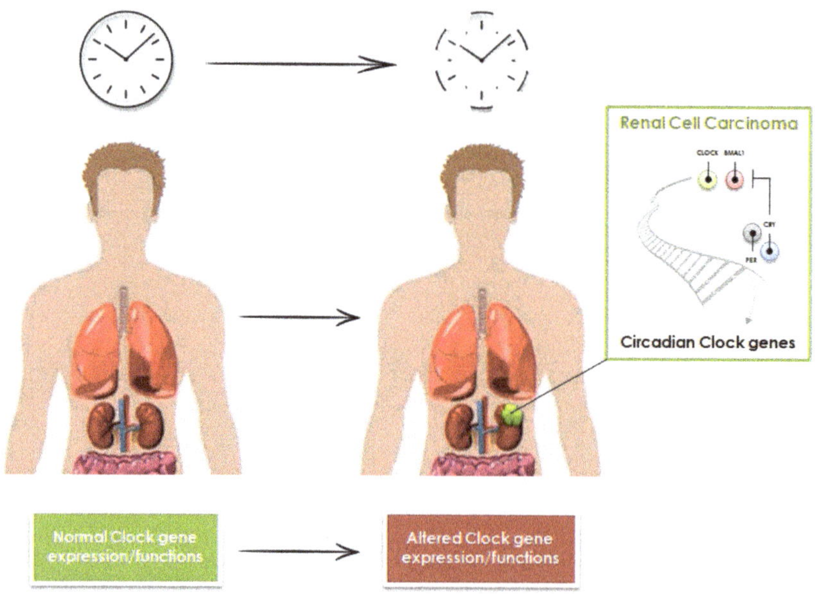

Figure 1. Role of Circadian Clock genes in the onset of Renal Cell Carcinoma.

Mazzocoli et al. [28] evaluated the expression of CC genes by DNA microarray assays and qRT-PCR in 11 RCC primary tumors and matched healthy tissues. They documented down-regulation of *PER2*, *TIMELESS* and *TIPIN* and up-regulation of *SERPINE1* genes. Furthermore, a statistically significant correlation between mRNA levels of *PER2* and *CSNKIE*, *PER2* and *TIPIN*, *PER2* and *SERPINE1*, *TIMELESS* and *TIPIN*, *TIMELESS* and *CSNKIE*, *TIPIN* and *CSNKIE* was reported [28].

CC system also shapes the activity of the mammalian target of rapamycin (mTOR) pathway, which is crucial for the development of RCC (Figure 2) [29].

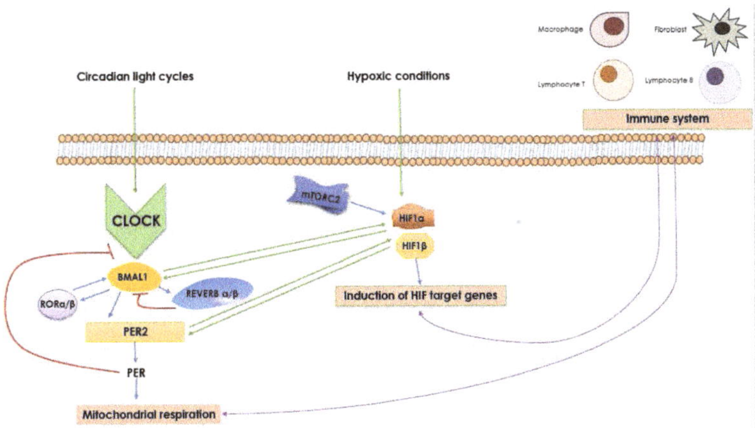

Figure 2. Relationship between HIF and CLOCK pathway. Under circadian light cycles, CLOK protein combines with BMAL1 protein to generate BMAL1/CLOCK heterodimer, which cause transcriptional activation of core clock genes (for example, transcription of *PER2*); this in turn inhibits BMAL1/CLOCK dimer activity. Meanwhile, the BMAL1/CLOCK dimers activate the transcription of the *Rev-Erb-α/β* and *RORα/β* genes, and the resulting translated proteins inhibit and promote the transcription of BMAL1, respectively. PER pathway regulates many transcriptional–translational processes influencing the whole cell metabolism and particularly mitochondrial activity. Together with the mTOR axis, hypoxia inducible factor (HIF) pathway orchestrates responses to oxygen and nutrient availability. HIF-1α can increase the amplitude of the PER2 circadian rhythm oscillation; colocalize with BMAL1 on E-box regions to increase expression of circadian and HIF target genes; CLOCK and BMAL1 can also express the *HIF-1α* gene. HIF patwhay activation and mitochondrial respiration can interplay with immune response. BMAL1 = Brain and Muscle ARNT-Like 1; REVERB *α*/*β* (REV-ERB-*α* and REV-ERB-*β*) = nuclear receptors that regulate the expression of genes involved in the control of circadian rhythm, metabolism and inflammatory responses; ROR*α*/*β* = RAR-related orphan recep-tor alpha/beta; CLOCK = circadian locomotor output cycles protein kaput; PER2 = period2; PER = period; mTORC2 = mammalian target of rapamycin complex-2; HIF1*α* and HIF*β* = hypoxia inducible factor-*α* and hypoxia inducible factor-*β*.

The expression levels of mTOR proteins show a 24-h rhythm in RCC tissues, mainly due to the activity of the ubiquitination factor Fbxw7, which is regulated by the circadian regulator D-site-binding protein. Of note is the fact that, the administration of everolimus, an oral mTOR inhibitor, improved survival in animal models during periods of elevated mTOR expression [30].

Together with the mTOR pathway, hypoxia inducible factor (HIF) pathway orchestrates responses to oxygen and nutrient availability. HIF plays a vital role in renal tumorigenesis, being constitutively activated by inactivation of the von Hippel-Lindau gene. Recently, the role of HIF as a potential therapeutic target in RCC patients has been supported by the positive results obtained by Belzutifan (MK-6482), a potent selective small molecule inhibitor of the HIF-2α subunit [31]. In Caki-2 RCC cell lines, an interaction between PER2 and HIF has been observed. Indeed, it has been shown that HIF-1α can increase the amplitude of the PER2 circadian rhythm oscillation by directly binding to the HIF-binding site located on the PER2 promoter [32].

As for the prognostic role of CC genes, the analysis of Cancer Genome Atlas (TCGA) data reported that the overexpression of *PER1-3*, *CRY2*, *CLOCK*, *NR1D2*, and *RORα*, as well as the under-expression of *TIMELESS* and *NPAS2*, were correlated with longer survival in patients with RCC [33]. Furthermore, the expression of *PER2*, *DBP*, *PER3*, *CRY2*, and *RORα*, has been shown to be significantly associated with kidney cancer prognosis [34]. Interestingly, in this study the expression of *PER2*, *DBP*, *PER3*, *CRY2*, and *RORα* genes was positively associated with the infiltration levels of CD4 and CD8 T cells [34].

4. Circadian Variations of Cytokines and Chemokines Involved in Renal Cell Carcinoma

Pro-inflammatory cytokines, such as Interleukin (IL)-1, IL-6 and tumor-necrosis factor (TNF)-α, and the expression of inflammatory chemokines (i.e., CXCL9 and CXCL10) has been associated with RCC tumor growth, angiogenesis, and response to therapy [35–37]. On the other hand, anti-inflammatory cytokines, including IL-10 and transforming growth factor (TGF)-β, and non-inflammatory chemokines (i.e., CCL17 and CCL24) produced by M2 phenotype tumor-associated macrophages (TAMs) promote tissue remodeling and angiogenesis in RCC microenvironment [38].

Moreover, the levels of several cytokines and chemokines involved in RCC onset and progression vary during the 24 h day. In this regard, it has been observed that the secretion of IL-6 varies during daytime, with two nadirs at about 8.00 and 21.00. Of note, IL-6 secretion is strictly associated with the sleep-wake rhythm and results high in disorders of excessive daytime sleepiness such as narcolepsy and obstructive sleep apnea [39].

Furthermore, TNF-α and TGF-β have been shown to regulate the transcription of the CC genes [40]. In turn, circadian oscillations of TNF-α gene expression are regulated by clock genes *BMAL1* and *CLOCK1* [41].

As for chemokines, the *PER2* gene regulates the levels of CCL5 (Rantes), a C-C chemokine secreted by T lymphocytes late after activation, fibroblasts, epithelial cells, and endothelial cells after stimulation by TNF –α and interleukin-1β [42].

5. Evidence on the Time-of-Day Influence on the Efficacy of Immunotherapy and Targeted Therapy

The time-of-day of treatment administration has emerged as a potential factor on the effectiveness of anticancer drugs. These chrono-pharmacological phenomena result of bot from the pharmacodynamics and pharmacokinetics of these agents, which are affected by the activity of CC genes. In this regard, Hori et al. reported that the delivery of anti-cancer drugs to tumor tissues varies following the circadian oscillations of blood flow in tumor tissues [43]. Similarly, studies in mice reported reduced tumor growth when cyclin-dependent kinase 4/6 (CDK4/6) drugs were administered in a time-dependent manner, showing higher efficacy in a morning dosing regimen compared to nighttime dosing [44].

Chronotherapy is defined as the administration of a treatment in coordination with the body's circadian rhythms; the aims of chronotherapy were the optimization of efficacy and minimization of adverse events. In the study led by Deprés-Brummer et al. in 1991 [45], ten patients with advanced RCC or melanoma were treated with 15–20 MU/m^2/day recombinant alpha-interferon-2b through a continuous 21-day intravenous schedule at circadian modulated rate. Compared with standard administration schedules, this circadian infusion schedule registered a significant increment in the total daily dose and dose intensity, with seven of the ten patients alive at a median follow-up of 15 months and two patients who continued chronotherapy for at least 9 and 13 months, respectively.

In 2009, Shiba et al. [46] investigated the expression of type-1 interferon receptor (IFNAR2) in peripheral blood mononuclear cells (PBMC) from RCC patients highlighting a peak at night, followed by a downregulation within 48 h from IFN-alpha administration and a subsequent recovery within further 48 h. These findings were in line with the results published in 1995 on the seven-day continuous infusion of IFN-α through a circadian schedule with maximum delivery between 6 p.m. and 3 a.m. [47].

Immunotherapies targeting immune checkpoint receptors or ligands such as Cytotoxic T-Lymphocyte-Associated protein-4 (CTLA-4), Programmed Death -1 (PD-1) and its ligand (PD-L1), are currently being widely exploited in clinical trials for multiple cancer types. Nonetheless, these agents are known to induce a significant inflammatory response and immune-related adverse events [48]. Since immune cell trafficking and inflammatory pathways are under CC control, applying a chronotherapy approach could help mitigate the associated toxicity issues [49]. The influence of time-of-day infusion on the effectiveness of immune checkpoint inhibitors has been recently investigated by Qian et al. in patients with advanced melanoma treated by ipilimumab, nivolumab, or pembrolizumab [50]. Specifically, patients 20% who had received at least of infusions after 4.30 pm (the cut off time delineating the onset of the evening) reported the worst overall survival (OS) [Hazard Ratio (HR) 1.31, 95% Confidence Interval (CI) 1.00–1.71, $p = 0.046$]. Accordingly, at a propensity score-matched analysis, median OS was shorter in patients who received at least 20% of immunotherapy infusions after 4.30 pm (4.8 years vs. not reached, HR 2.04, 95%CI: 1.08–2.98, $p = 0.023$) [50,51].

More recently, our group had observed that patients treated by dual immune checkpoint inhibitors or by the combination of immune checkpoint inhibitors with anti-VEGR TKIs before 4.30 pm had a significantly longer median PFS compared with those with the latest administration (12.3 vs. 5.6 months; HR 2.28: 95%CI 1.1–5.15; $p = 0.048$). Overall Survival data were not mature but demonstrated a trend toward improved survival for patients with earlier infusion administration (HR 2.33 $p = 0.16$) [52].

6. Emerging Strategies to Modulate Clock Genes in Patients with Cancer

Compounds targeting the CC have been developed as potential therapies for clock-related diseases, including cancer. Oshima et al. [53] identified by phosphoproteomics a potent and selective inhibitor of CK2 named GO289. This molecule can inhibit multiple phosphorylation sites on clock proteins, including PER2 S693, leading to decreased cancer cell growth. More recently, Borgo et al. [54] compared the efficacy and selectivity of CK2 inhibitors through a phosphoproteomics approach, concluding that both GO289 and CX4945 showed negligible off-target effects and high inhibitory efficacy against CK2.

Currently, an observational study (COMBOREIN, NCT03571438) investigating the treatment of RCC patients' cell cultures with the combination of CK2, and ATM inhibitors (compared to sunitinib, pazopanib, or temsirolimus) is ongoing, with the study completion date estimated at 30 September 2024.

Promising strategies to modulate clock proteins in patients with cancer consist of small molecules targeting biological clock, and synthetic anticancer chronobiotics directed against mammalian circadian clock components (i.e., CRYs, REV-ERBs, and RORs), as well as casein kinases [55].

7. Discussion

It is well known that CC genes and the cell cycle are tightly coupled, cooperating for proper cell functioning, and the dysregulation of the CC can significantly affect cell homeostasis and promote cancer development [3]. While the functions of the CC genes in normal physiology have been fully elucidated, studies of CC gene alterations in cancer are still lacking, leaving a gap in the clarity and description of their functions in neoplastic cells.

In RCC, the lack of validated biomarkers of efficacy or resistance of current therapeutic approaches strongly supports the necessity of exploring a spectrum of clinical and behavioral variables that could impact on patient outcome. Diet [56,57], fasting [58], concomitant medications [59–62], physical activity [63], and various other factors have been linked to immunotherapy efficacy in RCC. In this scenario, dosing time is an emerging element to be considered while managing RCC patients.

Including chronotherapy in daily therapy for RCC may offer a more effective and less toxic approach, although biomarkers for chronotherapy strategies' efficiency still need to be adequately defined [64]. Nevertheless, the results on the timing of immune checkpoint administration in patients with melanoma and RCC clearly indicate the clinical relevance of chronotherapy.

Well-designed, larger-size, and higher-quality cancer patient cohort studies are needed to investigate the precise impact of CC genes on the pathobiological behaviors of cancers. Additional in vitro and in vivo experiments and ad hoc clinical trials are warranted to better elucidate CC involvement in RCC biology, with the ultimate aim to improve patient prognosis and QoL. Finally, we need to plan dedicated clinical trials to assess: (1) the combination of emerging drugs regulating the CC with current standards; (2) the employment of CC targeting agents as maintenance treatment, (3) and as adjuvant treatment in patients with high-risk of relapse. In particular for mRCC, we could hypothesize to design a randomized trial aimed to assess the impact of treatment administration before and after 4.30 p.m.

8. Conclusions

Role of CC genes and circadian rhythm in patients with RCC is an intriguing and recent area of research, particularly considering the new findings on the impact of circadian rhythm in anticancer therapy. To date, evidence clearly supports the idea that CC genes and proteins may represent future therapeutic targets and the time of administration of immunotherapy drugs during the day is a nonnegligible factor.

Author Contributions: Conceptualization, M.S., G.S., J.M.-C. and E.G.; writing—original draft preparation, M.S., G.S. and S.B.; writing—review and editing, M.S., J.M.-C., G.S., E.T.L., F.M., V.M., G.M., B.L.R., E.G. and S.B; supervision, S.B. All authors have read and agreed to the published version of the manuscript.

Funding: This research received no external funding.

Conflicts of Interest: E.G. has received honoraria for speaker engagements, advisory roles or funding of continuous medical education from Adacap, AMGEN, Angelini, Astellas, Astra Zeneca, Bayer, Blueprint, Bristol Myers Squibb, Caris Life Sciences, Celgene, Clovis-Oncology, Eisai, Eusa Pharma, Genetracer, Guardant Health, HRA-Pharma, IPSEN, ITM-Radiopharma, Janssen, Lexicon, Lilly, Merck KGaA, MSD, Nanostring Technologies, Natera, Novartis, ONCODNA (Biosequence), Palex, Pharmamar, Pierre Fabre, Pfizer, Roche, Sanofi-Genzyme, Servier, Taiho, and Thermo Fisher Scientific. E.G. has received research grants from Pfizer, Astra Zeneca, Astellas, and Lexicon Pharmaceuticals. J.M.-C. declares consultant, advisory or speaker roles for IPSEN, Roche, Pfizer, Sanofi, Janssen, and BMS; he also received research grants from Pfizer, IPSEN and Roche. The other authors declare no conflict of interest.

References

1. Rosbash, M. Circadian Rhythms and the Transcriptional Feedback Loop (Nobel Lecture). *Angew. Chem. Int. Ed.* **2021**, *60*, 8650–8666. [CrossRef] [PubMed]
2. Reddy, P.; Zehring, W.A.; Wheeler, D.A.; Pirrotta, V.; Hadfield, C.; Hall, J.C.; Rosbash, M. Molecular Analysis of the Period Locus in Drosophila Melanogaster and Identification of a Transcript Involved in Biological Rhythms. *Cell* **1984**, *38*, 701–710. [CrossRef] [PubMed]
3. Takahashi, J.S. Transcriptional architecture of the mammalian circadian clock. *Nat. Rev. Genet.* **2017**, *18*, 164–179. [CrossRef] [PubMed]
4. Gu, F.; Gomez, E.C.; Chen, J.; Buas, M.F.; Schlecht, N.F.; Hulme, K.; Kulkarni, S.V.; Singh, P.K.; O'Connor, R.; Ambrosone, C.B.; et al. Genes Relevant to Tissue Response to Cancer Therapy Display Diurnal Variation in mRNA Expression in Human Oral Mucosa. *J. Circadian Rhythm.* **2021**, *19*, 8. [CrossRef] [PubMed]
5. Santoni, M.; Piva, F.; Porta, C.; Bracarda, S.; Heng, D.Y.; Matrana, M.R.; Grande, E.; Mollica, V.; Aurilio, G.; Rizzo, M.; et al. Artificial Neural Networks as a Way to Predict Future Kidney Cancer Incidence in the United States. *Clin. Genitourin. Cancer* **2021**, *19*, e84–e91. [CrossRef]
6. Allemani, C.; Matsuda, T.; Di Carlo, V.; Harewood, R.; Matz, M.; Nikšić, M.; Bonaventure, A.; Valkov, M.; Johnson, C.J.; Estève, J.; et al. Global surveillance of trends in cancer survival 2000-14 (CONCORD-3): Analysis of individual records for 37 513 025

patients diagnosed with one of 18 cancers from 322 population-based registries in 71 countries. *Lancet* **2018**, *391*, 1023–1075. [CrossRef]
7. Santoni, M.; Conti, A.; Porta, C.; Procopio, G.; Sternberg, C.N.; Basso, U.; De Giorgi, U.; Bracarda, S.; Rizzo, M.; Ortega, C.; et al. Sunitinib, pazopanib or sorafenib for the treatment of patients with late relapsing metastatic renal cell carcinoma. *J. Urol.* **2015**, *193*, 41–47. [CrossRef]
8. Conti, A.; Santoni, M.; Amantini, C.; Burattini, L.; Berardi, R.; Santoni, G.; Cascinu, S.; Muzzonigro, G. Progress of molecular targeted therapies for advanced renal cell carcinoma. *Biomed Res. Int.* **2013**, *2013*, 419176. [CrossRef]
9. Ciccarese, C.; Alfieri, S.; Santoni, M.; Santini, D.; Brunelli, M.; Bergamini, C.; Licitra, L.; Montironi, R.; Tortora, G.; Massari, F. New toxicity profile for novel immunotherapy agents: Focus on immune-checkpoint inhibitors. *Expert Opin. Drug Metab. Toxicol.* **2016**, *12*, 57–75. [CrossRef]
10. Rizzo, A.; Mollica, V.; Dall'Olio, F.G.; Ricci, A.D.; Maggio, I.; Marchetti, A.; Rosellini, M.; Santoni, M.; Ardizzoni, A.; Massari, F. Quality of life assessment in renal cell carcinoma Phase II and III clinical trials published between 2010 and 2020: A systematic review. *Future Oncol.* **2021**, *17*, 2671–2681. [CrossRef]
11. Hsieh, J.J.; Purdue, M.P.; Signoretti, S.; Swanton, C.; Albiges, L.; Schmidinger, M.; Heng, D.Y.; Larkin, J.; Ficarra, V. Renal cell carcinoma. *Nat. Rev. Dis. Primers* **2017**, *3*, 17009. [CrossRef] [PubMed]
12. Ruan, W.; Yuan, X.; Eltzschig, H.K. Circadian Rhythm as a Therapeutic Target. *Nat. Rev. Drug Discov.* **2021**, *20*, 287–307. [CrossRef] [PubMed]
13. Patke, A.; Young, M.W.; Axelrod, S. Molecular Mechanisms and Physiological Importance of Circadian Rhythms. *Nat. Rev. Mol. Cell Biol.* **2020**, *21*, 67–84. [CrossRef]
14. Shafi, A.A.; Knudsen, K.E. Cancer and the Circadian Clock. *Cancer Res.* **2019**, *79*, 3806–3814. [CrossRef] [PubMed]
15. Chen, S.T.; Choo, K.B.; Hou, M.F.; Yeh, K.T.; Kuo, S.J.; Chang, J.-G. Deregulated expression of the PER1, PER2 and PER3 genes in breast cancers. *Carcinogenesis* **2005**, *26*, 1241–1246. [CrossRef]
16. Liu, B.; Xu, K.; Jiang, Y.; Li, X. Aberrant expression of Per1, Per2 and Per3 and their prognostic relevance in non-small cell lung cancer. *Int. J. Clin. Exp. Pathol.* **2014**, *7*, 7863–7871.
17. Kang, T.H.; Sancar, A. Circadian regulation of DNA excision repair: Implications for chrono-chemotherapy. *Cell Cycle* **2009**, *8*, 1665–1667. [CrossRef]
18. Xue, X.; Liu, F.; Han, Y.; Li, P.; Yuan, B.; Wang, X.; Chen, Y.; Kuang, Y.; Zhi, Q.; Zhao, H. Silencing NPAS2 promotes cell growth and invasion in DLD-1 cells and correlated with poor prognosis of colorectal cancer. *Biochem. Biophys. Res. Commun.* **2014**, *450*, 1058–1062. [CrossRef]
19. Wang, L.; Chen, B.; Wang, Y.; Sun, N.; Lu, C.; Qian, R.; Hua, L. hClock gene expression in human colorectal carcinoma. *Mol. Med. Rep.* **2013**, *8*, 1017–1022. [CrossRef]
20. Wang, X.; Yan, D.; Teng, M.; Fan, J.; Zhou, C.; Li, D.; Qiu, G.; Sun, X.; Li, T.; Xing, T.; et al. Reduced expression of PER3 is associated with incidence and development of colon cancer. *Ann. Surg. Oncol.* **2012**, *19*, 3081–3088. [CrossRef]
21. Li, W.; Liu, L.; Liu, D.; Jin, S.; Yang, Y.; Tang, W.; Gong, L. Decreased circadian component Bmal1 predicts tumor progression and poor prognosis in human pancreatic ductal adenocarcinoma. *Biochem. Biophys. Res. Commun.* **2016**, *472*, 156–162. [CrossRef]
22. Hsu, C.M.; Lin, S.F.; Lu, C.T.; Lin, P.M.; Yang, M.Y. Altered expression of circadian clock genes in head and neck squamous cell carcinoma. *Tumor Biol.* **2012**, *33*, 149–155. [CrossRef] [PubMed]
23. Hwang-Verslues, W.W.; Chang, P.H.; Jeng, Y.M.; Kuo, W.H.; Chiang, P.H.; Chang, Y.C.; Hsieh, T.H.; Su, F.Y.; Lin, L.C.; Abbondante, S.; et al. Loss of corepressor PER2 under hypoxia up-regulates OCT1-mediated EMT gene expression and enhances tumor malignancy. *Proc. Natl. Acad. Sci. USA* **2013**, *110*, 12331–12336. [CrossRef] [PubMed]
24. Bellet, M.M.; Stincardini, C.; Costantini, C.; Gargaro, M.; Pieroni, S.; Castelli, M.; Piobbico, D.; Sassone-corsi, P.; Della-Fazia, M.A.; Romani, L.; et al. The Circadian Protein PER1 Modulates the Cellular Response to Anticancer Treatments. *Int. J. Mol. Sci.* **2021**, *22*, 2974. [CrossRef] [PubMed]
25. de Assis, L.V.M.; Kinker, G.S.; Moraes, M.N.; Markus, R.P.; Fernandes, P.A.; de Lauro Castrucci, A.M. Expression of the circadian clock gene BMAL1 positively correlates with antitumor immunity and patient survival in metastatic melanoma. *Front. Oncol.* **2018**, *8*, 185. [CrossRef]
26. Ukai, H.; Ueda, H.R. Systems Biology of Mammalian Circadian Clocks. *Annu. Rev. Physiol.* **2009**, *72*, 579–603. [CrossRef]
27. Ohashi, M.; Umemura, Y.; Koike, N.; Tsuchiya, Y.; Inada, Y.; Watanabe, H.; Tanaka, T.; Minami, Y.; Ukimura, O.; Miki, T.; et al. Disruption of circadian clockwork in in vivo reprogramming-induced mouse kidney tumors. *Genes Cells* **2018**, *23*, 60–69. [CrossRef]
28. Mazzoccoli, G.; Piepoli, A.; Carella, M.; Panza, A.; Pazienza, V.; Benegiamo, G.; Palumbo, O.; Ranieri, E. Altered expression of the clock gene machinery in kidney cancer patients. *Biomed. Pharmacother.* **2012**, *66*, 175–179. [CrossRef]
29. Santoni, M.; Pantano, F.; Amantini, C.; Nabissi, M.; Conti, A.; Burattini, L.; Zoccoli, A.; Berardi, R.; Santoni, G.; Tonini, G.; et al. Emerging strategies to overcome the resistance to current mTOR inhibitors in renal cell carcinoma. *Biochim. Biophys. Acta* **2014**, *1845*, 221–231. [CrossRef]
30. Okazaki, H.; Matsunaga, N.; Fujioka, T.; Okazaki, F.; Akagawa, Y.; Tsurudome, Y.; Ono, M.; Kuwano, M.; Koyanagi, S.; Ohdo, S. Circadian regulation of mTOR by the ubiquitin pathway in renal cell carcinoma. *Cancer Res.* **2014**, *74*, 543–551. [CrossRef]
31. Jonasch, E.; Donskov, F.; Iliopoulos, O.; Rathmell, W.K.; Narayan, V.K.; Maughan, B.L.; Oudard, S.; Else, T.; Maranchie, J.K.; Welsh, S.J.; et al. Belzutifan for Renal Cell Carcinoma in von Hippel-Lindau Disease. *N. Engl. J. Med.* **2021**, *385*, 2036–2046. [CrossRef]

32. Okabe, T.; Kumagai, M.; Nakajima, Y.; Shirotake, S.; Kodaira, K.; Oyama, M.; Ueno, M.; Ikeda, M. The impact of HIF1α on the Per2 circadian rhythm in renal cancer cell lines. *PLoS ONE* **2014**, *9*, e109693. [CrossRef] [PubMed]
33. Qiu, M.J.; Liu, L.P.; Jin, S.; Fang, X.F.; He, X.X.; Xiong, Z.F.; Yang, S.L. Research on circadian clock genes in common abdominal malignant tumors. *Chronobiol. Int.* **2019**, *36*, 906–918. [CrossRef]
34. Liu, S.; Cheng, Y.; Wang, S.; Liu, H. Circadian Clock Genes Modulate Immune, Cell Cycle and Apoptosis in the Diagnosis and Prognosis of Pan-Renal Cell Carcinoma. *Front. Mol. Biosci.* **2021**, *8*, 747629. [CrossRef]
35. Aggen, D.H.; Ager, C.R.; Obradovic, A.Z.; Chowdhury, N.; Ghasemzadeh, A.; Mao, W.; Chaimowitz, M.G.; Lopez-Bujanda, Z.A.; Spina, C.S.; Hawley, J.E.; et al. Blocking IL1 Beta Promotes Tumor Regression and Remodeling of the Myeloid Compartment in a Renal Cell Carcinoma Model: Multidimensional Analyses. *Clin. Cancer Res.* **2021**, *27*, 608–621. [CrossRef] [PubMed]
36. Ishibashi, K.; Koguchi, T.; Matsuoka, K.; Onagi, A.; Tanji, R.; Takinami-Honda, R.; Hoshi, S.; Onoda, M.; Kurimura, Y.; Hata, J.; et al. Interleukin-6 induces drug resistance in renal cell carcinoma. *Fukushima J. Med. Sci.* **2018**, *64*, 103–110. [CrossRef] [PubMed]
37. Santoni, M.; Bracarda, S.; Nabissi, M.; Massari, F.; Conti, A.; Bria, E.; Tortora, G.; Santoni, G.; Cascinu, S. CXC and CC chemokines as angiogenic modulators in nonhaematological tumors. *Biomed Res. Int.* **2014**, *2014*, 768758. [CrossRef] [PubMed]
38. Santoni, M.; Massari, F.; Amantini, C.; Nabissi, M.; Maines, F.; Burattini, L.; Berardi, R.; Santoni, G.; Montironi, R.; Tortora, G.; et al. Emerging role of tumor-associated macrophages as therapeutic targets in patients with metastatic renal cell carcinoma. *Cancer Immunol. Immunother.* **2013**, *62*, 1757–1768. [CrossRef]
39. Vgontzas, A.N.; Bixler, E.O.; Lin, H.M.; Prolo, P.; Trakada, G.; Chrousos, G.P. IL-6 and its circadian secretion in humans. *Neuroimmunomodulation* **2005**, *12*, 131–140. [CrossRef]
40. Ertosun, M.G.; Kocak, G.; Ozes, O.N. The regulation of circadian clock by tumor necrosis factor alpha. *Cytokine Growth Factor Rev.* **2019**, *46*, 10–16. [CrossRef]
41. Onoue, T.; Nishi, G.; Hikima, J.I.; Sakai, M.; Kono, T. Circadian oscillation of TNF-α gene expression regulated by clock gene, BMAL1 and CLOCK1, in the Japanese medaka (*Oryzias latipes*). *Int. Immunopharmacol.* **2019**, *70*, 362–371. [CrossRef]
42. Chen, X.; Hu, Q.; Zhang, K.; Teng, H.; Li, M.; Li, D.; Wang, J.; Du, Q.; Zhao, M. The clock-controlled chemokine contributes to neuroinflammation-induced depression. *FASEB J.* **2020**, *34*, 8357–8366. [CrossRef] [PubMed]
43. Hori, K.; Zhang, Q.H.; Li, H.C.; Saito, S.; Sato, Y. Timing of cancer chemotherapy based on circadian variations in tumor tissue blood flow. *Int. J. Cancer* **1996**, *65*, 360–364. [CrossRef]
44. Lee, Y.; Lahens, N.F.; Zhang, S.; Bedont, J.; Field, J.M.; Sehgal, A. G1/S Cell Cycle Regulators Mediate Effects of Circadian Dysregulation on Tumor Growth and Provide Targets for Timed Anticancer Treatment. *PLoS Biol.* **2019**, *17*, e3000228. [CrossRef] [PubMed]
45. Deprés-Brummer, P.; Levi, F.; Di Palma, M.; Beliard, A.; Lebon, P.; Marion, S.; Jasmin, C.; Misset, J.L. A phase I trial of 21-day continuous venous infusion of alpha-interferon at circadian rhythm modulated rate in cancer patients. *J. Immunother.* **1991**, *10*, 440–447. [CrossRef]
46. Shiba, M.; Nonomura, N.; Nakai, Y.; Nakayama, M.; Takayama, H.; Inoue, H.; Tsujimura, A.; Nishimura, K.; Okuyama, A. Type-I interferon receptor expression: Its circadian rhythm and downregulation after interferon-alpha administration in peripheral blood cells from renal cancer patients. *Int. J. Urol.* **2009**, *16*, 356–359. [CrossRef]
47. Iacobelli, S.; Garufi, C.; Irtelli, L.; Martino, M.T.; Santobuono, F.; Vicario, G.; Tinari, N.; Fiorentino, B.; Innocenti, P.; Natoli, C. A phase I study of recombinant interferon-alpha administered as a seven-day continuous venous infusion at circadian-rhythm modulated rate in patients with cancer. *Am. J. Clin. Oncol.* **1995**, *18*, 27–31. [CrossRef]
48. Puzanov, I.; Diab, A.; Abdallah, K.; Bingham, C.O.; Brogdon, C.; Dadu, R.; Hamad, L.; Kim, S.; Lacouture, M.E.; LeBoeuf, N.R.; et al. Managing Toxicities Associated with Immune Checkpoint Inhibitors: Consensus Recommendations from the Society forImmunotherapy of Cancer (SITC) Toxicity Management Working Group. *J. Immunother. Cancer* **2017**, *5*, 95. [CrossRef]
49. Deng, W.; Zhu, S.; Zeng, L.; Liu, J.; Kang, R.; Yang, M.; Cao, L.; Wang, H.; Billiar, T.R.; Jiang, J.; et al. The Circadian Clock Controls Immune Checkpoint Pathway in Sepsis. *Cell Rep.* **2018**, *24*, 366–378. [CrossRef]
50. Qian, D.C.; Kleber, T.; Brammer, B.; Xu, K.M.; Switchenko, J.M.; Janopaul-Naylor, J.R.; Zhong, J.; Yushak, M.L.; Harvey, R.D.; Paulos, C.M.; et al. Effect of immunotherapy time-of-day infusion on overall survival among patients with advanced melanoma in the USA (MEMOIR): A propensity score-matched analysis of a single-centre, longitudinal study. *Lancet Oncol.* **2021**, *22*, 1777–1786. [CrossRef]
51. Santoni, M.; Molina-Cerrillo, J.; Massari, F.; Montironi, R.; Grande, E. Re: Effect of Immunotherapy Time-of-day Infusion on Overall Survival among Patients with Advanced Melanoma in the USA (MEMOIR): A Propensity Score-matched Analysis of a Single-centre, Longitudinal Study. *Eur. Urol.* **2022**, *81*, 623–624. [CrossRef] [PubMed]
52. Molina-Cerrillo, J.; Ortego, I.; Pinto, A.; Alonso-Gordoa, T.; Massari, F.; Aurilio, G.; Buti, S.; Santoni, M.; Grande, E. Does timing of Immune checkpoint inhibitors (ICIs) administration in first line Metastatic Renal Cell Carcinoma (mRCC) have impact in survival outcomes? *J. Clin. Oncol.* **2022**, *40*, e16512. [CrossRef]
53. Oshima, T.; Niwa, Y.; Kuwata, K.; Srivastava, A.; Hyoda, T.; Tsuchiya, Y.; Kumagai, M.; Tsuyuguchi, M.; Tamaru, T.; Sugiyama, A.; et al. Cell-based screen identifies a new potent and highly selective CK2 inhibitor for modulation of circadian rhythms and cancer cell growth. *Sci. Adv.* **2019**, *5*, eaau9060. [CrossRef] [PubMed]
54. Borgo, C.; Cesaro, L.; Hirota, T.; Kuwata, K.; D'Amore, C.; Ruppert, T.; Blatnik, R.; Salvi, M.; Pinna, L.A. Comparing the efficacy and selectivity of Ck2 inhibitors. A phosphoproteomics approach. *Eur. J. Med. Chem.* **2021**, *214*, 113217. [CrossRef]

55. Rahman, S.; Wittine, K.; Sedić, M.; Markova-Car, E.P. Small Molecules Targeting Biological Clock; A Novel Prospective for Anti-Cancer Drugs. *Molecules* **2020**, *25*, 4937. [CrossRef] [PubMed]
56. Greathouse, K.L.; Wyatt, M.; Johnson, A.J.; Toy, E.P.; Khan, J.M.; Dunn, K.; Clegg, D.J.; Reddy, S. Diet-microbiome interactions in cancer treatment: Opportunities and challenges for precision nutrition in cancer. *Neoplasia* **2022**, *29*, 100800. [CrossRef]
57. Westheim, A.J.F.; Stoffels, L.M.; Dubois, L.J.; van Bergenhenegouwen, J.; van Helvoort, A.; Langen, R.C.J.; Shiri-Sverdlov, R.; Theys, J. Fatty Acids as a Tool to Boost Cancer Immunotherapy Efficacy. *Front. Nutr.* **2022**, *9*, 868436. [CrossRef]
58. Cortellino, S.; Raveane, A.; Chiodoni, C.; Delfanti, G.; Pisati, F.; Spagnolo, V.; Visco, E.; Fragale, G.; Ferrante, F.; Magni, S.; et al. Fasting renders immunotherapy effective against low-immunogenic breast cancer while reducing side effects. *Cell Rep.* **2022**, *40*, 111256. [CrossRef]
59. Santoni, M.; Massari, F.; Matrana, M.R.; Basso, U.; De Giorgi, U.; Aurilio, G.; Buti, S.; Incorvaia, L.; Rizzo, M.; Martignetti, A.; et al. Statin use improves the efficacy of nivolumab in patients with advanced renal cell carcinoma. *Eur. J. Cancer* **2022**, *172*, 191–198. [CrossRef]
60. Santoni, M.; Molina-Cerrillo, J.; Myint, Z.W.; Massari, F.; Buchler, T.; Buti, S.; Matrana, M.R.; De Giorgi, U.; Rizzo, M.; Zabalza, I.O.; et al. Concomitant Use of Statins, Metformin, or Proton Pump Inhibitors in Patients with Advanced Renal Cell Carcinoma Treated with First-Line Combination Therapies. *Target. Oncol.* **2022**, *17*, 571–581. [CrossRef]
61. Bersanelli, M.; Giannarelli, D.; De Giorgi, U.; Pignata, S.; Di Maio, M.; Clemente, A.; Verzoni, E.; Giusti, R.; Di Napoli, M.; Aprile, G.; et al. INfluenza Vaccine Indication During therapy with Immune checkpoint inhibitors: A multicenter prospective observational study (INVIDIa-2). *J. Immunother. Cancer* **2021**, *9*, e002619. [CrossRef] [PubMed]
62. Buti, S.; Bersanelli, M.; Perrone, F.; Bracarda, S.; Di Maio, M.; Giusti, R.; Nigro, O.; Cortinovis, D.L.; Aerts, J.G.J.V.; Guaitoli, G.; et al. Predictive ability of a drug-based score in patients with advanced non-small-cell lung cancer receiving first-line immunotherapy. *Eur. J. Cancer* **2021**, *150*, 224–231. [CrossRef] [PubMed]
63. Shaver, A.L.; Sharma, S.; Nikita, N.; Lefler, D.S.; Basu-Mallick, A.; Johnson, J.M.; Butryn, M.; Lu-Yao, G. The Effects of Physical Activity on Cancer Patients Undergoing Treatment with Immune Checkpoint Inhibitors: A Scoping Review. *Cancers* **2021**, *13*, 6364. [CrossRef] [PubMed]
64. Ozturk, N.; Ozturk, D.; Kavakli, I.H.; Okyar, A. Molecular Aspects of Circadian Pharmacology and Relevance for Cancer Chronotherapy. *Int. J. Mol. Sci.* **2017**, *18*, 2168. [CrossRef] [PubMed]

cancers

MDPI

Review

Cystic Clear Cell Renal Cell Carcinoma: A Morphological and Molecular Reappraisal

Giacomo Maria Pini [1], Roberta Lucianò [1] and Maurizio Colecchia [2,*]

[1] Department of Pathology, IRCCS San Raffaele Scientific Institute, 20132 Milan, Italy
[2] IRCCS San Raffaele Scientific Institute, Vita-Salute San Raffaele University, 20132 Milan, Italy
* Correspondence: colecchia.maurizio@hsr.it

Simple Summary: Renal cancer is a common malignant neoplasm. Indeed, not every cancer is created equal, as there are different entities with specific morphological and molecular features. These differences also lead to different clinical behaviors, ranging from benign to highly aggressive neoplasms. In renal cancer, it is not unusual to have cystic hollow spaces. Clear cell renal cell carcinoma is the most frequent type of renal cancer, and it can be cystic. Distinguishing it from other subtypes of renal carcinomas can, in some cases, be challenging.

Abstract: A wide variety of renal neoplasms can have cystic areas. These can occur for different reasons: some tumors have an intrinsic cystic architecture, while others exhibit pseudocystic degeneration of necrotic foci or they have cystically dilated renal tubules constrained by stromal neoplastic cells. Clear cell renal cell carcinoma (CCRCC), either solid or cystic, is the most frequent type of renal cancer. While pseudocysts are found in high-grade aggressive CCRCC, cystic growth is associated with low-grade indolent cases. The latter also form through a cyst-dependent molecular pathway, and they are more frequent in patients suffering from VHL disease. The differential diagnosis of multilocular cystic renal neoplasm of low malignant potential and clear cell papillary renal cell tumor can be especially hard and requires a focused macroscopical and microscopical pathological analysis. As every class of renal tumor includes cystic forms, knowledge of the criteria required for a differential diagnosis is mandatory.

Keywords: renal cell carcinoma; cystic renal neoplasm; differential diagnosis

Citation: Pini, G.M.; Lucianò, R.; Colecchia, M. Cystic Clear Cell Renal Cell Carcinoma: A Morphological and Molecular Reappraisal. *Cancers* **2023**, *15*, 3352. https://doi.org/10.3390/cancers15133352

Academic Editor: José I. López

Received: 31 May 2023
Revised: 15 June 2023
Accepted: 20 June 2023
Published: 26 June 2023

1. Introduction

Renal cancer is a common malignant neoplasm whose classification has been expanding over the decades [1]. There are, indeed, more frequent and rarer subtypes of tumors, some of which are molecularly defined [2,3]. Morphological analysis is, however, still the basis of pathological diagnosis, and cystic areas can be present in a wide variety of renal neoplasms (whether benign or malignant) as minor or dominant components [4,5]. They are estimated to be present in 5–15% of lesions, where they reflect an inherent architecture of the tumor [4,6]. They must be distinguished from the pseudocystic degeneration of necrotic foci: while cystic growth is associated with a more indolent behavior, tumoral necrosis is present in aggressive masses [5]. This is especially true in cystic clear cell renal cell carcinoma (CCRCC), which is also the most frequent cystic renal cancer [4]. Cystic CCRCC is more frequent in patients with Von Hippel–Lindau (VHL) syndrome, and different molecular patterns are also implicated in its development compared with solid cases [4]. Nevertheless, cystic CCRCC is not classified as a separate pathologic entity. By contrast, multilocular cystic renal neoplasm of low malignant potential (MCNLMP) is independently identified in the WHO classification, despite molecular overlaps with CCRCC [5]. Cystic areas can be present in non-renal-cell neoplasms of the kidney as well, further complicating the diagnostic process [4]. In this review, we address the main morphological and molecular features of cystic CCRCC, together with its main differential diagnoses.

2. Macroscopic and Microscopic Features of Cystic CCRCC

According to the 2019 Bosniak classification (BC), the term "cystic renal mass" can be applied to neoplasms with a predominant cystic pattern and less than 25% enhancing tissue [7,8]. This term has an agnostic character, as it can imply both benign and malignant lesions [7,8]. A distinction must be made between renal cysts, which are benign, and solid neoplasms with minor cystic components. The latter are more likely to be malignant with pseudocystic degenerative areas with tumoral necrosis [7,8]. Both cystic growth pattern and pseudocystic degeneration can occur in CCRCC [6]. Less than 5% of CCRCCs have multiple cysts as their predominant architecture [4]. The minimum amount of cystic architecture necessary to define cystic CCRCC varies in the literature. Some authors mirror BC, as they require cystic areas of at least 75% [9], while others lower the threshold to 50% [10]. Interestingly, both cutoffs have proven to discriminate CCRCCs associated with a better prognosis [9,10].

Macroscopically, cystic growths appear as variably sized hollow spaces filled with clear or hemorrhagic fluid, with a clear separation from adjacent solid neoplastic tissue. Cysts can be single or multiple, with or without internal septations. When multiple cysts are predominant, the neoplasm can, overall, resemble a multilocular cyst. Evident solid areas have instead the typical golden-yellow color, with reddish hemorrhagic foci. Pseudocystic degenerative areas contain, instead, darker, denser, hemorrhagic material with cellular debris. They are more frequently centrally located within the lesion, and they can be surrounded by soft, greyish necrotic tissue. Vital parts of the tumor can also have, apart from the typical colors, whitish areas, where sarcomatoid differentiation is present.

Microscopically, along with macrocysts, even solid regions of CCRCCs can reveal a microcystic growth pattern (Figure 1A–D). Microcysts arise within tumoral nests, and cystic spaces are usually filled with red blood cells. Cells at the border of these microcysts do not have significantly different histological and immunohistochemical (IHC) features compared with solid acini. They have clear cytoplasm and variably sized nucleoli. Nuclei are usually basally located, although occasional apical alignment can be present. Positive labeling is present for carbonic anhydrase IX (CAIX) in a diffuse, box-shaped fashion, together with CD10, RCC, Vimentin and pan-cytokeratin. High-molecular-weight cytokeratins (HMWCKs) and CK7 are usually negative.

Microscopical analysis of cystic CCRCC usually reveals bland-looking clear cells with a low grade of differentiation (i.e., G1–G2 WHO grading). The epithelial coating of cysts, different from solid areas, is more likely to be CK7-positive, a feature that can be misleading in small biopsy samples. Nevertheless, HMWCKs are negative. In cases with a marked predominance of cystic growth, sampling of the capsule and septation can reveal clear cell clusters exceeding a 20× (1 mm) microscopic field, which is sufficient for a diagnosis of cystic CCRCC. Another criterion is the presence of an expansile growth of clear cells large enough to alter the contours of the capsule/septum. Finally, necrosis or vascular invasion could be present. Cellular clusters below the 20×/1 mm cutoff without expansile growth, necrosis or vascular invasion allow instead a diagnosis of MCNLMP [5].

Pseudocystic degenerative areas are filled with nuclear and cytoplasmatic debris of necrotic cells, together with varying numbers of red blood cells (Figure 2A–D). No epithelial lining can be identified, and the surrounding tissue can be necrotic as well. Vital neoplastic cells are high-grade (i.e., G3–G4 WHO grading). Blandly eosinophilic cytoplasm and hyaline globules are commonly found in high-grade CCRCC, which can be misleading if clear cell areas cannot be identified. Moreover, CAIX tends to become positive near necrotic areas in different types of renal neoplasms as a hypoxia-induced factor [11,12]. As such, the diagnosis of high-grade pseudocystic CCRCC can be challenging and requires more extensive sampling.

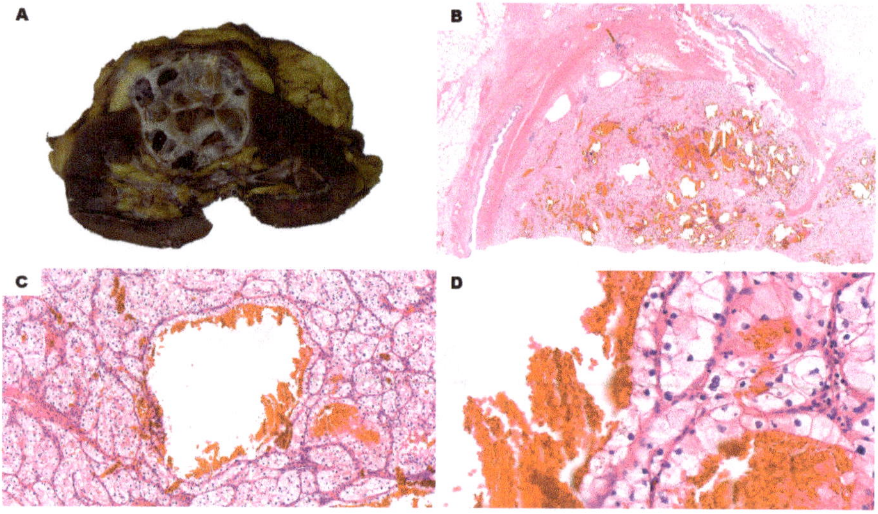

Figure 1. Cystic Clear Cell Renal Cell Carcinoma. (**A**) The gross specimen features a multiloculated, predominantly cystic nodule with variably thin walls and greyish solid areas. (**B**) (H&E) Low-power view of the lesion shows multiple, scattered, blood-filled cystic spaces, along with solid, whitish areas. (**C**) (H&E, 10×) Cysts are delimited by an epithelial lining with the same features of solid pericystic tissue. Around bigger cysts, higher magnification reveals the presence of microcystic spaces within neoplastic acini. No prominent nucleoli are evident at 10×. (**D**) (H&E, 40×) In this low-grade lesion, nucleoli are either very bland or absent, even with a high-power view.

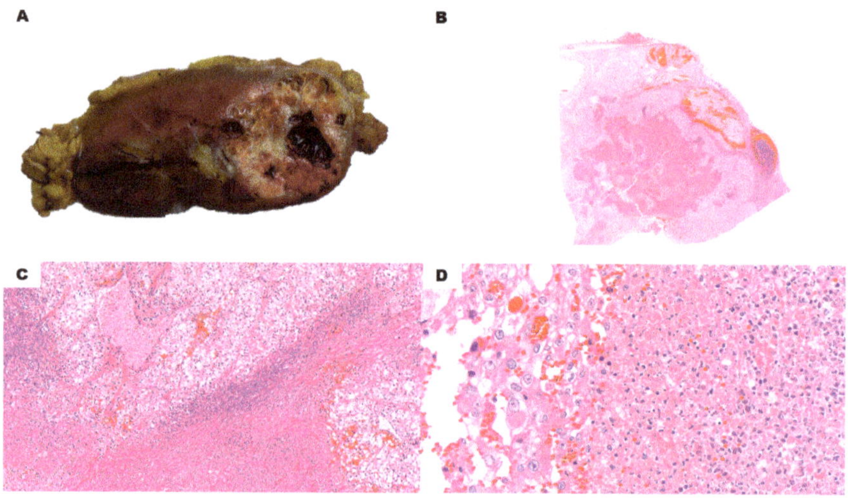

Figure 2. Pseudocystic Clear Cell Renal Cell Carcinoma. (**A**) Macroscopically, the nodule has a large hemorrhagic area surrounded by whitish and yellowish solid tissue. (**B**) (H&E) Low-power view shows a blood-filled area with blueish material at the border with the adjacent solid, whitish neoplastic tissue. (**C**) (H&E, 10×) The cystic area at the bottom of the picture, along with red blood cells, also contains blueish necrotic debris. Vital neoplastic cells at the top show prominent nucleoli already at this magnification, as it is a high-grade lesion. Other areas also showed rhabdoid cells. (**D**) (H&E, 40×) There is no clear-cut boundary between the necrotic debris on the right and vital solid neoplastic tissue on the left of the picture. This must be considered a pseudocyst rather than a cyst.

3. Molecular Features of Cystic CCRCC

In CCRCC, tumor-initiating molecular alterations involve the deletion of the 3p chromosome [13]. Specifically, loss of the 3p25 region is observed in 85% of CCRCCs [14]. As the *Von Hippel–Lindau (VHL)* tumor-suppressor gene is located in this area of the DNA, 3p25 deletion leads to the loss of one allele. The second *VHL* allele is, instead, inactivated either by mutation or methylation. Mutations of *VHL* are found in 64% of CCRCCs [14]. VHL protein is implicated in different molecular mechanisms, including microtubular stabilization for cilia formation and inhibition of the alpha subunit of hypoxia-inducible factor (HIF) [15,16]. When *VHL* is mutationally inactivated, the accumulation of HIFα upregulates vascular endothelial growth factor (VEGF), inducing angiogenesis. After initiating factors, other molecular events drive tumoral evolution towards different neoplastic subtypes [13,17]. For example, *BAP1* and *PBRM1* are two evolution-driver onco-suppressor genes (also located on chromosome 3p) mutated in 13% and 36% of CCRCCs, respectively [14]. Their mutations are mutually exclusive, leading to CCRCCs with different features. *BAP1*-mutated CCRCC is a high-grade neoplasm with poor vascularization, including renal cell carcinoma with sarcomatoid and rhabdoid features [18–20]. In these cases, also, *CDKN2A* deletions and increased expression of *MYC* transcriptional programs can be present [18]. Moreover, *BAP1*-mutated CCRCCs can be composed of large tumoral cells with abundant cytoplasm and a papillary architecture (reminiscent of RCC with MITF-family rearrangement), along with IHC positivity for racemase/AMACR and CK7 [19]. In addition, a rich T lymphocyte infiltration can be present. Such an immune-inflamed phenotype is characterized by immune activation and increased cytotoxic immune infiltration with upregulation of antigen presentation machinery genes and PD-L1 expression [18]. Infiltrated tumors are also enriched for chromosomal losses of 9p21.3 [21]. *PBRM1*-mutated CCRCC is a low-grade neoplasm with high levels of angiogenesis and lower levels of inflammation. Novel mutations can also be acquired by neoplasms during therapy with small molecules, giving rise to acquired drug resistance [22].

Different molecular patterns seem to be implied in the formation of cystic CCRCC, for which a cyst-dependent CCRCC progression pathway has been identified [4]. As previously mentioned, VHL contributes to cilia formation through microtubule stabilization. Loss of VHL is followed by an aberrant orientation of newly formed microtubules, which, in turn, hinders ciliogenesis. Such an effect upregulates the cell cycle, since cells without cilia cannot rest in the G0 phase, as differentiated cells would do. Therefore, cilia can be considered tumor-suppressor organelles, and their absence promotes the transition towards malignancy [23]. Loss of cilia is also associated with cyst development caused by impaired cellular signaling [15]. This process happens both in sporadic cystic CCRCC, as well as in inherited diseases, such as polycystic kidney disease (PKD) and VHL disease (VHLd) [24,25]. PKD and VHL diseases are therefore both considered among so-called ciliopathies [26]. The latter is an autosomal-dominant tumor syndrome: patients suffering from it develop renal cysts and CCRCC in 60% and 30% of cases, respectively [5,27]. Renal cancer in VHLd has been reported as early as 16 years of age, with a mean age of 37 years [28]. Renal cysts in VHLd are also potential precursors of CCRCC, as their epithelial linings can demonstrate dysplastic areas as well as loss of the remaining *VHL* non-genetically mutated allele [4]. It follows that CCRCC in VHLd is often both cystic and bilateral. Interestingly, just as *VHL* is an early cancer-initiator gene that requires further downstream molecular events, cyst formation cannot rely on VHL deficiency alone [23,29]. A critical role is played by GSK3β, a protein kinase that regulates cell proliferation, microtubule assembly, stability and dynamics [15]. Combined loss of VHL and GSK3β disrupts ciliary maintenance, and it is considered a key player in the cyst-dependent CCRCC progression pathway. The role of GSK3β is, however, yet to be fully elucidated, as evidence has also shown higher levels of expression both in PKD and in some CCRCCs [30,31]. According to these studies, its inhibition might actually be therapeutically useful to hinder cystic expansion and the progression of both PKD and CCRCCs [30,31].

4. Differential Diagnosis of Cystic CCRCC

As already mentioned, the range of renal neoplasms with cystic areas is wide. It encompasses every WHO group of tumors of the kidney (i.e., renal cell, metanephric, mixed epithelial and stromal, mesenchymal, embryonal and germ-cell tumors), including frequent and rare, adult and pediatric, and inherited and sporadic forms [5]. Attention must therefore be paid to patient age and the bilaterality of lesions. Pathological analysis must focus on the cellular lining of cysts, as well as the pericystic stroma and possible solid areas which can be focal.

Cystic areas in frequent renal neoplasms, such as chromophobe carcinoma, papillary carcinoma and oncocytoma, are possible but rather unusual [5]. Although rarer, the main differential diagnosis for cystic-predominant CCRCC is MNCLMP. Since the vast majority of CCRCCs harbor the VHL mutation, 3p copy number loss or both, tumors with clear cell histology lacking these alterations can often be reclassified as different established or emerging entities [32]. However, in the case of MNCLMP, there are molecular overlaps with cystic CCRCC, including deletion of the 3p chromosome and similar mutated genes which are part of the cyst-dependent pathway [5,33]. For this reason, MNCLMP might be considered a subtype of CCRCC, at the most indolent end of the spectrum. Nevertheless, it also has distinct clinical, morphological and molecular features that allow a separate classification [5,33,34]. MNCLMP accounts for less than 5% of renal tumors. It is usually incidentally detected as a monolateral lesion in patients slightly younger than CCRCC patients (median age: 55 vs. 62). The macroscopic appearance is entirely composed of variably sized cysts with a small total diameter (usually pT1, i.e., $\leq$7 cm) [5]. Neither solid nodules nor necrotic foci can be present. Even microscopical necrosis is not accepted, together with rhabdoid/sarcomatoid differentiation, lymphovascular invasion, frequent mitoses or any atypical mitosis. The epithelial lining of the cysts features one to a few layers of clear cells. Nuclei are randomly distributed, without a predilection for the apical portion of cells, and they must be low-grade (G1–G2 WHO grading). The capsule and septa are fibrous, and they can include clusters of clear cells, but they must be small (i.e., <1 mm or <20× microscopic area). When diagnostic criteria are strictly applied, tumors identified as MNCLMPs have a benign clinical behavior [5]. IHC analysis is not of aid in differential diagnosis with respect to CCRCCs, as they have the same profile [35]. Apart from the molecular similarities between MNCLMP and CCRCC, the former has also been shown to have a lower frequency of mutations. Six genes have been found significantly more frequently mutated in cystic CCRCC: SETD2, GIGYF2, FGFR3, BCR, KMT2C and TSC2 [36]. These are potential candidate genes that could help to elucidate the mechanisms in the development and progression of CCRCC, as well as in the differential diagnosis with MNCLMP [36].

Another benign renal cell tumor that can be nearly entirely cystic, featuring bland-looking clear cells, is clear cell papillary renal cell tumor (CCPRCT) (Figure 3A–D). Histologically, nuclei are oriented towards the luminal apex of the cells [37,38]. As in cystic CCRCC, CK7 is positive. However, CCPRCT also expresses HMWCKs (specifically, CK34βE12). CAIX signal has a cup-like pattern (i.e., with a missing luminal border), while CD10 is negative. Nevertheless, CCPRCT and low-grade CCRCC can have histologically identical areas, and unequivocal diagnosis of CCPRCT on needle biopsy may not be possible [5]. Molecularly, CCPRCTs have a distinct miRNA expression profile which also lacks the pattern typically associated with aggressive neoplastic behavior [39].

A cystic architecture combined with prominent nucleoli in epithelial cells can be found in WHO/ISUP category 5 neoplasms: tubulocystic RCC (TcRCC), acquired cystic disease-associated RCC (ACD-RCD), and eosinophilic solid and cystic RCC [2,5]. These neoplasms have a potentially misleading nucleolar appearance, as they look high-grade (equivalent to WHO grade 3) despite an indolent clinical behavior. They have eosinophilic cytoplasm, which distinguishes them from cystic CCRCCs. Moreover, TcRCC is composed of small cystic areas, which macroscopically reminds one of a sponge, rather than a multiloculated cyst. The cellular morphology ranges from flat to columnar, sometimes even with hobnail cells. Despite high-grade nucleoli, poorly differentiated or sarcomatoid areas must be absent, and the mitotic count is minimal. Differently from CCRCCs, CAIX is negative and racemase/AMACR is positive. ACD-RCDs are often multiple and bilateral solid masses in the setting of acquired cystic disease. As in VHLd, cysts are possible precursor lesions, and an ACD-RCD is often an intracystic mass. They are, however, derived from a history of long-term dialysis rather than an inherited gene mutation. Other than cystic areas, tubules lined by a multilayered epithelium with cytoplasmic vacuolation yield to a cribriform sieve-like pattern of growth. Other architectures might be present as well (e.g., papillary and solid). Oxalate crystals can be numerous within neoplastic tissue, and they are highlighted by polarized light. Both CD10 and racemase/AMACR are positive. In eosinophilic solid and cystic RCC, yellowish solid tissue is mixed with cystic spaces. Neoplastic cells are eosinophilic, but they also have basophilic intracytoplasmic inclusions surrounded by a clear halo (such inclusions are usually compared with Leishmania parasites). Binucleation and hobnail cellular profiles can also be present. Immunophenotypically, they are characterized by positive CD10 and racemase/AMACR, with a negative reaction for CAIX.

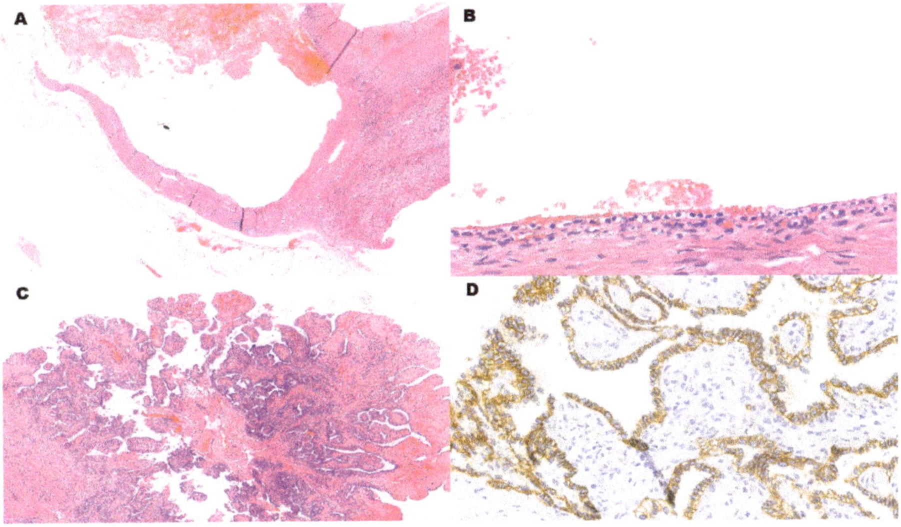

Figure 3. Clear Cell Papillary Renal Cell Tumor. (**A**) (H&E) Low-power view shows multiple blood-filled areas with fibrotic walls. (**B**) (H&E, 10×) The epithelial lining is composed of cuboidal to low-columnar clear cells. Nuclei in low-columnar cells tend to be oriented towards the cellular luminal apex. Nucleoli are not prominent. (**C**) (H&E) As the name of the tumor implies, papillary areas can also be present alongside cystic areas and they can protrude inside the cystic lumen. (**D**) (CAIX, 40×) Immunohistochemistry for CAIX signal has a cup-like pattern (i.e., with a missing luminal border). This pattern is typical of clear cell papillary renal cell tumors. The neoplasm is also CK7- and HMWCK-positive, while CD10 is negative.

While cystic CCRCCs have fibrotic septa and capsules, other neoplasms are biphasic with specific stromal proliferations. Angiomyolipoma with epithelial cysts (AMLEC) is a rare subtype of angiomyolipoma, a benign mesenchymal tumor of the kidney which is part of the perivascular epithelioid cell (PEC)/PEComa tumor family. The majority of AMLECs are sporadic lesions, mainly occurring in middle-aged females, but some may be part of tuberous sclerosis. The latter may be suspected in young patients, with no sex predilection. Along with solid areas predominantly composed of smooth muscle and blood vessels, in AMLEC, cystic spaces are present. They have a cuboidal-to-hobnail epithelium and a dense pericystic stroma, similar to the cambium layer in rhabdomyosarcoma. The epithelium is cytokeratin-positive, while the cambium-like stroma and solid areas are cytokeratin-negative and positive for melanocytic markers (HMB-45, melan-A and MiTF). Adult cystic nephromas (ACNs) and mixed epithelial and stromal tumors (MESTs) are two other closely related biphasic neoplasms that usually arise in women [40–42]. Their biphasic nature is embodied by a renal cell epithelial component, along with the proliferation of bland-looking spindle stromal cells (Figure 4A–C). The morphology recalls ovarian stroma, together with the expression of estrogen and progesterone receptors, as well as inhibin. While ACN is entirely cystic, MEST has solid, whitish areas with different patterns of growth (e.g., glandular, papillary and thyroid-like). Pediatric cystic nephroma (PCN) is a similar lesion, epidemiologically restricted to children (usually males) below 2 years of age, and is molecularly characterized by a DICER1 mutation [43]. While it can be cured by radical excision, it can also be part of DICER1 syndrome. The latter is characterized by an increased risk of developing benign and malignant diseases, including Sertoli–Leydig cell tumors, pleuropulmonary blastoma and embryonal rhabdomyosarcoma. If any immature nephroblastic element is present in a PCN-like lesion, the diagnosis switches to cystic partially differentiated nephroblastoma (CPDN) [5]. While nephroblastoma has a slight female preponderance and is a malignant neoplasm, CPDN is more frequent among males, and it is cured by surgery in stage I disease. CPDN lacks solid nodules both on gross and microscopic examination. Cystic septa are lined hobnail epithelial cells, and the walls contain primitive WT1-positive blastemal cells that differentiate into abortive tubules or glomerulus-like structures.

A

B

C

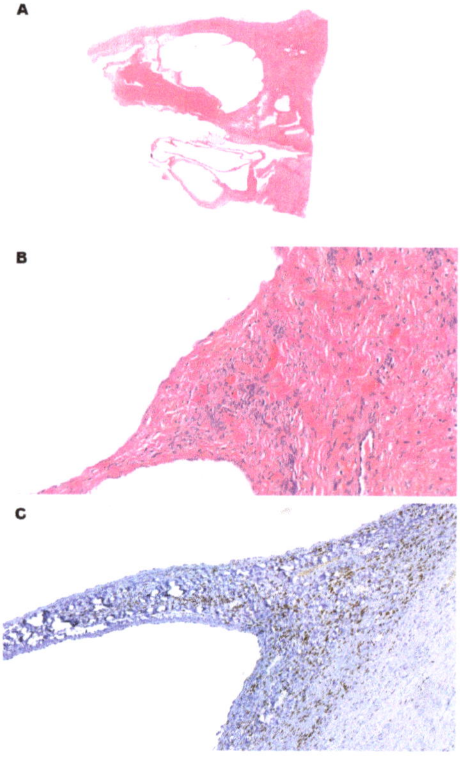

Figure 4. Mixed Epithelial and Stromal Tumor. (**A**) (H&E) Low-power view shows multiple empty cystic areas with fibrotic walls. (**B**) (H&E, 10×) The epithelial lining is flat and bland, while the pericystic stroma has foci with higher cellularity. Stromal cells are spindled and bland. These foci can be focal and hard to find. (**C**) (Estrogen Receptor, 40×) Immunohistochemistry for estrogen receptors is positive in stromal cells. They are also reactive for progesterone receptors and inhibin.

Metanephric stromal tumor is another pediatric renal neoplasm which can have cystic areas. Solid parts show a concentric peritubular growth of spindle cells expressing CD34 and with BRAF v600e mutation [5,44,45]. While tubules are more commonly unaltered by the encircling spindle cells, some become obstructed and therefore cystically dilated, rendering a cystic gross appearance. This is a third mechanism of cystic formation, alongside the aforementioned cystic architectural growth and pseudocystic degeneration in necrotic areas.

Renal teratomas are rare, most often cystic and mature, with mixed epithelial and stromal elements [5,46]. They can be pure or accompanied by a yolk-sac component. Microscopically, cystic spaces can be lined by a keratinizing squamous epithelium with skin adnexa or, alternatively, by a thick fibromuscular stroma without any lining. Generally speaking, considering the rarity of primary renal teratomas, more frequent diseases must always be ruled out, including renal metastasis from a distant germ-cell tumor, direct extension from a retroperitoneal germ-cell tumor and teratoid nephroblastoma [5,46].

5. Conclusions and Future Directions

Our knowledge about CCRCC pathogenesis and its molecular features has been increasing over the years. This is important not only for identifying different subtypes of CCRCC, but it also allows us to track novel therapeutic targets and diagnostic markers. While there are molecular overlaps between cystic CCRCC and MCNLMP, there are also some significant differences. The validation of such data and the implementation of molecular studies in daily pathology practice will be of great aid in challenging differential diagnoses. Moreover, further clarifying the role of GSK3β in the formation and progression of cystic renal lesions could lead to a targetable protein in the treatment of cystic CCRCC, as well as PKD.

Author Contributions: G.M.P. conception, writing; R.L.: conception; M.C. conception. All authors have read and agreed to the published version of the manuscript.

Funding: This research received no external funding.

Conflicts of Interest: The authors declare no conflict of interest.

Abbreviations

ACD-RCD	Acquired Cystic Disease-Associated Renal Cell Carcinoma
ACN	Adult Cystic Nephroma
AMLEC	Angiomyolipoma with Epithelial Cyst
BC	Bosniak Classification
CAIX	Carbonic Anhydrase IX
CCPRCT	Clear Cell Papillary Renal Cell Tumor
CCRCC	Clear Cell Renal Cell Carcinoma
CPDN	Cystic Partially Differentiated Nephroblastoma
HIF	Hypoxia-Inducible Factor
HMWCK	High-Molecular-Weight Cytokeratin
IHC	Immunohistochemical
MEST	Mixed Epithelial and Stromal Tumor
MiTF	Melanocyte-Inducing Transcription Factor
MCNLMP	Multilocular Cystic Renal Neoplasm of Low Malignant Potential
PEC	Perivascular Epithelioid Cell
PCN	Pediatric Cystic Nephroma
PKD	Polycystic Kidney Disease
TcRCC	Tubulo-Cystic Renal Cell Carcinoma
VEGF	Vascular Endothelial Growth Factor
VHL	Von Hippel–Lindau
VHLd	Von Hippel–Lindau disease

References

1. Caliò, A.; Marletta, S.; Brunelli, M.; Martignoni, G. WHO 2022 Classification of Kidney Tumors: What is relevant? An update and future novelties for the pathologist. *Pathologica* **2022**, *115*, 23–31. [CrossRef]
2. Alaghehbandan, R.; Siadat, F.; Trpkov, K. What's new in the WHO 2022 classification of kidney tumours? *Pathologica* **2022**, *115*, 8–22. [CrossRef]
3. Udager, A.M.; Mehra, R. Morphologic, Molecular, and Taxonomic Evolution of Renal Cell Carcinoma: A Conceptual Perspective with Emphasis on Updates to the 2016 World Health Organization Classification. *Arch. Pathol. Lab. Med.* **2016**, *140*, 1026–1037. [CrossRef]
4. Moch, H. Cystic renal tumors: New entities and novel concepts. *Adv. Anat. Pathol.* **2010**, *17*, 209–214. [CrossRef]
5. WHO Classification of Tumours Editorial Board. Urinary and male genital tumours. In *WHO Classification of Tumour Series*, 5th ed.; International Agency for Research on Cancer: Lyon, France, 2022; Volume 8.
6. Alrumayyan, M.; Raveendran, L.; Lawson, K.A.; Finelli, A. Cystic Renal Masses: Old and New Paradigms. *Urol. Clin. N. Am.* **2023**, *50*, 227–238. [CrossRef]
7. Silverman, S.G.; Pedrosa, I.; Ellis, J.H.; Hindman, N.M.; Schieda, N.; Smith, A.D.; Remer, E.M.; Shinagare, A.B.; Curci, N.E.; Raman, S.S.; et al. Bosniak Classification of Cystic Renal Masses, Version 2019: An Update Proposal and Needs Assessment. *Radiology* **2019**, *292*, 475–488. [CrossRef]
8. Krishna, S.; Schieda, N.; Pedrosa, I.; Hindman, N.; Baroni, R.H.; Silverman, S.G.; Davenport, M.S. Update on MRI of Cystic Renal Masses Including Bosniak Version 2019. *J. Magn. Reson. Imaging* **2021**, *54*, 341–356. [CrossRef]
9. Westerman, M.E.; Cheville, J.C.; Lohse, C.M.; Sharma, V.; Boorjian, S.A.; Leibovich, B.C.; Thompson, R.H. Long-Term Outcomes of Patients with Low Grade Cystic Renal Epithelial Neoplasms. *Urology* **2019**, *133*, 145–150. [CrossRef]
10. Tretiakova, M.; Mehta, V.; Kocherginsky, M.; Minor, A.; Shen, S.S.; Sirintrapun, S.J.; Yao, J.L.; Alvarado-Cabrero, I.; Antic, T.; Eggener, S.E.; et al. Predominantly cystic clear cell renal cell carcinoma and multilocular cystic renal neoplasm of low malignant potential form a low-grade spectrum. *Virchows Arch.* **2018**, *473*, 85–93. [CrossRef]
11. Pastorekova, S.; Gillies, R.J. The role of carbonic anhydrase IX in cancer development: Links to hypoxia, acidosis, and beyond. *Cancer Metastasis Rev.* **2019**, *38*, 65–77. [CrossRef]
12. Genega, E.M.; Ghebremichael, M.; Najarian, R.; Fu, Y.; Wang, Y.; Argani, P.; Grisanzio, C.; Signoretti, S. Carbonic anhydrase IX expression in renal neoplasms: Correlation with tumor type and grade. *Am. J. Clin. Pathol.* **2010**, *134*, 873–879. [CrossRef]
13. Jonasch, E.; Walker, C.L.; Rathmell, W.K. Clear cell renal cell carcinoma ontogeny and mechanisms of lethality. *Nat. Rev. Nephrol.* **2021**, *17*, 245–261. [CrossRef]
14. Bui, T.O.; Dao, V.T.; Nguyen, V.T.; Feugeas, J.P.; Pamoukdjian, F.; Bousquet, G. Genomics of Clear-cell Renal Cell Carcinoma: A Systematic Review and Meta-analysis. *Eur. Urol.* **2022**, *81*, 349–361. [CrossRef]
15. Thoma, C.R.; Frew, I.J.; Hoerner, C.R.; Montani, M.; Moch, H.; Krek, W. pVHL and GSK3beta are components of a primary cilium-maintenance signalling network. *Nat. Cell Biol.* **2007**, *9*, 588–595. [CrossRef] [PubMed]
16. Rechsteiner, M.P.; von Teichman, A.; Nowicka, A.; Sulser, T.; Schraml, P.; Moch, H. VHL gene mutations and their effects on hypoxia inducible factor HIFα: Identification of potential driver and passenger mutations. *Cancer Res.* **2011**, *71*, 5500–5511. [CrossRef]
17. Cremona, M.; Espina, V.; Caccia, D.; Veneroni, S.; Colecchia, M.; Pierobon, M.; Deng, J.; Mueller, C.; Procopio, G.; Lanzi, C.; et al. Stratification of clear cell renal cell carcinoma by signaling pathway analysis. *Expert Rev. Proteom.* **2014**, *11*, 237–249. [CrossRef]
18. Bakouny, Z.; Braun, D.A.; Shukla, S.A.; Pan, W.; Gao, X.; Hou, Y.; Flaifel, A.; Tang, S.; Bosma-Moody, A.; He, M.X.; et al. Integrative molecular characterization of sarcomatoid and rhabdoid renal cell carcinoma. *Nat. Commun.* **2021**, *12*, 808. [CrossRef]
19. Akgul, M.; Williamson, S.R. How New Developments Impact Diagnosis in Existing Renal Neoplasms. *Surg. Pathol. Clin.* **2022**, *15*, 695–711. [CrossRef]
20. Kapur, P.; Rajaram, S.; Brugarolas, J. The expanding role of BAP1 in clear cell renal cell carcinoma. *Hum. Pathol.* **2023**, *133*, 22–31. [CrossRef]
21. Braun, D.A.; Hou, Y.; Bakouny, Z.; Ficial, M.; Sant'Angelo, M.; Forman, J.; Ross-Macdonald, P.; Berger, A.C.; Jegede, O.A.; Elagina, L.; et al. Interplay of somatic alterations and immune infiltration modulates response to PD-1 blockade in advanced clear cell renal cell carcinoma. *Nat. Med.* **2020**, *26*, 909–918. [CrossRef]
22. Elgendy, M.; Fusco, J.P.; Segura, V.; Lozano, M.D.; Minucci, S.; Echeveste, J.I.; Gurpide, A.; Andueza, M.; Melero, I.; Sanmamed, M.F.; et al. Identification of mutations associated with acquired resistance to sunitinib in renal cell cancer. *Int. J. Cancer* **2019**, *145*, 1991–2001. [CrossRef]
23. Guinot, A.; Lehmann, H.; Wild, P.J.; Frew, I.J. Combined deletion of Vhl, Trp53 and Kif3a causes cystic and neoplastic renal lesions. *J. Pathol.* **2016**, *239*, 365–373. [CrossRef]
24. Kuehn, E.W.; Walz, G.; Benzing, T. Von hippel-lindau: A tumor suppressor links microtubules to ciliogenesis and cancer development. *Cancer Res.* **2007**, *67*, 4537–4540. [CrossRef] [PubMed]
25. Seeger-Nukpezah, T.; Geynisman, D.M.; Nikonova, A.S.; Benzing, T.; Golemis, E.A. The hallmarks of cancer: Relevance to the pathogenesis of polycystic kidney disease. *Nat. Rev. Nephrol.* **2015**, *11*, 515–534. [CrossRef]
26. Santoni, M.; Piva, F.; Cimadamore, A.; Giulietti, M.; Battelli, N.; Montironi, R.; Cosmai, L.; Porta, C. Exploring the Spectrum of Kidney Ciliopathies. *Diagnostics* **2020**, *10*, 1099. [CrossRef]

27. Louise MBinderup, M.; Smerdel, M.; Borgwadt, L.; Beck Nielsen, S.S.; Madsen, M.G.; Møller, H.U.; Kiilgaard, J.F.; Friis-Hansen, L.; Harbud, V.; Cortnum, S.; et al. von Hippel-Lindau disease: Updated guideline for diagnosis and surveillance. *Eur. J. Med. Genet.* **2022**, *65*, 104538. [CrossRef]
28. Chahoud, J.; McGettigan, M.; Parikh, N.; Boris, R.S.; Iliopoulos, O.; Rathmell, W.K.; Daniels, A.B.; Jonasch, E.; Spiess, P.E.; International VHL Surveillance Guidelines Consortium-Renal Committee. Evaluation, diagnosis and surveillance of renal masses in the setting of VHL disease. *World J. Urol.* **2021**, *39*, 2409–2415. [CrossRef]
29. Schönenberger, D.; Harlander, S.; Rajski, M.; Jacobs, R.A.; Lundby, A.K.; Adlesic, M.; Hejhal, T.; Wild, P.J.; Lundby, C.; Frew, I.J. Formation of Renal Cysts and Tumors in Vhl/Trp53-Deficient Mice Requires HIF1α and HIF2α. *Cancer Res.* **2016**, *76*, 2025–2036. [CrossRef]
30. Tao, S.; Kakade, V.R.; Woodgett, J.R.; Pandey, P.; Suderman, E.D.; Rajagopal, M.; Rao, R. Glycogen synthase kinase-3β promotes cyst expansion in polycystic kidney disease. *Kidney Int.* **2015**, *87*, 1164–1175. [CrossRef]
31. Bilim, V.; Ougolkov, A.; Yuuki, K.; Naito, S.; Kawazoe, H.; Muto, A.; Oya, M.; Billadeau, D.; Motoyama, T.; Tomita, Y. Glycogen synthase kinase-3: A new therapeutic target in renal cell carcinoma. *Br. J. Cancer* **2009**, *101*, 2005–2014. [CrossRef]
32. Favazza, L.; Chitale, D.A.; Barod, R.; Rogers, C.G.; Kalyana-Sundaram, S.; Palanisamy, N.; Gupta, N.S.; Williamson, S.R. Renal cell tumors with clear cell histology and intact VHL and chromosome 3p: A histological review of tumors from the Cancer Genome Atlas database. *Mod. Pathol.* **2017**, *30*, 1603–1612. [CrossRef]
33. Halat, S.; Eble, J.N.; Grignon, D.J.; Lopez-Beltran, A.; Montironi, R.; Tan, P.H.; Wang, M.; Zhang, S.; MacLennan, G.T.; Cheng, L. Multilocular cystic renal cell carcinoma is a subtype of clear cell renal cell carcinoma. *Mod. Pathol.* **2010**, *23*, 931–936. [CrossRef]
34. Gong, K.; Zhang, N.; He, Z.; Zhou, L.; Lin, G.; Na, Y. Multilocular cystic renal cell carcinoma: An experience of clinical management for 31 cases. *J. Cancer Res. Clin. Oncol.* **2008**, *134*, 433–437. [CrossRef]
35. Williamson, S.R.; Halat, S.; Eble, J.N.; Grignon, D.J.; Lopez-Beltran, A.; Montironi, R.; Tan, P.H.; Wang, M.; Zhang, S.; MacLennan, G.T.; et al. Multilocular cystic renal cell carcinoma: Similarities and differences in immunoprofile compared with clear cell renal cell carcinoma. *Am. J. Surg. Pathol.* **2012**, *36*, 1425–1433. [CrossRef] [PubMed]
36. Kim, S.H.; Park, W.S.; Chung, J. SETD2, GIGYF2, FGFR3, BCR, KMT2C, and TSC2 as candidate genes for differentiating multilocular cystic renal neoplasm of low malignant potential from clear cell renal cell carcinoma with cystic change. *Investig. Clin. Urol.* **2019**, *60*, 148–155. [CrossRef]
37. Diolombi, M.L.; Cheng, L.; Argani, P.; Epstein, J.I. Do Clear Cell Papillary Renal Cell Carcinomas Have Malignant Potential? *Am. J. Surg. Pathol.* **2015**, *39*, 1621–1634. [CrossRef]
38. Williamson, S.R. Clear cell papillary renal cell carcinoma: An update after 15 years. *Pathology* **2021**, *53*, 109–119. [CrossRef]
39. Munari, E.; Marchionni, L.; Chitre, A.; Hayashi, M.; Martignoni, G.; Brunelli, M.; Gobbo, S.; Argani, P.; Allaf, M.; Hoque, M.O.; et al. Clear cell papillary renal cell carcinoma: micro-RNA expression profiling and comparison with clear cell renal cell carcinoma and papillary renal cell carcinoma. *Hum. Pathol.* **2014**, *45*, 1130–1138. [CrossRef]
40. Mohanty, S.K.; Parwani, A.V. Mixed epithelial and stromal tumors of the kidney: An overview. *Arch. Pathol. Lab. Med.* **2009**, *133*, 1483–1486. [CrossRef]
41. Turbiner, J.; Amin, M.B.; Humphrey, P.A.; Srigley, J.R.; De Leval, L.; Radhakrishnan, A.; Oliva, E. Cystic nephroma and mixed epithelial and stromal tumor of kidney: A detailed clinicopathologic analysis of 34 cases and proposal for renal epithelial and stromal tumor (REST) as a unifying term. *Am. J. Surg. Pathol.* **2007**, *31*, 489–500. [CrossRef]
42. Zhou, M.; Kort, E.; Hoekstra, P.; Westphal, M.; Magi-Galluzzi, C.; Sercia, L.; Lane, B.; Rini, B.; Bukowski, R.; Teh, B.T. Adult cystic nephroma and mixed epithelial and stromal tumor of the kidney are the same disease entity: Molecular and histologic evidence. *Am. J. Surg. Pathol.* **2009**, *33*, 72–80. [CrossRef]
43. Vanecek, T.; Pivovarcikova, K.; Pitra, T.; Peckova, K.; Rotterova, P.; Daum, O.; Davidson, W.; Montiel, D.P.; Kalusova, K.; Hora, M.; et al. Mixed Epithelial and Stromal Tumor of the Kidney: Mutation Analysis of the DICER 1 Gene in 29 Cases. *Appl. Immunohistochem. Mol. Morphol.* **2017**, *25*, 117–121. [CrossRef] [PubMed]
44. Argani, P.; Beckwith, J.B. Metanephric stromal tumor: Report of 31 cases of a distinctive pediatric renal neoplasm. *Am. J. Surg. Pathol.* **2000**, *24*, 917–926. [CrossRef]
45. Kacar, A.; Azili, M.N.; Cihan, B.S.; Demir, H.A.; Tiryaki, H.T.; Argani, P. Metanephric stromal tumor: A challenging diagnostic entity in children. *J. Pediatr. Surg.* **2011**, *46*, e7–e10. [CrossRef]
46. Idrissi-Serhrouchni, K.; El-Fatemi, H.; El Madi, A.; Benhayoun, K.; Chbani, L.; Harmouch, T.; Bouabdellah, Y.; Amarti, A. Primary renal teratoma: A rare entity. *Diagn. Pathol.* **2013**, *8*, 107. [CrossRef]

cancers

MDPI

Review

Recent Advances in Single-Cell RNA-Sequencing of Primary and Metastatic Clear Cell Renal Cell Carcinoma

Adele M. Alchahin †, Ioanna Tsea † and Ninib Baryawno *

Childhood Cancer Research Unit, Department of Women's and Children's Health, Karolinska Institutet, 10000-19999 Stockholm, Sweden; adele.alchahin@ki.se (A.M.A.); ioanna.tsea@ki.se (I.T.)
* Correspondence: n.baryawno@ki.se
† These authors contributed equally to this work.

Simple Summary: In recent years, several therapeutic advances have been made in clear cell renal cell carcinoma (ccRCC) resulting in novel treatment regimens of increased effectiveness. These advances are largely due to breakthroughs in technologies, particularly in transcriptomics, such as single-cell RNA sequencing (scRNA-seq). Using this technology, we have gained a deeper understanding of the biology of ccRCC and revealed various cell populations and their interactions in disease progression. While localized ccRCC patients have shown promising responses to treatment, however, patients with advanced or metastatic disease remain a therapeutic challenge. To address this gap, recent studies have utilized scRNA-seq to investigate both primary and metastatic ccRCC in search of promising therapeutic targets. This review aims to summarize the current state of knowledge in the field, highlight available treatment options and underscore the critical steps needed to improve survival rates, especially for metastatic ccRCC patients.

Abstract: Over the past two decades, significant progress has been made in the treatment of clear cell renal cell carcinoma (ccRCC), with a shift towards adopting new treatment approaches ranging from monotherapy to triple-combination therapy. This progress has been spearheaded by fundamental technological advancements that have allowed a deeper understanding of the various biological components of this cancer. In particular, the rapid commercialization of transcriptomics technologies, such as single-cell RNA-sequencing (scRNA-seq) methodologies, has played a crucial role in accelerating this understanding. Through precise measurements facilitated by these technologies, the research community has successfully identified and characterized diverse tumor, immune, and stromal cell populations, uncovering their interactions and pathways involved in disease progression. In localized ccRCC, patients have shown impressive response rates to treatment. However, despite the emerging findings and new knowledge provided in the field, there are still patients that do not respond to treatment, especially in advanced disease stages. One of the key challenges lies in the limited study of ccRCC metastases compared to localized cases. This knowledge gap may contribute to the relatively low survival rates and response rates observed in patients with metastatic ccRCC. To bridge this gap, we here delve into recent research utilizing scRNA-seq technologies in both primary and metastatic ccRCC. The goal of this review is to shed light on the current state of knowledge in the field, present existing treatment options, and emphasize the crucial steps needed to improve survival rates, particularly in cases of metastatic ccRCC.

Keywords: renal cell carcinoma; single-cell RNA-seq; cell-of-origin; tumor microenvironment; stromal cells; immune cells; novel therapies

Citation: Alchahin, A.M.; Tsea, I.; Baryawno, N. Recent Advances in Single-Cell RNA-Sequencing of Primary and Metastatic Clear Cell Renal Cell Carcinoma. *Cancers* **2023**, *15*, 4734. https://doi.org/10.3390/cancers15194734

Academic Editor: José I. López

Received: 18 August 2023
Revised: 22 September 2023
Accepted: 22 September 2023
Published: 26 September 2023

1. Introduction

Renal cell carcinoma (RCC) encompasses a group of malignant tumors originating from the epithelium of the proximal part of the renal tubules. The most prevalent histological subtype is clear cell RCC (ccRCC), accounting for approximately 75–80% of all RCC cases, followed by papillary RCC (pRCC; 10–15%) and chromophobe RCC (5–10%) [1]. Among RCC patients, about one-third develop bone metastasis, with ccRCC being the most common subtype. Unfortunately, the 5-year survival rate for ccRCC bone metastasis is less than 10%, compared to 75% for non-metastatic ccRCC [2].

ccRCC is often characterized by the loss of chromosome 3p and a second-hit loss-of-function mutation in the *VHL* tumor suppressor gene located on chromosome 3 [3]. Additional chromosomal alterations commonly observed in ccRCC include loss of 14q and gain of 5q. The inactivation of the *VHL* protein leads to an increase in the transcription factor hypoxia-inducible factor (*HIF*), resulting in the transcriptional upregulation of hypoxia-inducible genes, such as vascular endothelial growth factor (*VEGF*) [3,4]. The elevated *VEGF* levels in the tumor microenvironment drive various downstream effects, including enhanced cell proliferation, angiogenesis, migration, and altered metabolism. Besides *VHL*, other frequently mutated genes in ccRCC include *PBRM1, BAP1, SETD2, UTX, ARID1a*, and *KDM5a*, which further contribute to the complex genomic landscape of ccRCC [5]. ccRCC tends to metastasize to the liver, lung, bone, brain, pancreas, skin and muscle [6]. One of the most aggressive metastatic sites is the bone. Around 30% of patients with ccRCC develop secondary tumors with a 5-year survival of less than 10% [7,8].

Early interpretations of the cellular landscape of ccRCC were achieved using bulk RNA-sequencing [9,10]. Although a powerful tool, bulk RNA-sequencing measures average gene expression across a population of heterogeneous cells and thus fails to distinguish subtle transcriptional differences or rare populations of cells. Following the emergence and commercialization of single-cell RNA-sequencing (scRNA-seq) a new avenue of unprecedented resolution was opened through which tumor heterogeneity could be unraveled on the individual cell level [11–13]. Through the use of single-cell transcriptomics, essential gaps of knowledge are being uncovered along with the rising technological possibilities as the discovery of the cell of origin of ccRCC, immune cell reprogramming post-treatment, and immune and stromal cell characterization of treatment naïve ccRCC patients [14–17]. Hence, this review aims to shed light on how the use of single-cell transcriptomics technology has assisted recent scientific research in unraveling the tumor microenvironment as well as the way it has supported current clinical trial targets and treatment modifications.

2. Methods

The narrative of the review was predetermined to cover the aspect of single-cell technologies. Over a period of four months, searches were made on scientific databases such as PubMed including keywords including "clear cell renal cell carcinoma", "single-cell RNA sequencing" and "transcriptomics". For the section describing spatial and transcriptomic evidence, a key inclusion criterion was that the study had used a single-cell and "spatial technology" when describing their findings.

3. Unveiling the ccRCC Cell of Origin

Two recent human single-cell transcriptomics studies have provided compelling evidence supporting proximal tubular epithelial cells (PTECs) as the cellular origin of ccRCC. These studies utilized single-cell transcriptomic analysis to compare the transcriptome of captured ccRCC PTECs with single or bulk normal and ccRCC transcriptomes [14,18]. The combined analysis of these studies revealed the expression of several genes, including carbonic anhydrase 9 (*CA9*), vascular cell adhesion molecule-1 (*VCAM1*), solute carrier family 17 member 3 (*SLC17A3*), intercellular adhesion molecule 1 (*ICAM1*), integrin subunit beta 8 (*ITGB8*), alpha kinase 2 (*ALPK2*), and vimentin (*VIM*), in ccRCC PTECs [14,18].

Of particular interest was the identification of *VCAM1* and *CA9*-positive PTECs in ccRCC patients' adjacent morphologically normal kidney tissue which were termed precursor PTECs, representing morphologically normal PTECs with $VHL^{+/-}$ mutation [14]. This finding suggests that identifiable transcriptomic alterations occur following genomic alteration in precursor PTECs, which precede morphological changes in ccRCC development. It supports the proposed concept of a transition from normal to precursor and ultimately malignant PTEC states [14]. A multi-omics study by Muto et al. [19], combining single-nucleus RNA sequencing (snRNA-seq) with single-nucleus ATAC sequencing (snATAC-seq) further investigated precursor PTECs expressing *VCAM1* and *CA9* and showed that they exhibit a transcriptomic similarity to inflamed PTECs characterized by *VCAM1* expression but without *CA9* expression [19]. Inflamed PTECs were defined by the expression of *VCAM1*, *ICAM1*, *CD24*, *CD133*, and *HAVCR1*, and were associated with the response to acute and/or chronic tubular injury. The transcriptomic profile of inflamed PTECs exhibited the strongest similarity to malignant PTECs in ccRCC. This observation suggests an alternative PTEC transition from normal to inflamed, then to precursor, and finally to malignant PTEC states during ccRCC development (Figure 1). The presence of common gene expression patterns in both inflamed and malignant PTECs suggests a potential link between tubular-injury-related inflammation and ccRCC pathogenesis.

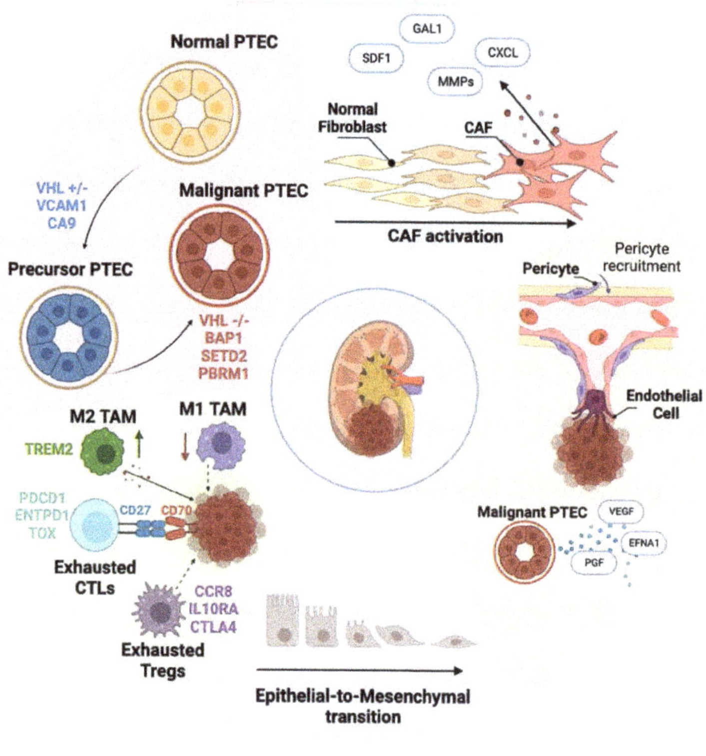

Figure 1. Summary of the biological findings validated with single-cell RNA sequencing studies in primary and metastatic ccRCC. These studies uncover the multilayer complexity of the disease and highlight. (1) The cell of origin of ccRCC as malignant PTECs. Normal PTECs transition towards Precursor PTECs harboring the $VHL^{+/-}$ mutation and finally to malignant PTECs harboring the $VHL^{-/-}$ mutation as well as additional genetic alterations. (2) The progressive dysfunction of the immune cell landscape is characterized by a simultaneous upregulation of dysfunctional M2−like TAMs and downregulation of M1−like TAMs as well as the emergence of terminally exhausted immune cell types including CTLs and Tregs. (3) The EMT transition process and its role in disease progression. (4) The role of non-immune TME cells in promoting angiogenesis and tumor cell invasion. In particular subsets of pro-angiogenic endothelial cells and capillary pericytes have been identified to aid ccRCC disease progression. (5) The role of CAFs in supporting tumor progression and metastasis through the secretion of molecules that drive immunosuppression, extracellular remodeling and EMT. Figure created in Biorender.com (accessed on 15 August 2023).

4. Elucidating the Transcriptomic Identity of Metastasizing Cells

In light of the increasing discoveries on the cell of origin of primary ccRCC, the field quickly proceeded toward the characterization of metastatic ccRCC using single-cell transcriptomic technologies. Indeed, a recent single-cell transcriptomics study analyzing ccRCC primary, locally invasive, and adjacent normal tissues identified enhanced extracellular matrix (ECM) remodeling by malignant PTECs in locally invasive lesions [20]. The findings suggest that while locally invasive ccRCC lesions may result from opportunistic extension into nearby vasculature, the extending malignant PTECs also depend on a supportive ECM. Similarly, metastatic ccRCC progression has been characterized by 17 metastasis-associated gene (MAG) markers identified in a single-cell transcriptomics study involving 121 single-cell samples [21,22]. These cells were captured from parental metastatic sites and patient-derived xenografts of primary and metastatic ccRCC samples [21,22]. The MAGs include chemokines (*CCL20* and *CXCL1*), as well as mitochondrial (*MT-ND3*, *MT-ND4*, and *MT-RNR2*) and cancer markers (*NDUFA5*, *NNMT*, *BHLHE41*, *ALDH1A1*, and *BNIP3*). Expression of these MAG markers correlate with a higher likelihood of ccRCC recurrence. Moreover, a single-cell transcriptomics study comparing treatment-naive primary tumor tissue matched adjacent normal kidney tissue and tumor samples collected from patients with bone metastases deduced a distinct transcriptional signature that is correlated with metastatic potential and patient survival [17]. Another study harnessed the power of multi-omics to highlight *SERPINE2*, a gene in metastatic RCC that could predict metastatic outcomes and further be targeted [23]. These studies provide novel insights into locally invasive and metastatic disease, which remain a therapeutic challenge and lead the field towards their etiology.

One of the major challenges in treating metastatic ccRCC is the significant intratumoral heterogeneity of tumors that is comprised of subclones with diverse genotypes [24]. Hence, recent single-cell studies have attempted to characterize the transcriptional identity of metastatic clones and found two distinct subpopulations able to separately stimulate VEGF- and Epithelial-to-Mesenchymal (EMT) related pathways [25]. Other studies focused on characterizing the process of EMT itself as a driver of tumor metastasis. In particular, a study accomplished to define an EMT metastatic program in ccRCC where they discover cells with an EMT high profile localized in the interface of the tumor and normal environment, which is the leading and migratory margins of a tumor [26].

5. Deciphering the Role of the TME in ccRCC Progression

In addition to cancer cells, the TME encompasses non-malignant cell types embedded within an altered ECM. The specific composition of the TME can vary among different tumor types, but it typically includes various cell types such as fibroblasts, adipocytes, neurons, endothelial cells, immune cells, and stem cells, along with secreted molecules like cytokines, chemokines, and growth factors[9]. The advent of innovative techniques like single-cell transcriptomic sequencing has facilitated a deeper understanding and cataloging of this context.

6. Immune TME in Primary and Metastatic ccRCC

Several single-cell transcriptomics studies have provided valuable information about immune cell populations captured from primary and metastatic ccRCC tumors. These studies shed light on the transcriptomic profiles of myeloid and lymphoid cell types, states, and their interactions in ccRCC, with particular emphasis on tumor-associated macrophages (TAMs) and CD8+ T cells, both of which play significant roles in tumor progression and evasion. More specifically, the high plasticity of TAM populations was highlighted in ccRCC, spanning a continuum from pro-inflammatory M1-like to anti-inflammatory M2-like states, with intermediate TAM subpopulations based on *HLA-DR* or interferon signaling gene expression levels [16,27]. One study has provided novel insights into the immune cell landscape and correlated its progressive dysfunction with disease stage in ccRCC patients by demonstrating a general shift in TAM states with ccRCC progression, with an increase in dysfunctional M2-like TAMs and a simultaneous decrease in M1-like TAMs [27]. An additional study correlated a *TREM2*-positive TAM population to lower survival in primary ccRCC patients [17,28]. Similarly, single-cell transcriptomic analysis of CD8+ T cells in ccRCC samples reveals a diverse and heterogeneous population, spanning a continuum that progresses to terminally exhausted clonotypes [28,29]. Several studies have unveiled distinct subsets of CD8+ T cells, including naïve, cytotoxic, exhausted, progenitor, and terminally exhausted states [15,27,28,30,31]. The identification of immune inhibitory markers on CD8+ T cells aligns with bulk RNA-seq studies, suggesting potential epigenetic reprogramming leading to exhaustive states through *TOX2* [32–35]. Within the exhausted CD8+ T cell population, the presence of progenitor and terminally exhausted subpopulations suggests a spectrum of exhausted states that may transition from a progenitor (*TCF7*) to a terminally exhausted (*ENTPD1*) state [15,27,36].

The role of immune cell infiltration in metastatic ccRCC is also gaining more attention as multiple single-cell transcriptomics studies suggest it might be affecting prognosis [27,30]. Two studies in metastatic ccRCC have characterized TAMs to have high expression of both HLA class I and II genes along with *IFI27*, *CTSL*, *CTSS*, *C1QA*, *C1QB*, *SERPING1*, *APOE*, and *PLTP* [27,30]. Moreover, inferred pseudotime trajectory analysis of CD8+ T cells in ccRCC, indicated a higher prevalence of exhausted CD8+ T cells in advanced and metastatic ccRCC compared to normal kidney tissues and peripheral blood [27,28,30,32]. Hence, the potential of immunotherapeutic strategies and immune-related pathways have been increasing as potential directors in cancer therapy care [37]. For instance, PD-1 has been implied to act negatively as an immunoregulatory molecule and to be involved in the regulation of cancer cell immune evasion [38]. However, a research study showed that standard pre-treatment T cell receptor (TCR) clonality could predict clinical response to anti-PD-1 therapy in ccRCC [39], while others observe a considerable variation of TCR clonality across disease stages of ccRCC [27].

While some single-cell studies have provided novel insights into the immune cell landscape of ccRCC, other studies focused on characterizing the complete TME of treatment-naive patients [17] as well as the consequences of ICB therapy in reprogramming the TME [15,31]. In metastatic RCC, one study integrated multi-omics analysis of bulk RNA-sequencing, scRNA-seq, ATAC-seq and 3D high-throughput chromosome conformation capture (Hi-C) to highlight the influence of the TME on the clinical responsiveness towards targeted therapy or immunotherapy [23]. Of particular clinical significance was the demonstration of the capacity of malignant PTECs to drive angiogenesis through the secretion of the *VEGFA*, *PGF* and *EFNA1* ligands and their interactions with receptors on macrophages, fibroblasts and endothelial cells [14,18,31,32]. This evidence indicated by numerous single-cell transcriptomics studies confirmed the existence of interactions between cancer cells and elements of the TME which could represent promising targets and spearheaded the clinical efforts for therapeutic targeting.

7. Non-Immune TME in Primary and Metastatic ccRCC

Nevertheless, while the use of antiangiogenic treatments has made VEGF targeting a favorable choice for ccRCC, these therapies often fail to sustain a long-term clinical response in patients. As a result, there has been increasing focus on non-malignant and non-immune stromal cells within the tumor microenvironment. Among these cells, cancer-associated fibroblasts (CAFs) have garnered attention due to their potential immunosuppressive functions within the ccRCC microenvironment. One study revealed that the immunosuppressive behavior mediated by CAFs is attributed to the secretion of Galectin-1 (*Gal1*) which induces apoptosis in cytotoxic CD8+ T cells in recurrent ccRCC [40]. It is suggested that the recruitment of CAFs into the ccRCC microenvironment occurs through interactions with malignant PTECs that upregulate *COL20A1*, *COL28A1*, and *TGFB1* [20]. Indeed, both Alchahin et al. and Shi et al. identified CAF-mediated extracellular matrix remodeling, which was associated with an increased gene signature for the EMT pathway in primary and locally invasive ccRCC, respectively [17,20]. Therefore, in both primary and recurrent ccRCC, the infiltration of CAFs should be considered as an additional critical cell type driving tumor progression and immunosuppression.

Endothelial cells, responsible for blood vessel formation and pericytes that surround and stabilize blood vessels exhibit distinct subpopulations within the ccRCC TME as revealed by scRNA-seq analysis. Multiple single-cell studies in primary ccRCC have identified endothelial cell subsets with differential expression of genes involved in angiogenesis, vascular stability, and immune modulation. These subsets were associated with immune cell infiltration, angiogenesis, and therapy response, highlighting their functional specialization and impact on the ccRCC TME [14,17,41]. Alchahin et al. highlighted the enrichment of pro-angiogenic capillary pericytes in treatment-naive ccRCC coupled with the reduction in vascular smooth muscle cells, known to maintain blood vessel integrity, thus showing the active remodeling of the TME in ccRCC progression [17].

8. Treatment of ccRCC and ccRCC Metastasis

A localized ccRCC tumor is still resected through partial or radical nephrectomy as the standard of care [42]. Even if there are signs of a simultaneous formation of micrometastasis, surgical resection is proven to be efficient in preventing the early steps of metastasis [43]. Although over the past two decades the treatment approaches have changed [44], particularly in metastatic ccRCC, the ongoing advancements in modern technology will continue to provide insights into the complexities of cancer and its metastasis, which may inspire further changes in treatment approaches. For metastatic disease, the treatment now comprises multiple targets that have been developed to block the activity in signaling pathways of mammalian target of rapamycin (*mTOR*), vascular endothelial growth factor (*VEGF*) pathways and platelet-derived growth factor (*PDGF*). They have demonstrated an involvement in angiogenesis and metastasis further promoting the development and progression of ccRCC [45–49]. Despite the efficacy of the initial treatments used, the median time for the patients to obtain drug resistance is around 6–15 months, differing based on therapeutic schedule and intratumor heterogeneity (ITH) [50,51]. Hence, further research is required to overcome this therapeutic hurdle.

The somatic *VHL* mutation known in ccRCC was one of the earlier discoveries [52] that resulted in the idea of preventing tumor angiogenesis via targeting the *VEGF* pathway, including its receptors, where mainly *VEGFR2* is targeted [53]. ccRCC tumors are a group of epithelial tumors that present with elevated expression of *VEGFA* and are therefore an understandable target of the disease [54]. With scRNA-seq technology, it was possible to identify cell populations in ccRCC that normally express *VEGF*-related pathway genes including subsets of endothelial cells [14,17]. Several studies have shown the binding of *VEGF-VEGFR2* to significantly enhance tumor development, contributing to its progression and expansion [55–57]. One of the first agents developed for targeting and inhibiting both *VEGF* and *PDGFR* in metastatic RCC was sunitinib, a tyrosine kinase inhibitor (TKI) which at the time presented significant progression-free survival of 11 months in comparison with 5 months with the earlier broadly used interferon alfa [58,59]. Along with the targeting of angiogenesis, there is a particular interest in inhibiting the *mTOR* pathway in ccRCC because of the known involvement of the regulatory effects on *HIF2a* production, but also its role in regulating cell proliferation and survival processes [60]. It is the *HIF2a* that is involved in the upregulation of *VEGF* further promoting angiogenesis [60]. When dual therapy was introduced by combining an *mTOR* inhibitor everolimus with levantinib (*VEGF* inhibitor), it demonstrated prolonged progression-free survival [61].

As dual therapy emerged, showing promising effects in patients with advanced ccRCC, the evolving single-cell transcriptomic knowledge of immune cell infiltration and dysfunction harbored the new therapeutic era [14,17,27,62]. The discovery that ccRCC tumors are highly infiltrated by T cells with the exhaustive phenotype [9] led to modified therapeutic approaches by adding immune checkpoint blockade (ICB) as an additional alternative in combination with antiangiogenic drugs [9,63–65]. When comparing monotherapy of antiangiogenic agents to a combination with ICB in clinical trials, a significantly improved overall survival was observed [61,64–67]. The response rates of combination therapy ranged between 42–71% [61,64–67]. Immunotherapy has therefore surpassed clinical expectations where both monotherapies of anti-*PD-1* agents and dual use of *PD-1* (programmed cell death protein 1) and *CTLA-4* (cytotoxic T lymphocyte–associated protein 4) inhibition show effective therapy responses of a median of 24 months in patients diagnosed with non-ccRCC and metastatic ccRCC [66,68,69]. These novel immunotherapies are capable of reviving T cell exhaustion to re-initiate tumor-killing effects [70]. Recent utilization of scRNA-seq has effectively uncovered the existence of a highly immunosuppressive microenvironment in both primary and metastatic ccRCC and provides a basis for therapeutic targeting using the aforementioned approaches and combinations [17]. Furthermore, combining immunotherapy with antiangiogenic agents has significantly improved progression-free survival indicating the important role of the TME and how it can be manipulated for new treatment strategies in metastatic ccRCC [64,65,71]. In patients with metastatic ccRCC that had previously been treated with antiangiogenic agents, the *PD-1* inhibitor nivolumab showed better overall survival compared to everolimus [68,72].

Despite promising and improved treatment results, patients with metastatic ccRCC may become unmanageable and the disease may recur. The latest guidelines specified by the European Association of Urology (Arnhem, The Netherlands) [42] concluded that partial nephrectomy remains a superior approach when the disease is localized. However, in metastatic ccRCC, surgical resection is not recommended as inhibition of *VEGFR* and *PDGFR* with sunitinib did not present worse outcomes compared with nephrectomy [73].

Following the idea of multiple targeting for metastatic disease, the latest therapeutic approach involves the triple combination of nivolumab (anti-*PD-1*), ipilimumab (anti-*CTLA-4*) and cabozantinib (TKI and VEGF inhibitor) covering both immune infiltration and known pathways involved in ccRCC progression. Interestingly, it demonstrated clinical efficiency in advanced RCC patients that are treatment naïve [74,75]. Nonetheless, as the results are encouraging, clinical analyses are being assessed. These, and more trials, will shape new therapeutic strategies after retrospective data representation in advanced ccRCC (Table 1). In the ongoing effort to manage advanced ccRCC, the question being explored is whether targeting multiple factors simultaneously offers improved treatment outcomes, as multiple cellular compartments and processes are simultaneously involved in cancer progression [14,17,27,30,62].

Table 1. Single-cell RNA sequencing studies that have supportive evidence of the targets in clinical trials of ccRCC and metastatic ccRCC.

Clinical Trials in ccRCC and Metastatic ccRCC	Target Involved in Trial	Single-Cell RNA Seq Studies Supporting the Trial
NCT05468190 **NCT05420519** **NCT00944905**	CD70	[17,76]
NCT03905889	CDK4/6	[11]
NCT03945773	Combination therapy with VEGF	[14,16]
NCT03473730	CD38	[16]
NCT03987698 **NCT05239728** **NCT04518046** **NCT03729245** **NCT03937219**	PD-1	[20,27,39,65,68]
NCT04518046 **NCT03937219**	CTLA-4, PD-1, MET	[17,27]
NCT04691375	TREM2	[17,30,77]
NCT05103722	IL-6	[25]
NCT04338269 **NCT03141177**	RTK, PD-1, CTLA-4, TKI c-MET, VEGFR2	[17]

9. Future Perspectives

Early treatment strategies already distinguished the importance of immunotherapy by T cell receptor proliferation cytokine IL-2 and interferon a2b [78,79]. Today, the recent standard of care is to use immune checkpoint inhibitors which have changed the paradigm of ccRCC therapy [64,80]. Despite these successful treatment strategies, a subset of renal cancer patients still do not respond to treatment, and those who do eventually progress [81,82]. The aforementioned single-cell studies have demonstrated a novel understanding of the cellular landscape of ccRCC. However, by also focusing scRNA-seq on T cells using 5′-sequencing and recombined V(D)J region of T cell receptor, researchers show in ccRCC that tumor-infiltrating T cells harbor a different expression and transcriptional pattern when it is compared to the normal renal tissue and peripheral blood [28], indicating a transcriptomic heterogeneity. Moreover, a single-cell study in melanoma achieved to connect a subset of melanoma-infiltrating lymphocytes to certain antigens of T cell receptors, implying that the expression level of intratumoral CD8+ T cells may be controlled by particular tumor phenotypes [83]. Similarly, in ccRCC, cells existing in a CD8+ T cell receptor (TCR) clonotype were shown to be controlled by their level of exhaustion [26]. This restriction of clonotypes based on phenotype may not depend on environmental factors, but instead on chronological mutations as individual tumors carry clonotypes through different states [84–86]. Thus, when cells infiltrate a tumor and undergo changes from an active to a dysfunctional state, they stay in the tissue depending on the phenotype of the tumor tissue they reside in [26].

Manipulating the adaptive immune responses by using ICB has shown us improved survival [65,68], emphasizing the role of the immune microenvironment in ccRCC. Hence, to better understand the architecture, tumor-infiltrating immune cells, immunotherapy and the T cell receptor immune cell atlas, advanced single-cell technologies need to be applied. The single-cell technologies have revealed and confirmed to us the rich tumor microenvironment, including the immune microenvironment [14,17,21,87,88]. The generation of gene signatures deciphering the roles of certain immune subsets in cancer has functioned as a tool in guiding clinicians to select patients, calculating the probability of benefiting from immunotherapy and characterizing clinically relevant subpopulations. Immunotherapy has reached the status of treatment stamina when treating treatment-naïve and advanced ccRCC [64,89].

The massive data collection and production from single-cell studies generate enormous information and hypotheses to test. The comprehensive output has been able to confirm that the current suggested triple therapy may work by proving the pathways and gene expressions that are augmented in primary and metastatic ccRCC as *CTLA-4*, *PD-1*, *VEGFR2* and *MET* [17,74]. In theory, this combination targets four principal aspects influencing ccRCC disease development; immunosuppression, angiogenesis (vascular remodeling) and MET overexpression [17].

With explorative science, new knowledge will be provided, interpreted, and tested. Equally, as single-cell technologies have computationally and quantitatively mapped the TME [14,17,21,87,88], clinical trials are pursuing to evaluate the findings. An example is the recent research emphasizing the *CD70-CD27* axis in primary ccRCC development as a potential target [17]. The finding supports clinical trials that are currently testing agents targeting this axis by engineering CAR-T cells for instance, as *CD27* was expressed by exhausted cytotoxic T cells [17]. This approach is now being tested in clinical trials [76] (NCT05468190, NCT05420519, see Table 1). Hence, single-cell technologies are able to generate massive amounts of information which can subsequently be used to provide solid facts to support clinical trials.

Nevertheless, while scRNA-seq technology has become a state-of-the-art approach for unraveling the heterogeneity and complexity of different cell types, it has also brought to light certain methodological challenges, such as the "artificial transcriptional stress responses". In particular, the process of single-cell isolation has been found to trigger the expression of stress-related genes, which in turn can lead to artificial changes in cell transcription patterns [90,91]. Another challenge in scRNA-seq is dealing with variation between different datasets [92]. Data collected at different times or using different sequencing platforms can have significant batch effects [93]. While often not biologically meaningful, these batch effects can disrupt patterns in gene expression and potentially lead to incorrect conclusions. Therefore, it is crucial to correct these batch effects during analysis of single-cell transcriptomics studies. To this end, several algorithms have been proposed to correct batch effects although these methods can be computationally intensive and often require significant amounts of memory and time [94,95].

Perhaps the most significant limitation of scRNA-seq methods is the loss of histological information through the need to dissociate tissue into single-cell suspensions. Hence, novel spatial biological technologies have been emerging rapidly as powerful tools to add the additional layer of spatial information that single-cell technologies are currently missing. A recent study in ccRCC combined the use of single-cell transcriptomics with barcode-based spatial transcriptomics to study the effect of the MC5 lncRNA signature on immunotherapy response and the TME [96]. Two studies combined spatial transcriptomics with scRNA-seq to reveal cell types within the TME of ccRCC that correlate with ICB resistance in ccRCC patients [97,98]. An intriguing example of the power of multi-omics is offered by Wu et al. [99], where single-nucleus transcriptomics, epigenomics, and spatial transcriptomics are employed to identify a novel tumor cell signature correlating with reduced survival in ccRCC patients [99].

In conclusion, the paradigm shift in ccRCC disease progression has been enhanced in the past two decades in terms of biological and therapeutic understanding. As the survival of primary disease is satisfactory, the challenges and future work lie in tackling the metastatic outcome. The research community of ccRCC has a broad and huge data output of hypotheses with computational rationale and strength to test. The next steps are to envision the disease by combining multi-omics data to better understand the cellular profiles and their biological communication patterns with spatial and imaging techniques. This is where the fundamental efforts need to be put on, in parallel with the ongoing clinical trials to understand therapy responses.

10. Conclusions

ccRCC tumors are known for their hypoxic, immunogenic, and angiogenic characteristics. To fully comprehend ccRCC, it is essential to investigate these features not only within tumor cells but also in immune and non-immune stromal cells that infiltrate the ccRCC TME. Recent advancements in single-cell transcriptomics applied to primary and metastatic ccRCC tumor samples have significantly enriched our understanding of the diverse cell types and states present in ccRCC.

The discovery of PTECs as the ccRCC cell of origin as well as their inflamed state raises the possibility of an alternative transcriptomic pathway in ccRCC development. Moreover, with our increasing understanding of the transcriptomic identity of metastasizing clones, new therapeutic avenues are being unlocked for metastatic ccRCC. Multiple studies have provided evidence on the dysfunctional interactions of M2-like TAMs and exhausted CD8+ T cells, which may contribute significantly to the resistance observed in available immune checkpoint therapies.

Recent findings on immunosuppressive, angiogenic, and extracellular matrix remodeling activities by CAFs and pericytes as well as endothelial cells suggest that stromal cells play an additional elusive role in primary and metastatic ccRCC. As a result, a comprehensive understanding of PTECs, immune cells, and stromal cell types and states within the ccRCC microenvironment is shedding light on tumor progression and evasion across different stages of ccRCC. Emerging integrative approaches to single-cell technologies overcome the current limitations of the technique and pave the way towards new discoveries that will drive future clinical management, therapeutics, and prognostics for both primary and metastatic ccRCC cases.

Author Contributions: Conceptualization, A.M.A., I.T. and N.B.; writing—original draft preparation, A.M.A., I.T.; writing—review and editing, A.M.A., I.T. and N.B.; visualization, A.M.A., I.T. and N.B. All authors have read and agreed to the published version of the manuscript.

Funding: This research received no external funding.

Acknowledgments: Figures were constructed using BioRender.com (accessed on 29 June 2010).

Conflicts of Interest: The authors declare no conflict of interest.

References

1. Lopez-Beltran, A.; Carrasco, J.C.; Cheng, L.; Scarpelli, M.; Kirkali, Z.; Montironi, R. 2009 update on the classification of renal epithelial tumors in adults. *Int. J. Urol.* **2009**, *16*, 432–443. [CrossRef] [PubMed]
2. Kumar, A.; Kumari, N.; Gupta, V.; Prasad, R. Renal Cell Carcinoma: Molecular Aspects. *Indian J. Clin. Biochem.* **2018**, *33*, 246–254. [CrossRef] [PubMed]
3. Pavlovich, C.P.; Schmidt, L.S.; Phillips, J.L. The genetic basis of renal cell carcinoma. *Urol. Clin. N. Am.* **2003**, *30*, 437–454. [CrossRef] [PubMed]
4. Beroukhim, R.; Brunet, J.P.; Di Napoli, A.; Mertz, K.D.; Seeley, A.; Pires, M.M.; Linhart, D.; Worrell, R.A.; Moch, H.; Rubin, M.A.; et al. Patterns of gene expression and copy-number alterations in von-hippel lindau disease-associated and sporadic clear cell carcinoma of the kidney. *Cancer Res.* **2009**, *69*, 4674–4681. [CrossRef] [PubMed]
5. Furge, K.A.; Tan, M.H.; Dykema, K.; Kort, E.; Stadler, W.; Yao, X.; Zhou, M.; Teh, B.T. Identification of deregulated oncogenic pathways in renal cell carcinoma: An integrated oncogenomic approach based on gene expression profiling. *Oncogene* **2007**, *26*, 1346–1350. [CrossRef]
6. Bhindi, B.; Wallis, C.J.D.; Boorjian, S.A.; Thompson, R.H.; Farrell, A.; Kim, S.P.; Karam, J.A.; Capitanio, U.; Golijanin, D.; Leibovich, B.C.; et al. The role of lymph node dissection in the management of renal cell carcinoma: A systematic review and meta-analysis. *BJU Int.* **2018**, *121*, 684–698. [CrossRef]
7. Woodward, E.; Jagdev, S.; McParland, L.; Clark, K.; Gregory, W.; Newsham, A.; Rogerson, S.; Hayward, K.; Selby, P.; Brown, J. Skeletal complications and survival in renal cancer patients with bone metastases. *Bone* **2011**, *48*, 160–166. [CrossRef]
8. Heinzelmann, J.; Unrein, A.; Wickmann, U.; Baumgart, S.; Stapf, M.; Szendroi, A.; Grimm, M.O.; Gajda, M.R.; Wunderlich, H.; Junker, K. MicroRNAs with prognostic potential for metastasis in clear cell renal cell carcinoma: A comparison of primary tumors and distant metastases. *Ann. Surg. Oncol.* **2014**, *21*, 1046–1054. [CrossRef]
9. Şenbabaoğlu, Y.; Gejman, R.S.; Winer, A.G.; Liu, M.; Van Allen, E.M.; de Velasco, G.; Miao, D.; Ostrovnaya, I.; Drill, E.; Luna, A.; et al. Tumor immune microenvironment characterization in clear cell renal cell carcinoma identifies prognostic and immunotherapeutically relevant messenger RNA signatures. *Genome Biol.* **2016**, *17*, 231. [CrossRef]
10. Clark, D.J.; Dhanasekaran, S.M.; Petralia, F.; Pan, J.; Song, X.; Hu, Y.; da Veiga Leprevost, F.; Reva, B.; Lih, T.M.; Chang, H.Y.; et al. Integrated Proteogenomic Characterization of Clear Cell Renal Cell Carcinoma. *Cell* **2020**, *180*, 207. [CrossRef]
11. Jerby-Arnon, L.; Shah, P.; Cuoco, M.S.; Rodman, C.; Su, M.J.; Melms, J.C.; Leeson, R.; Kanodia, A.; Mei, S.; Lin, J.R.; et al. A Cancer Cell Program Promotes T Cell Exclusion and Resistance to Checkpoint Blockade. *Cell* **2018**, *175*, 984–997.e24. [CrossRef] [PubMed]
12. Puram, S.V.; Tirosh, I.; Parikh, A.S.; Patel, A.P.; Yizhak, K.; Gillespie, S.; Rodman, C.; Luo, C.L.; Mroz, E.A.; Emerick, K.S.; et al. Single-Cell Transcriptomic Analysis of Primary and Metastatic Tumor Ecosystems in Head and Neck Cancer. *Cell* **2017**, *171*, 1611–1624.e24. [CrossRef]
13. Tirosh, I.; Izar, B.; Prakadan, S.M.; Wadsworth, M.H., 2nd; Treacy, D.; Trombetta, J.J.; Rotem, A.; Rodman, C.; Lian, C.; Murphy, G.; et al. Dissecting the multicellular ecosystem of metastatic melanoma by single-cell RNA-seq. *Science* **2016**, *352*, 189–196. [CrossRef] [PubMed]
14. Young, M.D.; Mitchell, T.J.; Vieira Braga, F.A.; Tran, M.G.B.; Stewart, B.J.; Ferdinand, J.R.; Collord, G.; Botting, R.A.; Popescu, D.M.; Loudon, K.W.; et al. Single-cell transcriptomes from human kidneys reveal the cellular identity of renal tumors. *Science* **2018**, *361*, 594–599. [CrossRef]
15. Bi, K.; He, M.X.; Bakouny, Z.; Kanodia, A.; Napolitano, S.; Wu, J.; Grimaldi, G.; Braun, D.A.; Cuoco, M.S.; Mayorga, A.; et al. Tumor and immune reprogramming during immunotherapy in advanced renal cell carcinoma. *Cancer Cell* **2021**, *39*, 649–661.e5. [CrossRef] [PubMed]
16. Chevrier, S.; Levine, J.H.; Zanotelli, V.R.T.; Silina, K.; Schulz, D.; Bacac, M.; Ries, C.H.; Ailles, L.; Jewett, M.A.S.; Moch, H.; et al. An Immune Atlas of Clear Cell Renal Cell Carcinoma. *Cell* **2017**, *169*, 736–749.e18. [CrossRef]
17. Alchahin, A.M.; Mei, S.; Tsea, I.; Hirz, T.; Kfoury, Y.; Dahl, D.; Wu, C.L.; Subtelny, A.O.; Wu, S.; Scadden, D.T.; et al. A transcriptional metastatic signature predicts survival in clear cell renal cell carcinoma. *Nat. Commun.* **2022**, *13*, 5747. [CrossRef]
18. Zhang, Y.; Narayanan, S.P.; Mannan, R.; Raskind, G.; Wang, X.; Vats, P.; Su, F.; Hosseini, N.; Cao, X.; Kumar-Sinha, C.; et al. Single-cell analyses of renal cell cancers reveal insights into tumor microenvironment, cell of origin, and therapy response. *Proc. Natl. Acad. Sci. USA* **2021**, *118*, e2103240118. [CrossRef]
19. Muto, Y.; Wilson, P.C.; Ledru, N.; Wu, H.; Dimke, H.; Waikar, S.S.; Humphreys, B.D. Single cell transcriptional and chromatin accessibility profiling redefine cellular heterogeneity in the adult human kidney. *Nat. Commun.* **2021**, *12*, 2190. [CrossRef]

20. Shi, Y.; Zhang, Q.; Bi, H.; Lu, M.; Tan, Y.; Zou, D.; Ge, L.; Chen, Z.; Liu, C.; Ci, W.; et al. Decoding the multicellular ecosystem of vena caval tumor thrombus in clear cell renal cell carcinoma by single-cell RNA sequencing. *Genome Biol.* **2022**, *23*, 87. [CrossRef]
21. Kim, K.T.; Lee, H.W.; Lee, H.O.; Song, H.J.; da Jeong, E.; Shin, S.; Kim, H.; Shin, Y.; Nam, D.H.; Jeong, B.C.; et al. Application of single-cell RNA sequencing in optimizing a combinatorial therapeutic strategy in metastatic renal cell carcinoma. *Genome Biol.* **2016**, *17*, 80. [CrossRef] [PubMed]
22. Zhang, C.; He, H.; Hu, X.; Liu, A.; Huang, D.; Xu, Y.; Chen, L.; Xu, D. Development and validation of a metastasis-associated prognostic signature based on single-cell RNA-seq in clear cell renal cell carcinoma. *Aging* **2019**, *11*, 10183–10202. [CrossRef] [PubMed]
23. Chen, W.J.; Dong, K.Q.; Pan, X.W.; Gan, S.S.; Xu, D.; Chen, J.X.; Chen, W.J.; Li, W.Y.; Wang, Y.Q.; Zhou, W.; et al. Single-cell RNA-seq integrated with multi-omics reveals SERPINE2 as a target for metastasis in advanced renal cell carcinoma. *Cell Death Dis.* **2023**, *14*, 30. [CrossRef] [PubMed]
24. Lawson, D.A.; Kessenbrock, K.; Davis, R.T.; Pervolarakis, N.; Werb, Z. Tumour heterogeneity and metastasis at single-cell resolution. *Nat. Cell Biol.* **2018**, *20*, 1349–1360. [CrossRef]
25. Liu, K.; Gao, R.; Wu, H.; Wang, Z.; Han, G. Single-cell analysis reveals metastatic cell heterogeneity in clear cell renal cell carcinoma. *J. Cell. Mol. Med.* **2021**, *25*, 4260–4274. [CrossRef]
26. Li, R.; Ferdinand, J.R.; Loudon, K.W.; Bowyer, G.S.; Laidlaw, S.; Muyas, F.; Mamanova, L.; Neves, J.B.; Bolt, L.; Fasouli, E.S.; et al. Mapping single-cell transcriptomes in the intra-tumoral and associated territories of kidney cancer. *Cancer Cell* **2022**, *40*, 1583–1599.e10. [CrossRef]
27. Braun, D.A.; Street, K.; Burke, K.P.; Cookmeyer, D.L.; Denize, T.; Pedersen, C.B.; Gohil, S.H.; Schindler, N.; Pomerance, L.; Hirsch, L.; et al. Progressive immune dysfunction with advancing disease stage in renal cell carcinoma. *Cancer Cell* **2021**, *39*, 632–648.e8.
28. Borcherding, N.; Vishwakarma, A.; Voigt, A.P.; Bellizzi, A.; Kaplan, J.; Nepple, K.; Salem, A.K.; Jenkins, R.W.; Zakharia, Y.; Zhang, W. Mapping the immune environment in clear cell renal carcinoma by single-cell genomics. *Commun. Biol.* **2021**, *4*, 122. [CrossRef]
29. Kim, M.C.; Jin, Z.; Kolb, R.; Borcherding, N.; Chatzkel, J.A.; Falzarano, S.M.; Zhang, W. Updates on Immunotherapy and Immune Landscape in Renal Clear Cell Carcinoma. *Cancers* **2021**, *13*, 5856. [CrossRef]
30. Obradovic, A.; Chowdhury, N.; Haake, S.M.; Ager, C.; Wang, V.; Vlahos, L.; Guo, X.V.; Aggen, D.H.; Rathmell, W.K.; Jonasch, E.; et al. Single-cell protein activity analysis identifies recurrence-associated renal tumor macrophages. *Cell* **2021**, *184*, 2988–3005.e16. [CrossRef]
31. Krishna, C.; DiNatale, R.G.; Kuo, F.; Srivastava, R.M.; Vuong, L.; Chowell, D.; Gupta, S.; Vanderbilt, C.; Purohit, T.A.; Liu, M.; et al. Single-cell sequencing links multiregional immune landscapes and tissue-resident T cells in ccRCC to tumor topology and therapy efficacy. *Cancer Cell* **2021**, *39*, 662–677.e6. [CrossRef] [PubMed]
32. Hu, J.; Chen, Z.; Bao, L.; Zhou, L.; Hou, Y.; Liu, L.; Xiong, M.; Zhang, Y.; Wang, B.; Tao, Z.; et al. Single-Cell Transcriptome Analysis Reveals Intratumoral Heterogeneity in ccRCC, which Results in Different Clinical Outcomes. *Mol. Ther.* **2020**, *28*, 1658–1672. [CrossRef] [PubMed]
33. Becht, E.; Giraldo, N.A.; Beuselinck, B.; Job, S.; Marisa, L.; Vano, Y.; Oudard, S.; Zucman-Rossi, J.; Laurent-Puig, P.; Sautès-Fridman, C.; et al. Prognostic and theranostic impact of molecular subtypes and immune classifications in renal cell cancer (RCC) and colorectal cancer (CRC). *Oncoimmunology* **2015**, *4*, e1049804. [CrossRef] [PubMed]
34. Khan, O.; Giles, J.R.; McDonald, S.; Manne, S.; Ngiow, S.F.; Patel, K.P.; Werner, M.T.; Huang, A.C.; Alexander, K.A.; Wu, J.E.; et al. TOX transcriptionally and epigenetically programs CD8(+) T cell exhaustion. *Nature* **2019**, *571*, 211–218. [CrossRef]
35. Scott, A.C.; Dündar, F.; Zumbo, P.; Chandran, S.S.; Klebanoff, C.A.; Shakiba, M.; Trivedi, P.; Menocal, L.; Appleby, H.; Camara, S.; et al. TOX is a critical regulator of tumour-specific T cell differentiation. *Nature* **2019**, *571*, 270–274. [CrossRef]
36. Blank, C.U.; Haining, W.N.; Held, W.; Hogan, P.G.; Kallies, A.; Lugli, E.; Lynn, R.C.; Philip, M.; Rao, A.; Restifo, N.P.; et al. Defining 'T cell exhaustion'. *Nat. Rev. Immunol.* **2019**, *19*, 665–674. [CrossRef]
37. Zeng, Q.; Zhang, W.; Li, X.; Lai, J.; Li, Z. Bioinformatic identification of renal cell carcinoma microenvironment-associated biomarkers with therapeutic and prognostic value. *Life Sci.* **2020**, *243*, 117273. [CrossRef]
38. Massari, F.; Santoni, M.; Ciccarese, C.; Santini, D.; Alfieri, S.; Martignoni, G.; Brunelli, M.; Piva, F.; Berardi, R.; Montironi, R.; et al. PD-1 blockade therapy in renal cell carcinoma: Current studies and future promises. *Cancer Treat. Rev.* **2015**, *41*, 114–121. [CrossRef]
39. Au, L.; Hatipoglu, E.; Robert de Massy, M.; Litchfield, K.; Beattie, G.; Rowan, A.; Schnidrig, D.; Thompson, R.; Byrne, F.; Horswell, S.; et al. Determinants of anti-PD-1 response and resistance in clear cell renal cell carcinoma. *Cancer Cell* **2021**, *39*, 1497–1518.e11. [CrossRef]
40. Peng, Y.L.; Xiong, L.B.; Zhou, Z.H.; Ning, K.; Li, Z.; Wu, Z.S.; Deng, M.H.; Wei, W.S.; Wang, N.; Zou, X.P.; et al. Single-cell transcriptomics reveals a low CD8(+) T cell infiltrating state mediated by fibroblasts in recurrent renal cell carcinoma. *J. Immunother. Cancer* **2022**, *10*, e004206. [CrossRef]
41. Long, Z.; Sun, C.; Tang, M.; Wang, Y.; Ma, J.; Yu, J.; Wei, J.; Ma, J.; Wang, B.; Xie, Q.; et al. Single-cell multiomics analysis reveals regulatory programs in clear cell renal cell carcinoma. *Cell Discov.* **2022**, *8*, 68. [CrossRef] [PubMed]
42. Ljungberg, B.; Albiges, L.; Abu-Ghanem, Y.; Bedke, J.; Capitanio, U.; Dabestani, S.; Fernández-Pello, S.; Giles, R.H.; Hofmann, F.; Hora, M.; et al. European Association of Urology Guidelines on Renal Cell Carcinoma: The 2022 Update. *Eur. Urol.* **2022**, *82*, 399–410. [CrossRef]
43. Klein, C.A. Parallel progression of primary tumours and metastases. *Nat. Rev. Cancer* **2009**, *9*, 302–312. [CrossRef] [PubMed]

44. Albiges, L.; Powles, T.; Staehler, M.; Bensalah, K.; Giles, R.H.; Hora, M.; Kuczyk, M.A.; Lam, T.B.; Ljungberg, B.; Marconi, L.; et al. Updated European Association of Urology Guidelines on Renal Cell Carcinoma: Immune Checkpoint Inhibition Is the New Backbone in First-line Treatment of Metastatic Clear-cell Renal Cell Carcinoma. *Eur. Urol.* **2019**, *76*, 151–156. [CrossRef] [PubMed]
45. Motzer, R.J.; Michaelson, M.D.; Redman, B.G.; Hudes, G.R.; Wilding, G.; Figlin, R.A.; Ginsberg, M.S.; Kim, S.T.; Baum, C.M.; DePrimo, S.E.; et al. Activity of SU11248, a multitargeted inhibitor of vascular endothelial growth factor receptor and platelet-derived growth factor receptor, in patients with metastatic renal cell carcinoma. *J. Clin. Oncol.* **2006**, *24*, 16–24. [CrossRef] [PubMed]
46. Motzer, R.J.; Rini, B.I.; Bukowski, R.M.; Curti, B.D.; George, D.J.; Hudes, G.R.; Redman, B.G.; Margolin, K.A.; Merchan, J.R.; Wilding, G.; et al. Sunitinib in patients with metastatic renal cell carcinoma. *JAMA* **2006**, *295*, 2516–2524. [CrossRef]
47. Rini, B.I.; Pal, S.K.; Escudier, B.J.; Atkins, M.B.; Hutson, T.E.; Porta, C.; Verzoni, E.; Needle, M.N.; McDermott, D.F. Tivozanib versus sorafenib in patients with advanced renal cell carcinoma (TIVO-3): A phase 3, multicentre, randomised, controlled, open-label study. *Lancet Oncol.* **2020**, *21*, 95–104. [CrossRef]
48. Atkins, M.B.; Hidalgo, M.; Stadler, W.M.; Logan, T.F.; Dutcher, J.P.; Hudes, G.R.; Park, Y.; Liou, S.H.; Marshall, B.; Boni, J.P.; et al. Randomized phase II study of multiple dose levels of CCI-779, a novel mammalian target of rapamycin kinase inhibitor, in patients with advanced refractory renal cell carcinoma. *J. Clin. Oncol.* **2004**, *22*, 909–918. [CrossRef]
49. Gibney, G.T.; Aziz, S.A.; Camp, R.L.; Conrad, P.; Schwartz, B.E.; Chen, C.R.; Kelly, W.K.; Kluger, H.M. c-Met is a prognostic marker and potential therapeutic target in clear cell renal cell carcinoma. *Ann. Oncol.* **2013**, *24*, 343–349. [CrossRef]
50. Rini, B.I.; Atkins, M.B. Resistance to targeted therapy in renal-cell carcinoma. *Lancet Oncol.* **2009**, *10*, 992–1000. [CrossRef]
51. Wang, C.; Li, Y.; Chu, C.M.; Zhang, X.M.; Ma, J.; Huang, H.; Wang, Y.N.; Hong, T.Y.; Zhang, J.; Pan, X.W.; et al. Gankyrin is a novel biomarker for disease progression and prognosis of patients with renal cell carcinoma. *EBioMedicine* **2019**, *39*, 255–264. [CrossRef] [PubMed]
52. Nickerson, M.L.; Jaeger, E.; Shi, Y.; Durocher, J.A.; Mahurkar, S.; Zaridze, D.; Matveev, V.; Janout, V.; Kollarova, H.; Bencko, V.; et al. Improved identification of von Hippel-Lindau gene alterations in clear cell renal tumors. *Clin. Cancer Res.* **2008**, *14*, 4726–4734. [CrossRef] [PubMed]
53. Rini, B.I.; Small, E.J. Biology and clinical development of vascular endothelial growth factor-targeted therapy in renal cell carcinoma. *J. Clin. Oncol.* **2005**, *23*, 1028–1043. [CrossRef]
54. Jubb, A.M.; Pham, T.Q.; Hanby, A.M.; Frantz, G.D.; Peale, F.V.; Wu, T.D.; Koeppen, H.W.; Hillan, K.J. Expression of vascular endothelial growth factor, hypoxia inducible factor 1alpha, and carbonic anhydrase IX in human tumours. *J. Clin. Pathol.* **2004**, *57*, 504–512. [CrossRef] [PubMed]
55. Karaman, S.; Leppänen, V.M.; Alitalo, K. Vascular endothelial growth factor signaling in development and disease. *Development* **2018**, *145*, dev151019. [CrossRef]
56. Jonasch, E.; Gao, J.; Rathmell, W.K. Renal cell carcinoma. *BMJ* **2014**, *349*, g4797. [CrossRef]
57. Ferrara, N.; Gerber, H.P.; LeCouter, J. The biology of VEGF and its receptors. *Nat. Med.* **2003**, *9*, 669–676. [CrossRef]
58. Motzer, R.J.; Hutson, T.E.; Tomczak, P.; Michaelson, M.D.; Bukowski, R.M.; Rixe, O.; Oudard, S.; Negrier, S.; Szczylik, C.; Kim, S.T.; et al. Sunitinib versus interferon alfa in metastatic renal-cell carcinoma. *N. Engl. J. Med.* **2007**, *356*, 115–124. [CrossRef]
59. Escudier, B.; Eisen, T.; Stadler, W.M.; Szczylik, C.; Oudard, S.; Siebels, M.; Negrier, S.; Chevreau, C.; Solska, E.; Desai, A.A.; et al. Sorafenib in advanced clear-cell renal-cell carcinoma. *N. Engl. J. Med.* **2007**, *356*, 125–134. [CrossRef]
60. Hoefflin, R.; Harlander, S.; Schäfer, S.; Metzger, P.; Kuo, F.; Schönenberger, D.; Adlesic, M.; Peighambari, A.; Seidel, P.; Chen, C.Y.; et al. HIF-1α and HIF-2α differently regulate tumour development and inflammation of clear cell renal cell carcinoma in mice. *Nat. Commun.* **2020**, *11*, 4111. [CrossRef]
61. Motzer, R.; Alekseev, B.; Rha, S.Y.; Porta, C.; Eto, M.; Powles, T.; Grünwald, V.; Hutson, T.E.; Kopyltsov, E.; Méndez-Vidal, M.J.; et al. Lenvatinib plus Pembrolizumab or Everolimus for Advanced Renal Cell Carcinoma. *N. Engl. J. Med.* **2021**, *384*, 1289–1300. [CrossRef] [PubMed]
62. Braun, D.A.; Hou, Y.; Bakouny, Z.; Ficial, M.; Sant' Angelo, M.; Forman, J.; Ross-Macdonald, P.; Berger, A.C.; Jegede, O.A.; Elagina, L.; et al. Interplay of somatic alterations and immune infiltration modulates response to PD-1 blockade in advanced clear cell renal cell carcinoma. *Nat. Med.* **2020**, *26*, 909–918. [CrossRef]
63. Motzer, R.J.; Hutson, T.E.; McCann, L.; Deen, K.; Choueiri, T.K. Overall survival in renal-cell carcinoma with pazopanib versus sunitinib. *N. Engl. J. Med.* **2014**, *370*, 1769–1770. [CrossRef] [PubMed]
64. Motzer, R.J.; Penkov, K.; Haanen, J.; Rini, B.; Albiges, L.; Campbell, M.T.; Venugopal, B.; Kollmannsberger, C.; Negrier, S.; Uemura, M.; et al. Avelumab plus Axitinib versus Sunitinib for Advanced Renal-Cell Carcinoma. *N. Engl. J. Med.* **2019**, *380*, 1103–1115. [CrossRef] [PubMed]
65. Rini, B.I.; Plimack, E.R.; Stus, V.; Gafanov, R.; Hawkins, R.; Nosov, D.; Pouliot, F.; Alekseev, B.; Soulières, D.; Melichar, B.; et al. Pembrolizumab plus Axitinib versus Sunitinib for Advanced Renal-Cell Carcinoma. *N. Engl. J. Med.* **2019**, *380*, 1116–1127. [CrossRef]
66. Motzer, R.J.; Tannir, N.M.; McDermott, D.F.; Arén Frontera, O.; Melichar, B.; Choueiri, T.K.; Plimack, E.R.; Barthélémy, P.; Porta, C.; George, S.; et al. Nivolumab plus Ipilimumab versus Sunitinib in Advanced Renal-Cell Carcinoma. *N. Engl. J. Med.* **2018**, *378*, 1277–1290. [CrossRef]

67. Choueiri, T.K.; Powles, T.; Burotto, M.; Escudier, B.; Bourlon, M.T.; Zurawski, B.; Oyervides Juárez, V.M.; Hsieh, J.J.; Basso, U.; Shah, A.Y.; et al. Nivolumab plus Cabozantinib versus Sunitinib for Advanced Renal-Cell Carcinoma. *N. Engl. J. Med.* **2021**, *384*, 829–841. [CrossRef]
68. Motzer, R.J.; Escudier, B.; McDermott, D.F.; George, S.; Hammers, H.J.; Srinivas, S.; Tykodi, S.S.; Sosman, J.A.; Procopio, G.; Plimack, E.R.; et al. Nivolumab versus Everolimus in Advanced Renal-Cell Carcinoma. *N. Engl. J. Med.* **2015**, *373*, 1803–1813. [CrossRef]
69. McDermott, D.F.; Lee, J.L.; Ziobro, M.; Suarez, C.; Langiewicz, P.; Matveev, V.B.; Wiechno, P.; Gafanov, R.A.; Tomczak, P.; Pouliot, F.; et al. Open-Label, Single-Arm, Phase II Study of Pembrolizumab Monotherapy as First-Line Therapy in Patients With Advanced Non-Clear Cell Renal Cell Carcinoma. *J. Clin. Oncol.* **2021**, *39*, 1029–1039. [CrossRef]
70. Wei, S.C.; Duffy, C.R.; Allison, J.P. Fundamental Mechanisms of Immune Checkpoint Blockade Therapy. *Cancer Discov.* **2018**, *8*, 1069–1086.
71. Motzer, R.J.; Powles, T.; Atkins, M.B.; Escudier, B.; McDermott, D.F.; Alekseev, B.Y.; Lee, J.L.; Suarez, C.; Stroyakovskiy, D.; De Giorgi, U.; et al. Final Overall Survival and Molecular Analysis in IMmotion151, a Phase 3 Trial Comparing Atezolizumab plus Bevacizumab vs Sunitinib in Patients with Previously Untreated Metastatic Renal Cell Carcinoma. *JAMA Oncol.* **2022**, *8*, 275–280. [CrossRef] [PubMed]
72. Motzer, R.J.; Rini, B.I.; McDermott, D.F.; Redman, B.G.; Kuzel, T.M.; Harrison, M.R.; Vaishampayan, U.N.; Drabkin, H.A.; George, S.; Logan, T.F.; et al. Nivolumab for Metastatic Renal Cell Carcinoma: Results of a Randomized Phase II Trial. *J. Clin. Oncol.* **2015**, *33*, 1430–1437. [CrossRef] [PubMed]
73. Méjean, A.; Ravaud, A.; Thezenas, S.; Chevreau, C.; Bensalah, K.; Geoffrois, L.; Thiery-Vuillemin, A.; Cormier, L.; Lang, H.; Guy, L.; et al. Sunitinib Alone or After Nephrectomy for Patients with Metastatic Renal Cell Carcinoma: Is There Still a Role for Cytoreductive Nephrectomy? *Eur. Urol.* **2021**, *80*, 417–424. [CrossRef] [PubMed]
74. Apolo, A.B.; Powles, T.; Escudier, B.; Burotto, M.; Zhang, J.; Simsek, B.; Scheffold, C.; Motzer, R.J.; Choueiri, T.K. Nivolumab plus ipilimumab plus cabozantinib triplet combination for patients with previously untreated advanced renal cell carcinoma: Results from a discontinued arm of the phase III CheckMate 9ER trial. *Eur. J. Cancer* **2022**, *177*, 63–71. [CrossRef]
75. Fahey, C.C.; Shevach, J.W.; Flippot, R.; Albiges, L.; Haas, N.B.; Beckermann, K.E. Triplet Strategies in Metastatic Clear Cell Renal Cell Carcinoma: A Worthy Option in the First-Line Setting? *Am. Soc. Clin. Oncol. Educ. Book* **2023**, *43*, e389650. [CrossRef]
76. Adotévi, O.; Galaine, J. Antitumor CAR T-cell Screening Platform: Many Are Called, but Few Are Chosen. *Cancer Res.* **2022**, *82*, 2517–2519. [CrossRef]
77. Binnewies, M.; Pollack, J.L.; Rudolph, J.; Dash, S.; Abushawish, M.; Lee, T.; Jahchan, N.S.; Canaday, P.; Lu, E.; Norng, M.; et al. Targeting TREM2 on tumor-associated macrophages enhances immunotherapy. *Cell Rep.* **2021**, *37*, 109844. [CrossRef]
78. Koneru, R.; Hotte, S.J. Role of cytokine therapy for renal cell carcinoma in the era of targeted agents. *Curr. Oncol.* **2009**, *16* (Suppl. S1), S40–S44. [CrossRef]
79. Motzer, R.J.; Bander, N.H.; Nanus, D.M. Renal-cell carcinoma. *N. Engl. J. Med.* **1996**, *335*, 865–875. [CrossRef]
80. Dudani, S.; Graham, J.; Wells, J.C.; Bakouny, Z.; Pal, S.K.; Dizman, N.; Donskov, F.; Porta, C.; de Velasco, G.; Hansen, A.; et al. First-line Immuno-Oncology Combination Therapies in Metastatic Renal-cell Carcinoma: Results from the International Metastatic Renal-cell Carcinoma Database Consortium. *Eur. Urol.* **2019**, *76*, 861–867. [CrossRef]
81. Sharma, P.; Allison, J.P. Immune checkpoint targeting in cancer therapy: Toward combination strategies with curative potential. *Cell* **2015**, *161*, 205–214. [CrossRef] [PubMed]
82. Giraldo, N.A.; Becht, E.; Pagès, F.; Skliris, G.; Verkarre, V.; Vano, Y.; Mejean, A.; Saint-Aubert, N.; Lacroix, L.; Natario, I.; et al. Orchestration and Prognostic Significance of Immune Checkpoints in the Microenvironment of Primary and Metastatic Renal Cell Cancer. *Clin. Cancer Res.* **2015**, *21*, 3031–3040. [CrossRef] [PubMed]
83. Oliveira, G.; Stromhaug, K.; Klaeger, S.; Kula, T.; Frederick, D.T.; Le, P.M.; Forman, J.; Huang, T.; Li, S.; Zhang, W.; et al. Phenotype, specificity and avidity of antitumour CD8(+) T cells in melanoma. *Nature* **2021**, *596*, 119–125. [CrossRef] [PubMed]
84. Turajlic, S.; Xu, H.; Litchfield, K.; Rowan, A.; Horswell, S.; Chambers, T.; O'Brien, T.; Lopez, J.I.; Watkins, T.B.K.; Nicol, D.; et al. Deterministic Evolutionary Trajectories Influence Primary Tumor Growth: TRACERx Renal. *Cell* **2018**, *173*, 595–610.e11. [CrossRef] [PubMed]
85. Turajlic, S.; Xu, H.; Litchfield, K.; Rowan, A.; Chambers, T.; Lopez, J.I.; Nicol, D.; O'Brien, T.; Larkin, J.; Horswell, S.; et al. Tracking Cancer Evolution Reveals Constrained Routes to Metastases: TRACERx Renal. *Cell* **2018**, *173*, 581–594.e12. [CrossRef]
86. Mitchell, T.J.; Turajlic, S.; Rowan, A.; Nicol, D.; Farmery, J.H.R.; O'Brien, T.; Martincorena, I.; Tarpey, P.; Angelopoulos, N.; Yates, L.R.; et al. Timing the Landmark Events in the Evolution of Clear Cell Renal Cell Cancer: TRACERx Renal. *Cell* **2018**, *173*, 611–623 e617. [CrossRef]
87. Wu, T.D.; Madireddi, S.; de Almeida, P.E.; Banchereau, R.; Chen, Y.J.; Chitre, A.S.; Chiang, E.Y.; Iftikhar, H.; O'Gorman, W.E.; Au-Yeung, A.; et al. Peripheral T cell expansion predicts tumour infiltration and clinical response. *Nature* **2020**, *579*, 274–278. [CrossRef]
88. Zhou, W.; Yang, F.; Xu, Z.; Luo, M.; Wang, P.; Guo, Y.; Nie, H.; Yao, L.; Jiang, Q. Comprehensive Analysis of Copy Number Variations in Kidney Cancer by Single-Cell Exome Sequencing. *Front. Genet.* **2019**, *10*, 1379. [CrossRef]
89. Tannir, N.M.; Signoretti, S.; Choueiri, T.K.; McDermott, D.F.; Motzer, R.J.; Flaifel, A.; Pignon, J.C.; Ficial, M.; Frontera, O.A.; George, S.; et al. Efficacy and Safety of Nivolumab Plus Ipilimumab versus Sunitinib in First-line Treatment of Patients with Advanced Sarcomatoid Renal Cell Carcinoma. *Clin. Cancer Res.* **2021**, *27*, 78–86. [CrossRef]

90. van den Brink, S.C.; Sage, F.; Vértesy, Á.; Spanjaard, B.; Peterson-Maduro, J.; Baron, C.S.; Robin, C.; van Oudenaarden, A. Single-cell sequencing reveals dissociation-induced gene expression in tissue subpopulations. *Nat. Methods* **2017**, *14*, 935–936. [CrossRef]
91. Adam, M.; Potter, A.S.; Potter, S.S. Psychrophilic proteases dramatically reduce single-cell RNA-seq artifacts: A molecular atlas of kidney development. *Development* **2017**, *144*, 3625–3632. [CrossRef] [PubMed]
92. Tran, H.T.N.; Ang, K.S.; Chevrier, M.; Zhang, X.; Lee, N.Y.S.; Goh, M.; Chen, J. A benchmark of batch-effect correction methods for single-cell RNA sequencing data. *Genome Biol.* **2020**, *21*, 12. [CrossRef] [PubMed]
93. Goh, W.W.B.; Wang, W.; Wong, L. Why Batch Effects Matter in Omics Data, and How to Avoid Them. *Trends Biotechnol.* **2017**, *35*, 498–507. [CrossRef] [PubMed]
94. Hie, B.; Bryson, B.; Berger, B. Efficient integration of heterogeneous single-cell transcriptomes using Scanorama. *Nat. Biotechnol.* **2019**, *37*, 685–691. [CrossRef]
95. Hao, Y.; Hao, S.; Andersen-Nissen, E.; Mauck, W.M., 3rd; Zheng, S.; Butler, A.; Lee, M.J.; Wilk, A.J.; Darby, C.; Zager, M.; et al. Integrated analysis of multimodal single-cell data. *Cell* **2021**, *184*, 3573–3587.e29. [CrossRef]
96. Gui, C.P.; Wei, J.H.; Zhang, C.; Tang, Y.M.; Shu, G.N.; Wu, R.P.; Luo, J.H. Single-cell and spatial transcriptomics reveal 5-methylcytosine RNA methylation regulators immunologically reprograms tumor microenvironment characterizations, immunotherapy response and precision treatment of clear cell renal cell carcinoma. *Transl. Oncol.* **2023**, *35*, 101726. [CrossRef]
97. Davidson, G.; Helleux, A.; Vano, Y.A.; Lindner, V.; Fattori, A.; Cerciat, M.; Elaidi, R.T.; Verkarre, V.; Sun, C.M.; Chevreau, C.; et al. Mesenchymal-like Tumor Cells and Myofibroblastic Cancer-Associated Fibroblasts Are Associated with Progression and Immunotherapy Response of Clear Cell Renal Cell Carcinoma. *Cancer Res.* **2023**, *83*, 2952–2969. [CrossRef]
98. Raghubar, A.M.; Matigian, N.A.; Crawford, J.; Francis, L.; Ellis, R.; Healy, H.G.; Kassianos, A.J.; Ng, M.S.Y.; Roberts, M.J.; Wood, S.; et al. High risk clear cell renal cell carcinoma microenvironments contain protumour immunophenotypes lacking specific immune checkpoints. *NPJ Precis. Oncol.* **2023**, *7*, 88. [CrossRef]
99. Wu, Y.; Terekhanova, N.V.; Caravan, W.; Naser Al Deen, N.; Lal, P.; Chen, S.; Mo, C.K.; Cao, S.; Li, Y.; Karpova, A.; et al. Epigenetic and transcriptomic characterization reveals progression markers and essential pathways in clear cell renal cell carcinoma. *Nat. Commun.* **2023**, *14*, 1681. [CrossRef]

cancers

MDPI

Systematic Review

Comprehensive Systematic Review of Biomarkers in Metastatic Renal Cell Carcinoma: Predictors, Prognostics, and Therapeutic Monitoring

Komal A. Dani [1,*,†], Joseph M. Rich [1,†], Sean S. Kumar [2,3,4], Harmony Cen [5], Vinay A. Duddalwar [6,7,8] and Anishka D'Souza [9]

1 Keck School of Medicine, University of Southern California, Los Angeles, CA 90033, USA; jmrich@usc.edu
2 Eastern Virginia Medical School, Norfolk, VA 23507, USA; sshkumar@ucdavis.edu
3 Children's Hospital Los Angeles, Los Angeles, CA 90027, USA
4 Norris Comprehensive Cancer Center, University of Southern California, Los Angeles, CA 90033, USA
5 University of Southern California, Los Angeles, CA 90033, USA; hscen@usc.edu
6 Department of Radiology, Keck School of Medicine, University of Southern California, Los Angeles, CA 90033, USA; vinay.duddalwar@med.usc.edu
7 Institute of Urology, University of Southern California, Los Angeles, CA 90033, USA
8 Department of Biomedical Engineering, University of Southern California, Los Angeles, CA 90089, USA
9 Department of Medical Oncology, Norris Comprehensive Cancer Center, University of Southern California, Los Angeles, CA 90033, USA; anishkad@med.usc.edu
* Correspondence: kadani@usc.edu; Tel.: +1-(916)-779-9316
† These authors contributed equally to this work.

Simple Summary: This comprehensive systematic review provides valuable insights into the landscape of biomarkers in clear cell renal cell carcinoma (ccRCC) and their potential applications in the prediction of treatment response, prognosis, and therapeutic monitoring. One of the major challenges in ccRCC is determining the most effective treatment strategies and identifying patients who would benefit from adjuvant or neoadjuvant therapy. This review aims to provide a comprehensive overview of biomarkers in ccRCC and their utility in the prediction of treatment response, prognosis, and therapeutic monitoring in patients receiving systemic therapy for metastatic disease. The findings underscore the importance of incorporating biomarker assessment into clinical practice to guide treatment decisions and improve patient outcomes in ccRCC.

Abstract: Background: Challenges remain in determining the most effective treatment strategies and identifying patients who would benefit from adjuvant or neoadjuvant therapy in renal cell carcinoma. The objective of this review is to provide a comprehensive overview of biomarkers in metastatic renal cell carcinoma (mRCC) and their utility in prediction of treatment response, prognosis, and therapeutic monitoring in patients receiving systemic therapy for metastatic disease. Methods: A systematic literature search was conducted using the PubMed database for relevant studies published between January 2017 and December 2022. The search focused on biomarkers associated with mRCC and their relationship to immune checkpoint inhibitors, targeted therapy, and VEGF inhibitors in the adjuvant, neoadjuvant, and metastatic settings. Results: The review identified various biomarkers with predictive, prognostic, and therapeutic monitoring potential in mRCC. The review also discussed the challenges associated with anti-angiogenic and immune-checkpoint monotherapy trials and highlighted the need for personalized therapy based on molecular signatures. Conclusion: This comprehensive review provides valuable insights into the landscape of biomarkers in mRCC and their potential applications in prediction of treatment response, prognosis, and therapeutic monitoring. The findings underscore the importance of incorporating biomarker assessment into clinical practice to guide treatment decisions and improve patient outcomes in mRCC.

Keywords: renal cell carcinoma; predictive biomarkers; prognostic biomarkers; therapeutic monitoring; immune checkpoint inhibitors; VEGF inhibitors; metastatic disease

Citation: Dani, K.A.; Rich, J.M.; Kumar, S.S.; Cen, H.; Duddalwar, V.A.; D'Souza, A. Comprehensive Systematic Review of Biomarkers in Metastatic Renal Cell Carcinoma: Predictors, Prognostics, and Therapeutic Monitoring. *Cancers* **2023**, *15*, 4934. https://doi.org/10.3390/cancers15204934

Academic Editors: Jörg Ellinger and Leo J. Schouten

Received: 22 July 2023
Revised: 30 September 2023
Accepted: 9 October 2023
Published: 11 October 2023

1. Introduction

In 2023, there will be an estimated 76,080 new cases and 13,780 new deaths due to kidney cancer in the US [1]. Approximately 90–95% of these neoplasms are renal cell carcinoma (RCC), with 16% presenting with regional spread and another 16% presenting with distant metastasis [2,3]. Clear cell RCC (ccRCC) is the most common subtype, accounting for over 70–80% of RCC [4]. The cure rate is high for patients with early, localized disease, with 5-year survival at more than 90% [5]. In contrast, 5-year survival drops to 12% for patients with distant metastatic disease.

Historically used chemotherapy and radiation therapy lacked sensitivity in ccRCC [6,7], and therefore an improved understanding of the biochemistry and genetic molecular landscape in ccRCC has led to the discovery of novel therapeutic agents [8–10].

Despite the advances that have been made, there are many challenges to treating ccRCC. In particular, pure anti-angiogenic trials (which have been largely negative) and pure immune-checkpoint monotherapy trials have been applied (with one positive trial so far) to the adjuvant setting with continued uncertainty as to who would benefit from adjuvant therapy or neoadjuvant therapy and for how long [11]. The advent of molecular signatures brings forth the opportunity to better understand how to personalize therapy in ccRCC, enabling clinicians to improve patient outcomes. The purpose of this review is to systematically identify biomarkers that have the potential to diagnose, predict, prognose, and track therapeutic monitoring in ccRCC patients receiving systemic therapy with immune oncology (IO), tyrosine kinase inhibitors (TKI), VEGF inhibitors (VEGFi), or a combination for treatment of adjuvant, neoadjuvant, and metastatic disease. These terms are defined in Table 1 [12].

Table 1. Important Definitions.

Term	Definition
Biomarker	A measurable substance whose presence is indicative of disease, infection, or environmental exposure.
Diagnostic Biomarker	Detects or confirms the presence of a disease or condition of interest or identifies an individual with a subtype of the disease.
Predictive Biomarker	Predicts an individual or group of individuals more likely to experience a favorable or unfavorable effect from the exposure to a medical product or environmental agent
Prognostic Biomarker	Identifies the likelihood of a clinical event, disease recurrence, or disease progression in patients with a disease or medical condition of interest
Therapeutic Monitoring Biomarker	Assesses the status of a disease or medical condition for evidence of exposure to a medical product or environmental agent, or to detects an effect of a medical product or biological agent

Prior reviews have broached this topic but have not provided a comprehensive review of the vast variety of biomarkers that exist in ccRCC or have focused on a specific form of therapy. For example, Gulati et al. aim to evaluate validated biomarkers that are being utilized to guide treatment choices and help identify pathways of resistance in other tumor types, and do this by organizing their paper by biomarker (PD-L1, tumor mutational burden, VHL, PBRM1, BAP1, SETD2, a few genomic signatures) [13]. Similarly, Raimondi et al. focus on predictive markers for immunotherapy response [14]. Farber et al. did explore the various serum, urine, imaging, and immunohistological biomarkers that had diagnostic, prognostic, and predictive utility up until 2017. However, this review once again organized their finding by type of biomarker [15]. Our review strove to incorporate all of this content in one place by exploring all biomarkers that have been studied from January 2017 to December 2022 by treatment line.

2. Materials and Methods

The protocol of the present systematic review and meta-analysis was following the Preferred Reporting Items for Systematic Reviews and Meta-Analyses (PRISMA) guidelines. The PRISMA checklist was shown in Supplementary Figure S1. The Covidence platform (reference) was used for paper importing and screening to streamline the review process [16]. This systemic review was conducted on schedule without its registration in PROSPERO.

2.1. Data Sources and Searches

Two authors independently searched relevant studies from the PubMed Advanced Search Builder literature database. The retrieval time ended on 31 December 2022. The systematic searching was restricted to the English language and human subjects. The following search strategy was employed in PubMed: ((biomarker) OR (prognostic marker) OR (gene expression)) AND ((immunotherapy) OR (immune checkpoint inhibitor) OR (ICI) OR (targeted therapy) OR (tyrosine kinase inhibitor) OR (TKI) OR (VEGF inhibitor)) AND ((clear cell renal cell carcinoma)) OR (ccRCC) OR (clear cell RCC)) AND ((metastatic) OR (metastasis) OR (metastases) OR (stage 4) OR (stage IV) OR (adjuvant) OR (neoadjuvant)) NOT ("case report"[Title]) NOT ("case reports"[Publication Type]) NOT (review [Publication Type]).

2.2. Inclusion and Exclusion Criteria

Two authors independently screened the literature according to the following criteria: (1) studies including patients who were diagnosed with primary or metastatic RCC after cytology or pathology; (2) studies that were Phase I, II, III, or IV or retrospective studies; (3) studies that focused on systemic therapies to treat adjuvant, neoadjuvant, or metastatic disease; (4) studies that used IO, TKI, VEGFi, or a combination of these therapies; and (5) articles that were published in English. For repeatedly published studies, we only chose the latest literature or the literature with the largest sample size. The articles were excluded based on the following reasons: (1) without the normal control group data; (2) review articles, editorials, comments, letters, case reports, etc.; (3) duplicated data; and (4) animal experiment. Only articles published between January 2017 and December 2022 were included, given that RCC treatment protocols have frequently changed.

2.3. Data Extraction and Literature

A standardized data collection form was employed to extract the following information by 2 authors independently: the first author's name, publication year, title, DOI, abstract, paper methodology, sample size, marker name, marker purpose, marker class, type of maker, drug, drug class, major findings, and limitations. These data can be found in Supplementary Table S1. In case no direct HR or 95% CI was provided in the publications or could not be calculated through the existing data, we tried to contact the corresponding author to obtain the relevant data. If no response was received from the author, data were extracted from the survival curve. In case of disagreements during data extraction, a third author would participate in the discussion.

2.4. Quality Assessment

The Newcastle–Ottawa quality (NOS) assessment scale was used to evaluate the quality of the included studies by the 2 authors independently. The full mark of the scale was 9. Scores with 0–3, 4–6, and 7–9 were regarded as low quality, moderate quality, and high quality, respectively.

3. Results/Discussion

The purpose of this review is to systematically identify biomarkers that have the potential to diagnose, predict, prognose, and track therapeutic monitoring in ccRCC patients receiving systemic therapy with ICI, TKI, VEGFi, or a combination for treatment of adjuvant, neoadjuvant, and metastatic disease. This review organizes our findings first by marker class, then drug class, and finally paper methodology (Figure 1). Various types of biomarkers currently exist including immunologic, genomic, radiogenomic, and physiologic biomarkers. Other markers do not fit into any of these classes. A complete table of all biomarkers and associated data can be found in Supplementary Table S1. Additionally, the data collected in this paper can be visualized via Flourish at the following website: https://public.flourish.studio/visualisation/14189718/ (accessed on 22 July 2023).

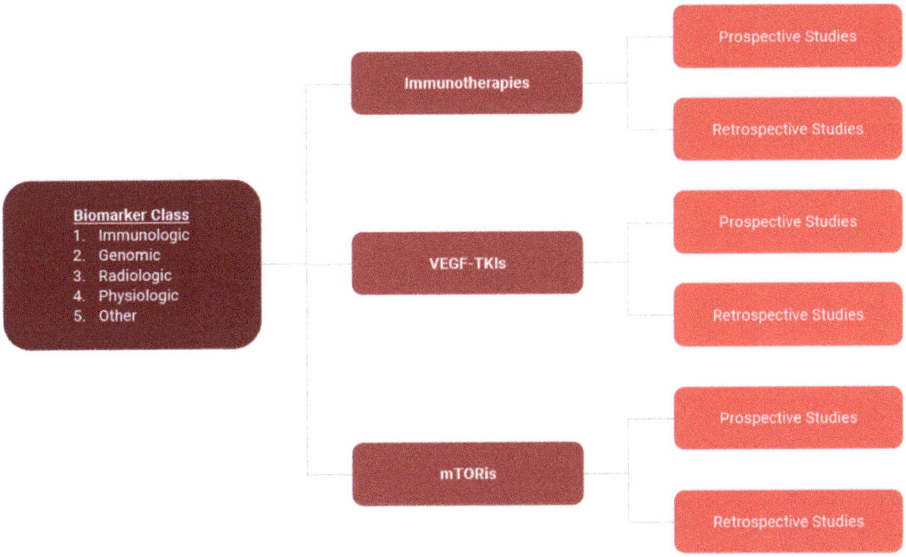

Figure 1. Organization of Review Article.

3.1. Immunologic Biomarkers

Immunologic markers are those that serve as surrogate markers for cellular activation and play an important role in the function of the immune system. These includes serum cytokines, chemokines, adipocytokines, soluble forms of cell receptors, and immune activation markers [17].

3.1.1. Immunotherapy

Immune checkpoint inhibitors are monoclonal antibodies that boost anticancer immune responses by targeting various immune receptors. These agents have been successfully used to treat RCC in clinical trials and/or are being studied in several prospective and retrospective studies.

3.1.2. Clinical Trials

There are few clinically proven predictive and prognostic biomarkers for ccRCC treated with first-line therapies, but novel drugs and biomarkers have been emerging from clinical trials. The safety and efficacy of rocapuldencel-T, a type of immunotherapy prepared from mature dendritic cells, was explored in a Phase III study in combination with standard of care sunitinib as a first-line therapy for mRCC. When compared to the standard of care, Rocapuldencel-T produced immune responses in 70% of patients and the strength of the response correlated with OS. High baseline numbers of T regulatory cells are associated with improved outcomes in patients treated with rocapuldencel-T, but were associated with worse outcomes in patients receiving SOC treatment [18].

Prospective Studies

PD-1 is a receptor found on T-cells which binds to PD-L1, a receptor often upregulated on cancer cells. This binding interaction prevents T-cells from destroying cancer cells, enabling the tumor to escape the immune system. Clinical trials have proven the efficacy of anti-PD1 therapies as a mainstay systemic treatment in ccRCC patients [9,19–21]. Therefore, it is natural that PD-1 expression levels can be predictive of treatment response to anti-PD1 therapies. Pignon et al. found that tumor cell PD-L1 expression in combination with PD-1 expression on CD8+ T-cells may predict outcome of nivolumab in mccRCC. Overall PD-L1 expression was not clinically relevant which suggests that in ccRCC, PD-L1 expression on T-cells (but not IC) drives immune evasion and can be reversed by anti-PD-1 therapies [22]. Atkins et al. found that PD-L1 expression has limited clinical utility in selecting patients for nivolumab monotherapy, but may have potential as a predictive biomarker of nivolumab monotherapy efficacy within a multifactorial predictive biomarker model [23].

By contrast, Mahoney et al. examined soluble PD-L1 (sPD-L1) as a biomarker within serum, rather than from tumor tissue. They found that sPD-L1 is a marker of nivolumab-refractory disease in RCC and is more complex than just a substitute for PD-L1. This is because aggressive disease produces sPD-L1, but there may be a distinct secondary pathway by which some patients with some or complete response to nivolumab produce sPD-L1. Therefore, a comparison of baseline and on-therapy sPD-L1 levels in RCC may be able to predict progressive disease in patients taking nivolumab [24]. Incorvaria et al. showed that the plasma levels of sPD-1, sPD-L1, and sBTN3A1 can predict improved response to second-line nivolumab. Incorvaria et al. analyzed the dynamic changes of plasma after nivolumab treatment and found a statistically significant decrease of sPD-1 after 28 days of therapy in the long-responder patients [25].

Ross-Macdonald et al. found that non-response to nivolumab was correlated with tumors highly infiltrated with T-cells. The degree of infiltration was measured at baseline and on day 28 by a T-cell receptor "CD3TCR" expression score using a gene set of the CD3-γ, CD3-δ, CD3-ε, CD3-ζ, TCR-α, and TCR-β subunits [26]. They also found that IL-18 mRNA was differentially upregulated in treatment responders [26]. Similarly, Chehrazi-Raffle found that patients with clinical benefit from immunotherapy had higher levels of interferon-γ and IL-12 [27].

Other predictive markers being explored include circulating endothelial cells. García-Donas et al. found that, when given antiangiogenic treatments (sunitinib and/or pazopanib), the detection of higher CEC levels, defined as DAPI+, CD105+, and CD45−, was associated with progression-free survival in ccRCC. The study also found that CEC levels did not change significantly despite tumor progression compared with baseline, which suggests that CECs could be determined at any time during treatment, remaining a stable predictive biomarker [28].

Prognostic liquid biopsy biomarkers have been explored prospectively as well. Bootsma et al. examined circulating tumor cell abundance and HLA I to PD-L1 (HP) ratio as prognostic markers in RCC and demonstrated that both markers can be used to monitor treatment response Specifically, the direction in which CTC enumeration changes was strongly associated with OS and increases in the HP ratio over time collated with worse outcomes [29]. Billon et al. suggested that the baseline level of plasmatic BTN2A1 could be an independent prognosis factor of PFS for second-line nivolumab after a TKI in patients with mRCC [30]. Incorvaria et al. also studied this biomarker, BTN2A1 was not significantly associated with OS or PFS [25]. However, Billon et al. found that patients with PFS > 18 months seemed to have lower levels of sBTN2A1 than patients with PFS < 18 months [30]. In a study that analyzed 106 immune cell populations in fresh blood, Carril et al. found that baseline unswitched memory B cells (NSwM B cells) were increased in responders and associated with improved OS and PFS in patients on second-line nivolumab following prior nivolumab therapy. They also found that BCA-1/CXCL13 and BAFF, which are chemokines strongly expressed in the secondary lymphoid organs, were both associated with worse OS and inversely correlated to NSwM B cells [31]. De Giorgi et al. studied the association between inflammation in general (and Body Mass Index, or BMI) with the clinical outcome and found that a normal BMI combined with inflammation tripled the risk of death, suggesting that these biomarkers are critical prognostic factors for OS in patients with RCC treated with nivolumab. In univariate analysis, markers such as systemic immune-inflammation index (SII), neutrophil to lymphocyte ratio (NLR), and platelet to lymphocyte ratio (PLR) were able to predict outcome; and in multivariate analyses, SII $\geq$ 1375, BMI < 25 kg/m^2, and age $\geq$ 70 years independently predicted overall survival. SII changes at 3 months also predicted OS [32].

A study by Saal et al. aimed to evaluate the prognostic value of the modified Glasgow prognostic score (mGPS) in patients with mRCC treated with ICIs, specifically in the context of the IMmotion151 trial, which compared atezolizumab plus bevacizumab to sunitinib. The mGPS assigns points based on elevated serum C-reactive protein (CRP) and decreased serum albumin levels. Patients are categorized as low, intermediate, or high risk. The results demonstrated that the mGPS outperformed the International Metastatic Renal Cell Carcinoma Database Consortium (IMDC) score, the current standard for risk stratification in mRCC. The mGPS had a higher concordance index and identified a larger proportion of patients as low risk while maintaining comparable survival rates. This study suggests that the mGPS could replace the IMDC score as a simple and effective prognostic tool in the era of immuno-oncology for mRCC patients [33].

Retrospective Studies

Retrospective studies disclose even more unique biomarkers that have both predictive and prognostic value when studying effects of immunologic treatments. Many biomarkers have been studied, the most common of which are C-reactive protein (CRP), NLR, and immune-related adverse events (irAEs). Many other markers were also studied and can be found in Supplementary Table S2.

One of the most commonly studied predictive and prognostic markers is CRP, which is a nonspecific marker of inflammation found in blood. Kankkunen et al., Roussel et al., and Ito et al. all studied elevated baseline CRP as a marker of poor prognosis in RCC, and found that this marker can predict worse OS and PFS on nivolumab [34–36]. Kankkunen also found on multivariate analyses that patients with elevated baseline and on-treatment CRP had shorter OS and PFS than patients with normal CRP [34]. Similarly, Ito et al. found that pretreatment C-reactive protein >3.0 mg/dl was an independent predictor for PFS [36].

Abuhelwa et al. also studied CRP as a prognostic maker in ccRCC. They found that CRP could be used to stratify patients into four levels, with CRP levels inversely correlated with OS and PFS [37]. Fukuda et al. studied the CRP flare response by categorizing patients as CRP flare responders, CRP responders, and non-CRP responders depending on if CRP levels more than double compared with baseline within 1 month after initiation of nivolumab (flare) and then decreased below baseline within 3 months, if CRP levels decreased by $\geq$30% within 3 months without "flare", or if CRP levels were minimally changed or unchanged, respectively. They found that the CRP-flare response was associated with significant tumor shrinkage and improved survival outcomes [38]. Takamatsu et al. found that serum CRP higher than 0.5 mg/dl in RCC patients after first-line treatment termination could be a marker of poor prognosis in intermediate-risk patients treated with second-line treatment, given that the median OS of elevated and non-elevated CRP group was 11.5 and 29.4, respectively [39].

Others studied CRP as a predictor of treatment response in RCC. Noguchi et al. found that CRP level at 1 month after treatment with nivolumab may be clinically useful in quickly predicting treatment effect [40]. Fukuda et al. and Klümper et al. both studied the predictive value of the CRP flare response described above [38,41]. Fukuda et al. suggested that the

CRP flare response in nivolumab treatment may be related to changes caused by the cancer-immune system within the tumor microenvironment [38]. Klümper et al. expanded on this idea and found that CRP responders, especially CRP flare responders, had significantly prolonged progression-free survival (PFS) compared with non-CRP responders and long-term response ($\geq$12 months) to first-line IO combination therapy [41]. Beyond CRP, several cytokines are being studied as predictive and prognostic biomarkers for immunotherapies, including PD-1, PD-L1, IL6, IL-8, and IL-12. Details regarding these studies can be found in Supplementary Table S2.

Another commonly studied predictive and prognostic biomarker is NLR, which is a biomarker that links the two parts of the immune system: the innate immune response, mediated by neutrophils, and adaptive immunity, supported by lymphocytes [42]. There is some conflicting evidence on the clinical significance of this marker. Roussel et al., Lalani et al., and Jeyakumar et al. found that higher baseline NLR predicted worse OS and PFS on nivolumab in mccRCC patients [35,43,44], while Nishiyama et al. found that baseline NLR was not associated with OS or PFS. Specifically, Jeyakumar et al. concluded that NLR $\geq$3 prior to initiating ICI therapy was an independent predictor of OS and PFS [44], while Ito et al. and Nishiyama et al. instead found that NLR of $\geq$3 at 4 weeks of nivolumab therapy was an independent predictor of OS and PFS [36,45]. Lalani's and Ikarashi's work supports Nishiyama's findings as they both also found that higher NLR at 6 weeks was a significantly stronger predictor of all three outcomes than baseline NLR [43,46]. Zahoor et al. studied NLR as a predictor of progressive disease (i.e., metastasis), and found that the risk of progressive disease was elevated with higher baseline NLR [47]. Beyond NLR, there are several immune cell populations being studied as predictive and prognostic biomarkers for immunotherapies, including lymphocyte counts, neutrophil counts, PLR, neutrophil to eosinophil ratio (NER), monocyte to eosinophil ratio (NER), monocyte to lymphocyte ratio (MLR). Details regarding these studies can be found in Supplementary Table S2.

irAEs, a known side effect associated with ICIs, are a set of autoimmune conditions that can affect any organ in the body, which makes them difficult to diagnose and manage [48]. Clinicians have theorized that the presence of irAEs may have some implications for mRCC prognosis but the relationship is fairly unclear. Kankkunen et al. and Martini et al. found that irAEs, particularly thyroid irAEs, had significantly improved clinical outcomes in mRCC patients treated with ICIs as second-line therapy [34,49]. Ikeda et al. and Ishihara et al. both studied the relationship between irAEs and PFS and OS in patients treated with nivolumab. Ikeda et al. found that irAE development was significantly associated with PFS but not with OS [50], while Ishihara et al. found that irAE development was significantly associated with both PFS and OS [51]. These findings suggest that irAEs may be used as a clinical biomarker predicting treatment outcome in mRCC patients treated with ICIs.

3.1.3. VEGFR TKIs

TKIs are a class of drugs that inhibit signal transduction of protein kinases which in turn critically disrupt cellular signaling [52]. One group of growth factor receptors affected by TKIs are VEGF receptors (VEGFRs), which have been known to play a key role in the angiogenesis caused by cancers, including RCC [53].

VEGFR TKIs form a class of adjuvant therapy used in RCC. Drugs that are approved by the US FDA include sorafenib, sunitinib, axitinib, and pazopanib [53]. These agents have been successfully used to treat RCC in clinical trials and biomarkers related to VEGFR TKI efficacy and prognosis are being studied in several prospective and retrospective studies.

Prospective Studies

Prospective studies have studied both the predictive and prognostic abilities of angiogenesis markers as well as plasma cytokines in patients treated with VEGFR TKIs. Mauge et al., Jilaveanu et al., Oudard et al., and Xu et al. all studied the prognostic abilities of various angiogenesis markers [54–57] in patients receiving first-line sunitinib or pazopanib. In this multicenter, prospective, open-label, phase II trial, Mauge et al. included treatment-naïve patients with mccRCC who had received two cycles of sunitinib before nephrectomy, and found that baseline values of angiogenesis markers (found in parentheses) were significantly associated with a change in primary renal tumor size (endothelial progenitor cells), PFS (vascular endothelial growth factor-A, stromal cell-derived factor(SDF)-1, soluble VEGF receptors (sVEGFR) 1 and 2), and OS (SDF-1 and sVEGFR1). They also found that changes in the following markers (found in parentheses) during treatment were significantly associated with a change in primary renal tumor size (SDF-1 and platelet-derived growth factor-BB), PFS (sVEGFR2), and OS (SDF-1 and sVEGFR1) [54]. Jilaveanu et al. focused on the prognostic of changes in microvessel density (MVD) in patients taking adjuvant sunitinib and sorafenib. They found that, on both univariate and multivariate analyses, high MVD (defined as above the median) was associated with increased OS as compared to patients receiving placebo, and there was a less significant association when comparing populations of patients treated with sunitinib or sorafenib [55]. Oudard et al. also looked at VEGFR-1 when they studied the prognostic value of subpopulations of pro-angiogenic monocytes and found that a more than 20% reduction from baseline value of VEGFR-1+CD14 monocytes at 20 weeks after starting sunitinib or pazopanib was associated with a significant increase in PFS and OS. They also found that more than 20% reduction from baseline value of Tie2 + CD14 cells monocytes was associated with a significant increase in OS [56]. Similarly, Xu et al. found that sVEGFR-2 decreased after both 4 and 6 weeks of treatment on sunitinib or sorafenib and that sFLT-1 decreased after 4 weeks on sunitinib and 6 weeks on sorafenib [57].

Jilaveanu et al. also studied the predictive value of MVD. They also found that high MVD significantly correlated with Fuhrman grade 1–2, clear cell histology, and absence of necrosis but not with gender, age, sarcomatoid features, lymphovascular invasion, or tumor size, and therefore concluded that MVD was a better potential as a prognostic marker, rather than a predictive marker [55]. Hakimi et al. also studied the predictive value of angiogenesis markers, as well as mutation profiles and macrophage infiltration, in patients receiving received first-line sunitinib or pazopanib and found that patients with higher angiogenesis scores had a superior outcome independent of the IMDC risk category. They also attributed the predictive capabilities of angiogenesis markers to upregulation and suppression of angiogenesis observed with loss-of-function mutations in PBRM1 and BAP1, respectively [58].

Other groups have prospectively looked at the predictive and prognostic abilities of various plasma cytokines in patients treated with VEGFR TKIs. IL-6 is a plasma cytokine that is known to play a pathologic role in chronic inflammation as well as RCC [59,60]. Pilskog et al. studied the ability of IL-6 to prognose PFS and OS and predict response to sunitinib [61,62] while Chehrazi-Raffle et al. focused on IL-6′s predictive abilities [27]. In Pilskog et al.'s first study, they evaluate both the prognostic and predictive ability of interleukin-6 receptor α (pIL6Rα) in mRCC patients treated with sunitinib, and found that low tumor expression of IL6Rα, which is directly correlated with expression of some angiogenesis markers, was significantly associated with improved response to sunitinib [61]. Pilskog et al.'s second study focused on the predictive ability of plasma interleukin-6 (pIL6), pIL6Rα, and interleukin-6 signal transducer (pIL6ST) in mRCC patients treated with sunitinib. In this study, they found that low pIL6 at baseline was also significantly associated with improved PFS and response to sunitinib. Furthermore, patients with high pIL6ST at baseline showed significantly improved OS. This signified that pIL6 may have both prognostic and predictive potential in mRCC. Chehrazi-Raffle et al. confirmed this finding when they found that patients treated with VEGF-TKIs had lower pretreatment levels of interleukin-6 (IL-6) (as well as IL-1RA and granulocyte CSF). This study, however, did support IL-6′s ability to predict treatment response and instead found that clinical benefit from VEGF TKIs could be monitored with decreases in IL-13 and granulocyte macrophage CSF as well as increases in VEGF at one month [27].

Like Chehrazi-Raffle et al., Bellmunt et al. and Xu et al. all looked at the predictive and prognostic value of plasma cytokines other than IL-6. Bellmunt et al. studied the predictive value of IL-10 levels in mRCC patients receiving second-line pazopanib after failure of a prior TKI. They found that lower circulating levels of IL-10 were observed in responding patients at 8 weeks after treatment [63]. Xu et al. investigated the effects of adjuvant VEGFR TKIs on circulating cytokines and found that when on sunitinib and sorafenib, CXCL10 elevated at 4 and 6 weeks was associated with worse DFS [57].

Zizzari et al. and Montemagno et al. both looked at soluble forms of PD-L1 and PD-1 as prognostic and predictive markers of sunitinib efficacy. Zizzari et al. identified soluble PD-L1 and PD-1 as two of seven soluble immune molecules (IFNγ, sPDL2, sHVEM, sPD1, sGITR, sPDL1, and sCTLA4) as well as CD3 + CD8 + CD137+ and CD3 + CD137 + PD1 + T-cell populations as markers modulated by TKI therapy [64]. Montemagno specifically studied first-line sunitinib and bevacizumab, and confirmed that levels of soluble PD-L1 and PD-1 correlated with PFS with sunitinib only. They found that sunitinib treated patients with high baseline plasmatic levels of sPD-L1 and sPD-1 had a shorter PFSm [65]. There are several other retrospective studies that have been conducted on PD-L1 and PD-1, and can be found in Supplementary Table S3.

Retrospective Studies

Retrospective studies identified even more unique potential biomarkers that have both predictive and prognostic value when studying the effects of immunologic treatments. Many biomarkers have been studied, the most common of which are CRP and NLR. Many other markers were also studied and can be found in Supplementary Table S3.

Several retrospective studies have investigated the prognostic value of CRP levels in mRCC patients and their response to TKIs. Takamatsu et al. investigated the prognostic value of baseline CRP level in intermediate-risk mRCC patients treated with first-line VEGFi therapy. They reported that higher baseline CRP levels were associated with inferior OS and PFS outcomes [66]. They then performed a second study in which they found that elevated baseline serum CRP levels prior to second-line treatment in intermediate-risk mRCC patients was also associated with poor prognosis [39]. Both studies showed that intermediate-risk mRCC patients could be divided into two prognostic subgroups [39,66]. Similarly, Wang et al. identified dynamic changes in the systemic inflammatory response, including CRP, that could be prognostic indicators in mRCC. They observed that patients with elevated CRP levels at baseline and during treatment had significantly worse OS and PFS rates compared to those with normal CRP levels [67]. Takagi et al. also confirmed this finding with their study aiming to identify prognostic markers, including CRP, for refined stratification of intermediate-risk ccRCC patients treated with first-line TKI therapy. They also found that elevated CRP levels were associated with worse OS and PFS outcomes in this patient population [68]. Furthermore, Teishima et al. observed that normalization of CRP levels following cytoreductive nephrectomy in mRCC patients treated with TKIs was associated with improved overall survival. They found that patients who achieved CRP normalization after surgery had better OS rates compared to those who did not [69].

Other studies aimed to show CRP's efficacy as a predictive marker. Yasuda et al. demonstrated that early response of CRP levels could predict survival in patients with mRCC undergoing TKI treatment. They found that patients who achieved a significant decrease in CRP levels within one month of TKI therapy initiation had improved OS and PFS [70]. Klümper et al. explored the CRP flare response as a predictor of long-term efficacy in first-line anti-PD-1-based combination therapy for mRCC. They observed that patients who exhibited a CRP flare response had significantly prolonged PFS and higher rates of long-term response [41]. In addition to these studies, Erdogan et al. investigated the association between early changes in systemic inflammatory markers, including CRP, and treatment response in patients receiving pazopanib. They observed that patients who showed a significant decrease in CRP levels within one month of pazopanib treatment initiation had improved clinical outcomes [71]. Collectively, these studies highlight the importance of CRP as a prognostic and predictive marker and highlight the role it may play in mRCC patients receiving TKIs in the future. Beyond CRP, there are several cytokines being studied as predictive and prognostic biomarkers for immunotherapies, including PD-1, PD-L1, HIF-1α, IL-8, and FGFR2. Details regarding these studies can be found in Supplementary Table S3.

3.1.4. Combination Therapy

Combination therapy involves the use of multiple drug classes to enhance the efficacy of anti-cancer treatments. This approach potentially reduces drug resistance, while reducing tumor growth, mitotically active and cancer stem cells, and metastatic potential [72]. In RCC, the current gold standard is antiangiogenic agents combined with tyrosine kinase, mTOR, or immune checkpoint inhibitors [73]. These agents have been successfully used to treat RCC in clinical trials and biomarkers related to combination therapy efficacy and prognosis are being studied in several prospective and retrospective studies.

Prospective Studies

Several studies have provided valuable insights into the clinical activity and molecular correlates of response to immunotherapy in renal cell carcinoma (RCC) patients. McDermott et al. found that atezolizumab–bevacizumab combination therapy showed improved clinical activity and higher overall response rates compared to sunitinib, providing molecular characteristics of each treatment option. The study also identified specific molecular markers associated with response to immunotherapy, such as high tumor mutational burden and PD-L1 expression. These findings provide important information for patient selection and treatment decision making [74].

In a study by Martini et al. the focus was on angiogenic and immune-related biomarkers in patients receiving axitinib/pembrolizumab treatment. The researchers observed that specific angiogenic and immune-related biomarkers, including angiopoietin-2 and vascular endothelial growth factor, were associated with treatment response and outcomes. This highlights the potential of these biomarkers as predictive factors for therapy response and the importance of considering both angiogenesis and immune pathways in RCC treatment strategies [75].

Msaouel et al. conducted a phase 1–2 trial evaluating sitravatinib and nivolumab in clear cell RCC patients who had progressed on antiangiogenic therapy. The results demonstrated promising clinical activity, with a significant proportion of patients experiencing tumor shrinkage and disease control. The study also highlighted the potential of sitravatinib to overcome resistance to antiangiogenic therapy and enhance the efficacy of immunotherapy. These findings suggest that the combination of sitravatinib and nivolumab could be a promising treatment approach for refractory RCC patients [76].

Retrospective Studies

In a study by Kamai et al. the researchers investigated the expression of adenosine 2A receptors (A2AR) in metastatic renal cell carcinoma (RCC) and its impact on treatment response and patient survival. The study revealed that increased expression of A2AR was associated with poorer response to anti-VEGF agents and anti-PD-1/anti-CTLA4 antibodies. Patients with higher A2AR expression levels demonstrated shorter overall survival rates, indicating the potential of A2AR as a prognostic biomarker in metastatic RCC. These findings highlight the significance of A2AR in the immunotherapy response and suggest that targeting A2AR signaling may improve treatment outcomes in RCC patients [77]. The study contributes to the understanding of the molecular mechanisms underlying immunotherapy resistance and provides insights for the development of novel therapeutic strategies in metastatic RCC. Beyond A2AR, there are several markers being studied for their predictive and prognostic effects in patients taking immunotherapies. Details regarding these studies can be found in Supplementary Table S4.

3.2. Genomic Biomarkers

Genomic markers are DNA or RNA sequences that are known to cause disease or increase susceptibility to disease. This marker class is the basis of genetic testing, which identifies changes in chromosomes, genes, and proteins [78], as well as companion diagnostics that match patients to a specific drug or therapy based on identified genetic changes or genomic markers.

3.2.1. Immunotherapy

Prospective Studies

Several studies have investigated predictive biomarkers for immunotherapy response in metastatic renal cell carcinoma (mRCC) patients. Incorporvaia et al. and Epaillard et al. specifically studied PD-1 and PD-L1 as genomic biomarkers that may predict response to nivolumab. Incorporvaia et al. identified a "Lymphocyte MicroRNA Signature" as a potential predictive biomarker of immunotherapy response and plasma PD-1/PD-L1 expression levels in mRCC patients, suggesting a role for epigenetic reprogramming in treatment outcomes. They found that patients with higher levels of specific microRNAs had better response rates and improved overall survival [79]. Epaillard et al. also investigated the efficacy of nivolumab and ipilimumab in treatment-naïve patients with metastatic kidney cancer. This phase 2 biomarker-driven trial found that patients with PD-L1 positive tumors exhibited higher response rates and improved progression-free survival when treated with nivolumab and ipilimumab combination therapy [80].

Miao et al. and Ross-Macdonald et al. studied other genomic biomarkers that may predict response to nivolumab. Miao et al. explored the genomic correlates of response to immune checkpoint therapies in clear cell RCC and identified specific genomic features associated with treatment response, including mutations in the PBRM1 gene. They observed that patients with PBRM1 mutations had a higher treatment response rate and longer progression-free survival [81]. Ross-Macdonald et al. investigated molecular correlates of response to nivolumab at baseline and during treatment and identified several gene expression signatures associated with clinical response. They suggested that these gene signatures could serve as potential predictive biomarkers for nivolumab therapy [26]. Kim et al. explored the potential of circulating tumor DNA as a predictor of therapeutic responses to immune checkpoint blockades in mRCC. They observed that the detection of specific mutations in circulating tumor DNA was associated with better treatment responses and improved progression-free survival [82].

Retrospective Studies

Several genomic markers are being studied as predictive and prognostic potential in immunotherapies. Details regarding these studies can be found in Supplementary Table S5.

3.2.2. VEGF TKIs

Prospective Studies

Several studies have investigated the role of biomarkers in predicting treatment response and prognosis in advanced renal cell carcinoma (RCC) patients receiving sunitinib therapy. In a study by Dietz et al. molecular alterations specific to individual patients with metastatic ccRCC were identified, and these alterations were found to be associated with disease progression despite TKI therapy. The results underscore the need for personalized treatment approaches that consider the unique molecular profiles of patients to improve therapeutic outcomes [83]. Examining the hypoxia-inducible factor (HIF) pathway and c-Myc as potential biomarkers, Maroto et al. assessed their predictive value in response to sunitinib treatment in metastatic ccRCC. Their findings indicated that the status of these biomarkers could help identify patients who are more likely to respond favorably to sunitinib therapy, enabling a more targeted approach to treatment selection [84]. Wierzbicki et al. evaluated the prognostic significance of several biomarkers, including VHL, HIF1A, HIF2A, VEGFA, and p53, in ccRCC patients treated with sunitinib as a first-line therapy.

The study revealed that the expression levels of these biomarkers were associated with clinical outcomes, providing valuable prognostic information [85]. Nayak et al. investigated the role of circulating tumor cells (CTCs) as biomarkers in metastatic ccRCC. They found that the presence of CTCs was associated with disease status and provided valuable information regarding treatment response. The study suggested that CTCs could serve as non-invasive markers for monitoring disease progression and guiding treatment decisions in ccRCC [86].

Other studies have looked at the role of biomarkers in predicting treatment response and prognosis in patients receiving sorafenib therapy. Gudkov et al. conducted a study aiming to develop a gene expression-based signature capable of predicting the response to sorafenib in kidney cancer patients. The findings demonstrated that the signature successfully predicted the efficacy of sorafenib treatment, providing valuable insights for personalized therapeutic decision making in renal cell carcinoma (RCC) patients [87]. Crona et al. investigated genetic variants of VEGFA and FLT4 as determinants of survival in RCC patients treated with sorafenib. The study revealed that specific genetic variations in these genes were associated with patient survival outcomes, highlighting the potential of genetic profiling as a prognostic tool to identify individuals who may benefit the most from sorafenib therapy [88].

Bevacizumab was also a commonly explored drug; in a study by Dorff et al. the efficacy of bevacizumab, either alone or in combination with TRC105, was evaluated in patients with refractory metastatic RCC. The results demonstrated the potential clinical benefit of bevacizumab-based therapy in this patient population, suggesting its relevance as a treatment option for refractory metastatic RCC [89]. Bamias et al. conducted a clinical and biomarker study assessing the combination of bevacizumab and temsirolimus as a second-line therapy in advanced RCC patients who had received prior anti-VEGF treatment. The study revealed promising results, with improved clinical outcomes and biomarker profiles observed in patients treated with combination therapy, indicating its potential as an effective therapeutic approach for advanced RCC [90].

Retrospective Studies

Several genomic markers were studied as predictive and prognostic potential in VEGF-TKI therapies. Details regarding these studies can be found in Supplementary Table S6.

3.2.3. mTORi

mTOR inhibitors (mTORi) are chemotherapy drugs that deactivate the PI3K/AKT/mTOR signaling pathway, which in turn prevents tumor angiogenesis and downregulates the expression of hypoxia-inducible factors in RCC. Currently, two mTOR inhibitors are approved for use in metastatic RCC: temsirolimus and everolimus. These agents have been successfully used to treat RCC in clinical trials and biomarkers related to mTORi efficacy and prognosis are being studied in several prospective and retrospective studies, and are summarized in Supplementary Table S7.

Prospective Studies

Several studies have focused on identifying prognostic and predictive biomarkers for mTOR inhibitors. Palomero et al. conducted a study investigating the role of EVI1, a transcription factor involved in cell proliferation and differentiation, as a prognostic and predictive biomarker in ccRCC. They found that higher EVI1 expression was associated with worse prognosis and resistance to mTORi therapy, indicating its potential as a valuable biomarker for guiding treatment decisions in ccRCC patients [91].

Zeuschner et al. aimed to identify predictive biomarkers for everolimus treatment in second-line metastatic ccRCC. Through their research, they identified thrombospondin-2 and lactate dehydrogenase (LDH) as potential predictive biomarkers associated with treatment response to everolimus. This suggests that assessing the levels of these biomarkers could aid in patient selection for everolimus therapy, allowing for more personalized and effective treatment strategies.

In addition, Voss et al. investigated the correlation between PTEN expression and treatment outcome in RCC patients receiving everolimus. They found that PTEN expression, rather than mutation status in TSC1, TSC2, or mTOR, was significantly associated with treatment response to everolimus. This highlights the importance of assessing PTEN expression levels as a potential predictive biomarker for everolimus therapy in RCC [92].

Retrospective Studies

Flaifel et al. conducted an analysis of the randomized clinical trials METEOR and CABOSUN to investigate the relationship between programmed death-ligand 1 (PD-L1) expression and clinical outcomes in mRCC patients treated with cabozantinib, everolimus, and sunitinib. The study aimed to determine whether PD-L1 expression levels could serve as predictive biomarkers for treatment response. The analysis revealed that PD-L1 expression was not significantly associated with clinical outcomes in terms of progression-free survival and overall survival in patients receiving any of the three treatments. These findings suggest that PD-L1 expression may not be a reliable biomarker for predicting response to cabozantinib, everolimus, or sunitinib in mRCC patients [93].

Several other genomic markers were studied as predictive and prognostic potential in mTORi therapies. Details regarding these studies can be found in Supplementary Table S7.

3.2.4. Combined Biomarkers

Prospective Studies

Both Motzer et al. and McDermott et al. studied gene expression signatures as a biomarker in mRCC by comparing atezolizumab alone or in combination with bevacizumab versus sunitinib. Motzer et al.'s molecular analysis revealed that immune-related gene expression signatures were associated with better outcomes for the combination therapy, while angiogenesis-related signatures showed better outcomes with sunitinib [94]. McDermott et al.'s results revealed that the combination therapy of atezolizumab plus bevacizumab demonstrated improved clinical outcomes, including overall survival, compared to sunitinib alone. Molecular analysis further identified immune-related gene expression signatures associated with better responses to the combination therapy, emphasizing the importance of molecular profiling for guiding treatment decisions in RCC patients.

Retrospective Studies

Several genomic markers were studied as predictive and prognostic potential in combination therapies. Details regarding these studies can be found in Supplementary Table S8.

3.3. Radiologic Biomarkers

Radiomics is an emerging field of study in which advanced imaging strategies provide structural and phenotypic biomarkers related to key disease processes. Radiomics-based biomarkers are being used to deeply analyze pathophysiologic processes and have provided insights to better diagnose, classify, stratify, and prognosticate tumors, and to assess their response to therapy [95].

3.3.1. ICI

Prospective Studies

In recent years, there has been growing interest in identifying radiographic biomarkers that can predict clinical outcomes in mRCC patients receiving immune checkpoint inhibitors (ICIs). Tabei et al. conducted a study to assess the early predictive value of 18F-2-fluoro-2-deoxyglucose positron emission tomography/computed tomography (18F-FDG PET/CT) in patients with ccRCC who received nivolumab, an immune checkpoint inhibitor. The study aimed to determine whether metabolic response assessed by 18F-FDG PET/CT after two cycles of nivolumab treatment could predict short-term outcomes. The findings revealed that patients treated with nivolumab had significantly longer progression-free survival and overall survival if they showed a metabolic response on 18F-FDG PET/CT. This suggests that early assessment with 18F-FDG PET/CT could serve as a valuable tool for predicting treatment response and patient outcomes in ccRCC patients receiving nivolumab therapy [96].

Drljevic-Nielsen et al. investigated the potential of spectral dual-layer detector CT parameters as imaging biomarkers in patients with metastatic renal cell carcinoma (mRCC) who were treated with TKIs and immunotherapies. Their first study aimed to assess whether early reduction in spectral dual-layer CT parameters could serve as favorable prognostic indicators. They found that early reduction iodine concentration and effective atomic number predicted significantly longer OS and PFS. Their second study continued the focus on CT parameters, corroborating the predictive value of iodine concentration and adding CT parameters, such as virtual monochromatic images at specific energy levels, as a predictive biomarker. These findings suggest that both pretreatment and spectral dual-layer CT parameters have prognostic utility and could potentially aid in risk stratification and treatment decision making in mRCC patients treated with different therapies [97,98].

Retrospective Studies

Martini et al. conducted a study to evaluate the potential of body composition variables as radiographic biomarkers in this patient population. They found that lower skeletal muscle index and higher visceral adipose tissue area were associated with worse clinical outcomes, including shorter overall survival and progression-free survival. They also found that lower skeletal muscle index was associated with higher levels of systemic inflammation, as indicated by elevated CRP levels. In addition, higher visceral adipose tissue area was linked to greater tumor burden and increased tumor inflammation. These findings suggest that body composition may influence treatment response and prognosis through its impact on systemic inflammation and tumor characteristics [99].

Malone et al. conducted a study to develop a predictive radiomics signature for treatment response to nivolumab in patients with advanced renal cell carcinoma (RCC). Using radiomics analysis, they extracted quantitative imaging features from pre-treatment computed tomography (CT) scans and developed a radiomics signature. The results showed that the radiomics signature was significantly associated with treatment response to nivolumab. This finding suggests that the radiomics signature has the potential to serve as a non-invasive predictive biomarker for treatment response in patients with advanced RCC undergoing nivolumab therapy [100].

Mittlmeier et al. investigated the utility of 18F-PSMA-1007 PET/CT for response assessment in patients with metastatic renal cell carcinoma undergoing TKI or checkpoint inhibitor therapy. The study evaluated the preliminary results of using PSMA-based positron emission tomography/computed tomography (PET/CT) imaging to assess treatment response. The findings indicated that 18F-PSMA-1007 PET/CT showed promising potential in monitoring response to therapy in patients with metastatic RCC. This suggests that PSMA-based PET/CT imaging may serve as a valuable tool for response assessment in RCC patients receiving TKI or checkpoint inhibitor therapy [101].

Zheng et al. compared radiological tumor response assessment based on immune Response Evaluation Criteria in Solid Tumors (iRECIST) and Response Evaluation Criteria in Solid Tumors (RECIST) 1.1 in metastatic clear-cell renal cell carcinoma (ccRCC) patients treated with programmed cell death-1 (PD-1) inhibitor therapy. The study aimed to determine the agreement between the two response evaluation criteria and assess their correlation with overall survival. The results showed that iRECIST provided a more accurate response assessment and better prediction of overall survival compared to RECIST 1.1 in metastatic ccRCC patients treated with PD-1 inhibitor therapy. These findings highlight the importance of using iRECIST as a more reliable and clinically relevant tool for response evaluation in this patient population [102].

Park et al. conducted a study to evaluate the use of computed tomography (CT) texture analysis as a predictive tool for clinical outcomes in patients with mRCC treated with immune checkpoint inhibitors (ICIs). They analyzed CT images of mRCC patients and extracted various texture features to assess tumor heterogeneity. The study found that certain CT texture features were significantly associated with treatment response, progression-free survival, and overall survival in patients receiving ICIs. These findings suggest that CT texture analysis could serve as a non-invasive and promising tool for predicting clinical outcomes and guiding treatment decisions in mRCC patients undergoing ICI therapy [103].

3.3.2. VEGF TKIs

Prospective Studies

Udayakumar et al. conducted a study aimed at deciphering the intratumoral molecular heterogeneity in ccRCC using a radiogenomics platform. They integrated radiomic and genomic data to gain insights into the diverse molecular subtypes within ccRCC tumors. The study demonstrated that the radiogenomics approach could identify distinct molecular subtypes based on imaging features, providing valuable information for personalized treatment strategies. The findings highlight the potential of utilizing radiogenomics to improve the understanding of intratumoral heterogeneity in ccRCC and guide precision medicine approaches [104].

Nakaigawa et al. investigated the predictive value of fluorodeoxyglucose positron emission tomography/computed tomography (FDG PET/CT) after the first molecular targeted therapy in patients with renal cell carcinoma (RCC). The study aimed to determine whether FDG PET/CT could serve as a prognostic tool to predict patient survival. The results demonstrated that FDG PET/CT findings after the initial targeted therapy were significantly associated with overall survival in RCC patients. This suggests that FDG PET/CT imaging can serve as a valuable non-invasive tool for predicting patient outcomes and guiding treatment decisions in RCC [105].

Drljevic-Nielsen et al. investigated the potential of spectral dual-layer detector CT parameters as imaging biomarkers in patients with mRCC who were treated with TKIs and immunotherapies. Their first study aimed to assess whether early reduction in spectral dual-layer CT parameters could serve as favorable prognostic indicators. The results demonstrated that patients who showed early reduction in CT parameters, such as iodine concentration and effective atomic number, had significantly longer progression-free survival and overall survival. Drljevic-Nielsen et al.'s second study assessed various CT parameters to determine their association with patient outcomes. This study revealed that certain CT parameters, such as iodine concentration and virtual monochromatic images at specific energy levels, were significantly associated with progression-free survival and overall survival in mRCC patients receiving different therapies. These findings suggest that both pretreatment and spectral dual-layer CT parameters have prognostic utility and could potentially aid in risk stratification and treatment decision making in mRCC patients treated with different therapies [97,98].

Retrospective Studies

In a study by Hall et al. the researchers investigated the prognostic significance of radiological response heterogeneity in mRCC patients treated with VEGF-targeted therapy. The study aimed to determine whether the variation in radiological response within individual patients could predict their overall survival. The findings revealed that higher radiological response heterogeneity was associated with poorer outcomes and shorter survival in mRCC patients. These results highlight the importance of considering the heterogeneity of radiological response as a prognostic factor when evaluating treatment response and predicting patient outcomes in mRCC patients undergoing VEGF-targeted therapy [106].

Go et al. conducted a study aimed at developing a response classifier for VEGFR TKI treatment in mRCC. They analyzed the expression of various biomarkers, including VEGFR and other molecular factors, and developed a response classifier to predict treatment response to VEGFR-TKI therapy. The findings provide insights into the molecular characteristics associated with treatment response, potentially enabling the identification of patients who are more likely to benefit from VEGFR-TKI therapy in mRCC [107].

Mytsyk et al. investigated the usefulness of the apparent diffusion coefficient (ADC) derived from diffusion-weighted magnetic resonance imaging (DW-MRI) in predicting early therapeutic response in mRCC patients. By analyzing the ADC values before and after systemic treatment, they found that changes in ADC values were correlated with treatment response. These results suggest that DW-MRI, specifically the ADC parameter, can serve as a useful tool for predicting early therapeutic response in mRCC patients, facilitating treatment monitoring and decision making [108].

In another study, Wu et al. aimed to assess the response to anti-angiogenic targeted therapy in pulmonary mRCC using the R2* value as a predictive biomarker. They utilized magnetic resonance imaging (MRI) to measure the R2* value, which reflects the concentration of deoxyhemoglobin and serves as an indicator of tumor angiogenesis and hypoxia. The study demonstrated that the baseline R2* value was associated with treatment response and overall survival in mRCC patients receiving anti-angiogenic therapy. These findings suggest that the R2* value could be a predictive biomarker for assessing treatment response in pulmonary metastatic RCC patients undergoing anti-angiogenic targeted therapy [109].

Mittlmeier et al. investigated the utility of 18F-PSMA-1007 PET/CT for response assessment in patients with metastatic renal cell carcinoma undergoing TKI or checkpoint inhibitor therapy. The study evaluated the preliminary results of using PSMA-based positron emission tomography/computed tomography (PET/CT) imaging to assess treatment response. The findings indicated that 18F-PSMA-1007 PET/CT showed promising potential in monitoring response to therapy in patients with metastatic RCC. This suggests that PSMA-based PET/CT imaging may serve as a valuable tool for response assessment in RCC patients receiving TKI or checkpoint inhibitor therapy [101].

3.3.3. Combined Biomarkers

Retrospective Studies

Navani et al. conducted a study to assess the imaging response in patients with mRCC treated with contemporary immuno-oncology combination therapies. The study analyzed radiographic imaging data, including computed tomography (CT) scans, of mRCC patients undergoing combination immunotherapy. They evaluated treatment response based on radiographic assessments such as tumor size reduction, stability, or progression. The results showed that patients who achieved a complete or partial response on imaging had significantly improved progression-free survival compared to those with stable disease or disease progression. This study highlights the importance of radiographic imaging in monitoring treatment response and predicting outcomes in mRCC patients receiving combination immuno-oncology therapies [110].

3.4. Physiologic Biomarkers

3.4.1. ICI

Prospective Studies

Only one prospective study by De Giorgi et al. studies the role of inflammation and BMI on clinical outcome. They found that a normal BMI combined with inflammation tripled the risk of death, suggesting that these biomarkers are critical prognostic factors for OS in patients with RCC treated with nivolumab. In univariate analysis, markers such as SII, NLR, and PLR were able to predict outcome, and on multivariate analyses, SII $\geq$ 1375, BMI < 25 kg/m^2, and age $\geq$ 70 years independently predicted overall survival. SII changes at 3 months also predicted OS [32].

Retrospective Studies

Other retrospective studies have also investigated the relationship between BMI and treatment outcomes in RCC patients receiving immune-checkpoint inhibition therapy. Labadie et al. conducted a study to evaluate the association between BMI, irAEs, and gene expression signatures with resistance to immune-checkpoint inhibitors and patient outcomes in RCC. They found that higher BMI was significantly associated with poorer response to therapy and reduced overall survival. Additionally, specific gene expression signatures were identified that correlated with resistance to immune-checkpoint inhibition. These findings highlight the potential role of BMI and gene expression profiling as predictive biomarkers for immune-checkpoint inhibition therapy in RCC patients, providing insights into patient stratification and treatment optimization [111].

In a related study, Herrmann et al. explored the prognostic value of BMI and sarcopenia, a condition characterized by muscle loss, in predicting outcomes for RCC patients treated with the immune-checkpoint inhibitor nivolumab. They analyzed the variations in BMI and sarcopenia before and after treatment and assessed their impact on treatment response and overall survival. The study revealed that patients with higher baseline BMI and those who experienced a decrease in BMI during treatment had significantly better outcomes, including improved response rates and prolonged survival. Moreover, the presence of sarcopenia at baseline was associated with worse treatment outcomes. These findings suggest that BMI and sarcopenia can serve as potential prognostic indicators for RCC patients undergoing immune-checkpoint inhibitor therapy, aiding in treatment decision making and patient management [112].

Ueki et al. also conducted a retrospective study to investigate the association between sarcopenia, as determined by the psoas muscle index (PMI), and the response to nivolumab in patients with mRCC. The PMI, which measures the cross-sectional area of the psoas muscle normalized to the patient's height, is a recognized indicator of muscle mass and overall body composition. The study findings revealed that, patients with a lower PMI, indicating sarcopenia, had a significantly poorer response to nivolumab treatment. These results suggest that sarcopenia, as assessed by the PMI, could serve as a potential biomarker for predicting the response to nivolumab in mRCC patients [113].

3.4.2. VEGF TKIs

Retrospective Studies

Several retrospective studies have also investigated the relationship between body mass and treatment outcomes in renal cell carcinoma (RCC) patients receiving TKI therapy. McKay et al. conducted a study to investigate the effect of weight change during treatment with targeted therapy in patients with mRCC. The researchers assessed the impact of weight loss or gain on treatment outcomes and survival. The study revealed that weight loss during treatment was associated with worse overall survival and progression-free survival in patients with mRCC. On the other hand, weight gain was not significantly associated with survival outcomes. These findings suggest that weight change during targeted therapy may serve as a prognostic factor in mRCC, emphasizing the importance of monitoring and managing weight during treatment [114]. Similarly, Ishihara et al. aimed to evaluate the effect of changes in skeletal muscle mass on oncological outcomes in patients with mRCC receiving first-line sunitinib therapy. The study investigated the association between skeletal muscle mass and treatment response, overall survival, and progression-free survival. The results demonstrated that a decrease in skeletal muscle mass during sunitinib therapy was associated with worse oncological outcomes, including shorter overall survival and progression-free survival. These findings highlight the potential role of skeletal muscle mass as a prognostic indicator in mRCC patients undergoing targeted therapy [115].

Others looked at the predictive and prognostic role of the De Ritis ratio. Janisch et al. conducted a study to assess the predictive value of the De Ritis ratio in mRCC patients treated with TKIs. The De Ritis ratio, calculated as the ratio of aspartate transaminase (AST) to alanine transaminase (ALT), is an indicator of liver function. The researchers found that a higher De Ritis ratio at baseline was associated with poorer survival outcomes in mRCC patients treated with TKIs. This suggests that the De Ritis ratio could serve as a potential predictive biomarker for treatment response and prognosis in mRCC patients receiving TKIs [116]. Kang et al. aimed to evaluate the prognostic impact of the pretreatment aspartate transaminase/alanine transaminase (AST/ALT) ratio in patients treated with first-line systemic TKI therapy for mRCC. The study analyzed the association between the AST/ALT ratio and survival outcomes, including overall survival and progression-free survival. The findings revealed that a higher pretreatment AST/ALT ratio was significantly associated with worse survival outcomes in mRCC patients treated with first-line TKIs. This suggests that the AST/ALT ratio may serve as a simple and readily available prognostic marker in guiding treatment decisions and predicting outcomes in mRCC [117].

Zhang et al. investigated the impact of renal impairment on survival outcomes in patients with mRCC treated with TKIs. The study assessed the association between renal impairment, as determined by estimated glomerular filtration rate (eGFR), and overall survival in mRCC patients. The results demonstrated that patients with impaired renal function had significantly worse overall survival compared to those with normal renal function. This highlights the importance of considering renal impairment in the management and treatment of mRCC patients receiving TKIs [118].

Aktepe et al. aimed to evaluate the impact of the albumin-to-globulin ratio (AGR) on survival outcomes of patients with mRCC. The study assessed the association between AGR, calculated as the ratio of albumin to globulin levels, and overall survival in mRCC patients. The findings revealed that a higher AGR was significantly associated with better overall survival in mRCC patients. This suggests that the AGR could serve as a potential prognostic biomarker for survival outcomes in mRCC, providing additional information for risk stratification and treatment decisions [119].

3.5. Miscellaneous Biomarkers

Many studies looked at measures of tumor progression and spread as predictive and prognostic markers in ccRCC. Several other markers are also being studied for predictive and prognostic potential, including race, TKI-induced hypertension, RDW levels, and hemoglobin levels. Details regarding these studies can be found in Supplementary Table S14.

3.5.1. VEGF TKIs

Retrospective Studies

Roussel et al. and Martini et al. aimed to find mechanisms to predict response to therapy. Roussel et al. conducted a study aimed at understanding the molecular mechanisms underlying the glandular tropism observed in metastatic ccRCC. By investigating the gene expression profiles of ccRCC tumors with glandular tropism, they identified specific molecular signatures associated with this phenotype. These findings have important therapeutic implications as they provide insights into the potential targets for developing novel treatment strategies specifically tailored to ccRCC with glandular tropism [120]. Martini et al. developed a novel risk-scoring system for mRCC patients treated with cabozantinib, a tyrosine kinase inhibitor. They integrated various clinical and laboratory parameters to develop a predictive model that can identify patients with different risk profiles and provide prognostic information. This risk-scoring system has the potential to assist clinicians in individualizing treatment approaches and optimizing patient outcomes in mRCC [121].

Several studies examined tumor physiology as a biomarker. Shirotake et al. conducted a single institutional study to evaluate early tumor shrinkage as a predictive factor for mRCC in patients undergoing molecular targeted therapy. They found that patients who achieved early tumor shrinkage demonstrated improved progression-free survival and overall survival. Early tumor shrinkage could serve as an early indicator of treatment response, enabling timely modifications in the treatment regimen to optimize outcomes in mRCC patients [122]. Kammerer-Jacquet et al. investigated hilar fat infiltration as a prognostic factor in metastatic ccRCC patients receiving first-line sunitinib treatment. Their study revealed that the presence of hilar fat infiltration was associated with poorer overall survival and progression-free survival. The identification of hilar fat infiltration as a prognostic factor provides valuable information for risk stratification and treatment decision making in metastatic ccRCC patients [123]. Pieretti et al. investigated the association between tumor diameter response and overall survival in patients with metastatic ccRCC. They found that greater tumor diameter response was significantly associated with improved overall survival. The study highlights the importance of tumor diameter response as a potential predictive factor and emphasizes the need for monitoring tumor size changes during the course of treatment in metastatic ccRCC patients [124]. Shi et al. evaluated the prognostic value of the ratio of maximum-to-minimum diameter of the primary tumor in metastatic ccRCC. They found that a higher ratio was associated with worse overall survival. The ratio of maximum-to-minimum diameter of the primary tumor could serve as a simple and accessible prognostic factor in metastatic ccRCC, aiding in risk stratification and treatment decision making [125]. Yildiz et al. examined prognostic factors for survival in mRCC patients with brain metastases receiving targeted therapy. They identified several factors, including Karnofsky performance status, number of brain metastases, and time to brain metastases, as independent predictors of survival. These findings contribute to the understanding of prognostic factors specific to mRCC patients with brain metastases, aiding in treatment planning and patient management [126].

In summary, these studies provide valuable insights into various aspects of mRCC, ranging from molecular underpinnings and therapeutic implications to prognostic factors and predictive biomarkers. The identification of molecular signatures associated with glandular tropism in ccRCC offers potential targets for personalized treatment strategies. Early tumor shrinkage and tumor diameter response serve as predictive factors for treatment response and overall survival, emphasizing the importance of monitoring tumor size changes during therapy. Prognostic factors such as hilar fat infiltration, the ratio of maximum-to-minimum diameter of the primary tumor, and specific clinical parameters contribute to risk stratification and treatment decision making in mRCC patients. These findings contribute to refining treatment approaches, improving patient outcomes, and advancing our understanding of the complex nature of metastatic renal cell carcinoma.

Artificial Intelligence and Multi-Omics Approaches

There has been a recent focus on the ability to apply artificial intelligence and multi-omics approaches to studying a range of cancers and other clinical conditions. Outside of ccRCC, tumor microarrays followed by image analysis and unsupervised learning with machine learning models has shown the ability to predict bladder cancer chemotherapy response [127]. Quantitative imaging followed up with computational analysis of immunohistochemistry revealed characteristic cancer cell profiling in pancreatic cancer [128]. Imaging mass cytometry with follow-up uniform manifold approximation and projection shows the ability to identify tumor biomarkers and predict patient outcomes for ccRCC [129]. Supervised machine learning approaches including convolutional neural networks, logistic regression, and support vector machine also show the ability to predict survival in ccRCC through hematoxylin and eosin histopathological image analysis [130,131]. Combining pathomic and genomic data also demonstrated the ability to stratify risk of ccRCC patients [132]. Due to the nascency of these methods, there is little research done applying these methods in involving prospective clinical trials or in response to specific forms of systemic therapy, but these methods will likely play a large role in shaping personalized and precision medicine.

4. Conclusions

The studies reviewed in this article provide important insights into the prognostic factors, treatment response, and response assessment in mRCC patients undergoing targeted therapies. These studies highlight the significance of various clinical, pathological, and radiological factors in predicting patient outcomes and tailoring treatment strategies. Factors such as C-reactive protein levels, lymphocyte microRNA signature, adenosine 2A receptor expression, EVI1 expression, and radiological response heterogeneity have been identified as potential prognostic markers and predictors of treatment response. Additionally, the use of advanced imaging techniques, such as PSMA-based PET/CT, shows promise in assessing treatment response in mRCC patients. The collective findings underscore the importance of a comprehensive approach to patient evaluation, incorporating multiple factors and innovative tools, to optimize the management of mRCC and improve patient outcomes. Further research and validation of these findings are necessary to refine risk stratification, guide treatment decisions, and ultimately enhance the care of patients with metastatic renal cell carcinoma.

While the studies discussed provide valuable insights into prognostic factors, treatment response, and response assessment in mRCC patients, certain limitations should be acknowledged. Firstly, the studies mentioned are based on retrospective analyses or small sample sizes, which may introduce selection bias and limit the generalizability of the findings. Larger prospective studies are needed to validate and further explore the identified prognostic factors and response predictors. Secondly, the studies predominantly focus on specific targeted therapies, such as VEGF-targeted therapy, TKIs, and immune checkpoint inhibitors. The results may not fully represent the entire spectrum of treatment options available for mRCC, including emerging therapies and combination approaches. Therefore, the findings should be interpreted within the context of the specific treatments studied. Additionally, while the studies assess various factors and biomarkers, the complex nature of mRCC and its heterogeneity suggest that multiple factors likely contribute to treatment response and patient outcomes. It is important to consider the interplay between different biomarkers, clinical characteristics, and tumor biology to obtain a comprehensive understanding of prognostic factors in mRCC.

Furthermore, the studies primarily rely on retrospective data or imaging modalities for response assessment, which may have inherent limitations. Variability in radiological assessments, lack of standardized criteria, and potential discrepancies between radiological and clinical response could impact the accuracy and reliability of response evaluation. Lastly, the studies reviewed have varying follow-up periods, and long-term outcomes and survival data may not be fully captured. Longer follow-up durations are necessary to assess the durability of treatment response and evaluate overall survival in mRCC patients. Considering these limitations, further research with larger cohorts, prospective designs, longer follow-up periods, and comprehensive evaluation of multiple factors is warranted to enhance our understanding of mRCC prognosis, treatment response, and response assessment.

5. Future Directions

The studies discussed in the previous paragraphs provide valuable insights into prognostic factors, treatment response, and response assessment in mRCC patients. Moving forward, there are several potential directions for future research in this field.

1. Conducting large-scale prospective studies with standardized protocols and longer follow-up periods can provide more robust evidence on prognostic factors, treatment response, and survival outcomes in mRCC patients. These studies can help validate the findings from retrospective analyses and further explore additional factors that may impact patient outcomes.
2. Continued research is needed to identify and validate novel biomarkers that can predict treatment response, prognosis, and therapeutic resistance in mRCC. The integration of genomic, proteomic, and immunological markers may offer a comprehensive approach to understand the underlying mechanisms and develop personalized treatment strategies.

3. Investigating the efficacy and safety of combination therapies, including targeted agents, immunotherapies, and other emerging treatment modalities, is crucial. Studying the synergistic effects of different therapeutic approaches and identifying predictive markers for optimal treatment selection can improve patient outcomes.
4. Exploring novel response assessment techniques, such as functional imaging modalities (e.g., dynamic contrast-enhanced MRI, diffusion-weighted imaging) and liquid biopsies (e.g., circulating tumor DNA, exosomes), may provide more accurate and timely evaluation of treatment response. Developing standardized criteria and guidelines for response assessment in mRCC can enhance comparability and facilitate clinical decision making.
5. Supplementing traditional clinical trials with real-world evidence from diverse patient populations and clinical settings can provide a more comprehensive understanding of treatment outcomes and enable personalized treatment decisions. Large-scale observational studies and data registries can help assess the effectiveness and safety of therapies in real-world clinical practice.
6. Including patient-reported outcomes, quality-of-life assessments, and patient preferences in research studies can provide a holistic perspective on treatment efficacy and impact on patients' lives. Understanding the patient experience and incorporating patient-centered endpoints can guide treatment decisions and improve the overall care of mRCC patients.

By focusing on these future directions, researchers can continue to advance our knowledge of mRCC, refine treatment strategies, and ultimately improve patient outcomes in this challenging disease.

Supplementary Materials: The following supporting information can be downloaded at: https://www.mdpi.com/article/10.3390/cancers15204934/s1, Supplementary Figure S1: PRISMA checklist; Supplementary Table S1: Detailed characteristics of included studies; Supplementary Table S2: Summary of studies investigating the use of immunologic biomarkers in patients on immunotherapy; Supplementary Table S3: Summary of studies investigating the use of immunologic biomarkers in patients on TKI therapy; Supplementary Table S4: Summary of studies investigating the use of immunologic biomarkers in patients on combination therapy; Supplementary Table S5: Summary of studies investigating the use of genomic biomarkers in patients on immunotherapy therapy; Supplementary Table S6: Summary of studies investigating the use of genomic biomarkers in patients on TKI therapy; Supplementary Table S7: Summary of studies investigating the use of genomic biomarkers in patients on mTOR inhibitors; Supplementary Table S8: Summary of studies investigating the use of genomic biomarkers in patients on combination therapy; Supplementary Table S9: Summary of studies investigating the use of radiologic biomarkers in patients on immunotherapy; Supplementary Table S10: Summary of studies investigating the use of radiologic biomarkers in patients on TKIs; Supplementary Table S11: Summary of studies investigating the use of radiologic biomarkers in patients on combination therapy; Supplementary Table S12: Summary of studies investigating the use of physiologic biomarkers in patients on immunotherapy; Supplementary Table S13: Summary of studies investigating the use of physiologic markers biomarkers in patients on TKIs; and Supplementary Table S14: Summary of studies investigating the use of novel biomarkers.

Author Contributions: Conceptualization, K.A.D., J.M.R., V.A.D. and A.D.; Methodology, K.A.D. and J.M.R.; Investigation, K.A.D. and J.M.R.; Resources, K.A.D. and J.M.R.; Data Curation, K.A.D., J.M.R., S.S.K. and H.C.; Writing—Original Draft Preparation, K.A.D., J.M.R. and S.S.K.; Writing—Review and Editing, K.A.D., J.M.R., V.A.D. and A.D.; Visualization, K.A.D., J.M.R., S.S.K. and H.C.; Supervision, K.A.D., J.M.R., V.A.D. and A.D.; Project Administration, V.A.D. and A.D. All authors have read and agreed to the published version of the manuscript.

Funding: Sean S. Kumar was a fellow in the USC/CHLA Summer Oncology Research Fellowship Program, supported in part by a National Cancer Institute R25 grant CA225513, the Norris Comprehensive Cancer Center in Los Angeles, Children's Hospital Los Angeles, Concern Foundation for Cancer Research, and Tri Delta.

Conflicts of Interest: Vinay A. Duddalwar reports consulting fees from Radmetrix, Cohere, and Westat; medical advisory board activity for DeepTek Inc; and grant funding from Samsung Healthcare. The other authors have no relevant financial interests or disclosures.

References

1. Siegel, R.L.; Miller, K.D.; Jemal, A. Cancer Statistics, 2020. *CA Cancer J. Clin.* **2020**, *70*, 7–30. [CrossRef] [PubMed]
2. US Department of Health and Human Services. *Kidney and Renal Pelvis Cancer—Cancer Stat Facts*; US Department of Health and Human Services: Washington, DC, USA, 2021.
3. Wong, M.C.S.; Goggins, W.B.; Yip, B.H.K.; Fung, F.D.H.; Leung, C.; Fang, Y.; Wong, S.Y.S.; Ng, C.F. Incidence and Mortality of Kidney Cancer: Temporal Patterns and Global Trends in 39 Countries. *Sci. Rep.* **2017**, *7*, 15698. [CrossRef] [PubMed]
4. Protzel, C.; Maruschke, M.; Hakenberg, O.W. Epidemiology, Aetiology, and Pathogenesis of Renal Cell Carcinoma. *Eur. Urol. Suppl.* **2012**, *11*, 52–59. [CrossRef]
5. National Cancer Institute. *Cancer Stat Facts: Kidney and Renal Pelvis Cancer*; National Cancer Institute: Bethesda, MD, USA, 2023.
6. Goyal, R.; Gersbach, E.; Yang, X.J.; Rohan, S.M. Differential Diagnosis of Renal Tumors with Clear Cytoplasm: Clinical Relevance of Renal Tumor Subclassification in the Era of Targeted Therapies and Personalized Medicine. *Arch. Pathol. Lab. Med.* **2013**, *137*, 467–480. [CrossRef] [PubMed]
7. Rini, B.I.; Campbell, S.C.; Escudier, B. Renal Cell Carcinoma. *Lancet* **2009**, *373*, 1119–1132. [CrossRef]
8. Bedke, J.; Gauler, T.; Grünwald, V.; Hegele, A.; Herrmann, E.; Hinz, S.; Janssen, J.; Schmitz, S.; Schostak, M.; Tesch, H.; et al. Systemic Therapy in Metastatic Renal Cell Carcinoma. *World J. Urol.* **2017**, *35*, 179–188. [CrossRef]
9. Atkins, M.B.; Tannir, N.M. Current and Emerging Therapies for First-Line Treatment of Metastatic Clear Cell Renal Cell Carcinoma. *Cancer Treat. Rev.* **2018**, *70*, 127–137. [CrossRef]
10. Zahoor, H.; Duddalwar, V.; D'Souza, A.; Merseburger, A.S.; Quinn, D.I. What Comes After Immuno-Oncology Therapy for Kidney Cancer? *Kidney Cancer* **2019**, *3*, 93–102. [CrossRef]
11. Haas, N.B.; Shevach, J.; Davis, I.D.; Eisen, T.; Gross-Gupil, M.; Kapoor, A.; Master, V.A.; Ryan, C.W.; Schimdinger, M. Chapter 12: Neoadjuvant and Adjuvant Therapy for Renal Cell Carcinoma. In *2nd ICUD-WUOF International Consultation: Management of Kidney Cancer*; Société Internationale d'Urologie: Montreal, QC, Canada, 2022; pp. 416–417.
12. Califf, R.M. Biomarker Definitions and Their Applications. *Exp. Biol. Med.* **2018**, *243*, 213–221. [CrossRef]
13. Gulati, S.; Vogelzang, N.J. Biomarkers in Renal Cell Carcinoma: Are We There Yet? *Asian J. Urol.* **2021**, *8*, 362–375. [CrossRef]
14. Raimondi, A.; Sepe, P.; Zattarin, E. Predictive Biomarkers of Response to Immunotherapy in Metastatic Renal Cell Cancer. *Front. Oncol.* **2020**, *10*, 1644. [CrossRef]
15. Farber, N.J.; Kim, C.J.; Modi, P.K.; Hon, J.D.; Sadimin, E.T.; Singer, E.A. Renal Cell Carcinoma: The Search for a Reliable Biomarker. *Transl. Cancer Res.* **2017**, *6*, 620–632. [CrossRef]
16. Covidence systematic review software, Veritas Health Innovation, Melbourne, Australia. Available online: http://www.covidence.org/ (accessed on 29 September 2023).
17. Aziz, N.; Detels, R.; Quint, J.J.; Gjertson, D.; Ryner, T.; Butch, A.W. Biological Variation of Immunological Blood Biomarkers in Healthy Individuals and Quality Goals for Biomarker Tests. *BMC Immunol.* **2019**, *20*, 33. [CrossRef] [PubMed]
18. Figlin, R.A.; Tannir, N.M.; Uzzo, R.G.; Tykodi, S.S.; Chen, D.Y.; Master, V.; Kapoor, A.; Vaena, D.; Lowrance, W.T.; Bratslavsky, G.; et al. Results of the ADAPT Phase 3 Study of Rocapuldencel-T in Combination with Sunitinib as First-Line Therapy in Patients with Metastatic Renal Cell Carcinoma. *Clin. Cancer Res.* **2020**, *26*, 2327–2336. [CrossRef] [PubMed]
19. Tang, B.; Yan, X.; Sheng, X.; Si, L.; Cui, C.; Kong, Y.; Mao, L.; Lian, B.; Bai, X.; Wang, X.; et al. Safety and Clinical Activity with an Anti-PD-1 Antibody JS001 in Advanced Melanoma or Urologic Cancer Patients. *J. Hematol. Oncol.* **2019**, *12*, 7. [CrossRef] [PubMed]
20. Velev, M.; Dalban, C.; Chevreau, C.; Gravis, G.; Negrier, S.; Laguerre, B.; Gross-Goupil, M.; Ladoire, S.; Borchiellini, D.; Geoffrois, L.; et al. Efficacy and Safety of Nivolumab in Bone Metastases from Renal Cell Carcinoma: Results of the GETUG-AFU26-NIVOREN Multicentre Phase II Study. *Eur. J. Cancer* **2023**, *182*, 66–76. [CrossRef]
21. Xu, J.X.; Maher, V.E.; Zhang, L.; Tang, S.; Sridhara, R.; Ibrahim, A.; Kim, G.; Pazdur, R. FDA Approval Summary: Nivolumab in Advanced Renal Cell Carcinoma After Anti-Angiogenic Therapy and Exploratory Predictive Biomarker Analysis. *Oncologist* **2017**, *22*, 311–317. [CrossRef]
22. Pignon, J.-C.; Jegede, O.; Shukla, S.A.; Braun, D.A.; Horak, C.E.; Wind-Rotolo, M.; Ishii, Y.; Catalano, P.J.; Grosha, J.; Flaifel, A.; et al. irRECIST for the Evaluation of Candidate Biomarkers of Response to Nivolumab in Metastatic Clear Cell Renal Cell Carcinoma: Analysis of a Phase II Prospective Clinical Trial. *Clin. Cancer Res.* **2019**, *25*, 2174–2184. [CrossRef]
23. Atkins, M.B.; Jegede, O.A.; Haas, N.B.; McDermott, D.F.; Bilen, M.A.; Stein, M.; Sosman, J.A.; Alter, R.; Plimack, E.R.; Ornstein, M.; et al. Phase II Study of Nivolumab and Salvage Nivolumab/Ipilimumab in Treatment-Naive Patients with Advanced Clear Cell Renal Cell Carcinoma (HCRN GU16-260-Cohort A). *J. Clin. Oncol.* **2022**, *40*, 2913–2923. [CrossRef]
24. Mahoney, K.M.; Ross-Macdonald, P.; Yuan, L.; Song, L.; Veras, E.; Wind-Rotolo, M.; McDermott, D.F.; Hodi, F.S.; Choueiri, T.K.; Freeman, G.J. Soluble PD-L1 as an Early Marker of Progressive Disease on Nivolumab. *J. Immunother. Cancer* **2022**, *10*, e003527. [CrossRef]

25. Incorvaia, L.; Fanale, D.; Badalamenti, G.; Porta, C.; Olive, D.; De Luca, I.; Brando, C.; Rizzo, M.; Messina, C.; Rediti, M.; et al. Baseline Plasma Levels of Soluble PD-1, PD-L1, and BTN3A1 Predict Response to Nivolumab Treatment in Patients with Metastatic Renal Cell Carcinoma: A Step toward a Biomarker for Therapeutic Decisions. *OncoImmunology* **2020**, *9*, 1832348. [CrossRef]
26. Ross-Macdonald, P.; Walsh, A.M.; Chasalow, S.D.; Ammar, R.; Papillon-Cavanagh, S.; Szabo, P.M.; Choueiri, T.K.; Sznol, M.; Wind-Rotolo, M. Molecular Correlates of Response to Nivolumab at Baseline and on Treatment in Patients with RCC. *J. Immunother. Cancer* **2021**, *9*, e001506. [CrossRef] [PubMed]
27. Chehrazi-Raffle, A.; Meza, L.; Alcantara, M.; Dizman, N.; Bergerot, P.; Salgia, N.; Hsu, J.; Ruel, N.; Salgia, S.; Malhotra, J.; et al. Circulating Cytokines Associated with Clinical Response to Systemic Therapy in Metastatic Renal Cell Carcinoma. *J. Immunother. Cancer* **2021**, *9*, e002009. [CrossRef] [PubMed]
28. García-Donas, J.; Leon, L.A.; Esteban, E.; Vidal-Mendez, M.J.; Arranz, J.A.; Garcia Del Muro, X.; Basterretxea, L.; González Del Alba, A.; Climent, M.A.; Virizuela, J.A.; et al. A Prospective Observational Study for Assessment and Outcome Association of Circulating Endothelial Cells in Clear Cell Renal Cell Carcinoma Patients Who Show Initial Benefit from First-Line Treatment. The CIRCLES (CIRCuLating Endothelial cellS) Study (SOGUG-CEC-2011-01). *Eur. Urol. Focus* **2017**, *3*, 430–436. [CrossRef] [PubMed]
29. Bootsma, M.; McKay, R.R.; Emamekhoo, H.; Bade, R.M.; Schehr, J.L.; Mannino, M.C.; Singh, A.; Wolfe, S.K.; Schultz, Z.D.; Sperger, J.; et al. Longitudinal Molecular Profiling of Circulating Tumor Cells in Metastatic Renal Cell Carcinoma. *J. Clin. Oncol.* **2022**, *40*, 3633–3641. [CrossRef]
30. Billon, E.; Chanez, B.; Rochigneux, P.; Albiges, L.; Vicier, C.; Pignot, G.; Walz, J.; Chretien, A.-S.; Gravis, G.; Olive, D. Soluble BTN2A1 Is a Potential Prognosis Biomarker in Pre-Treated Advanced Renal Cell Carcinoma. *Front. Immunol.* **2021**, *12*, 670827. [CrossRef]
31. Carril-Ajuria, L.; Desnoyer, A.; Meylan, M.; Dalban, C.; Naigeon, M.; Cassard, L.; Vano, Y.; Rioux-Leclercq, N.; Chouaib, S.; Beuselinck, B.; et al. Baseline Circulating Unswitched Memory B Cells and B-Cell Related Soluble Factors Are Associated with Overall Survival in Patients with Clear Cell Renal Cell Carcinoma Treated with Nivolumab within the NIVOREN GETUG-AFU 26 Study. *J. Immunother. Cancer* **2022**, *10*, e004885. [CrossRef]
32. De Giorgi, U.; Procopio, G.; Giannarelli, D.; Sabbatini, R.; Bearz, A.; Buti, S.; Basso, U.; Mitterer, M.; Ortega, C.; Bidoli, P.; et al. Association of Systemic Inflammation Index and Body Mass Index with Survival in Patients with Renal Cell Cancer Treated with Nivolumab. *Clin. Cancer Res.* **2019**, *25*, 3839–3846. [CrossRef]
33. Saal, J.; Bald, T.; Hölzel, M.; Ritter, M.; Brossart, P.; Ellinger, J.; Klümper, N. In the Phase III IMmotion151 Trial of Metastatic Renal Cell Carcinoma the Easy-to-Implement Modified Glasgow Prognostic Score Predicts Outcome More Accurately than the IMDC Score. *Ann. Oncol.* **2022**, *33*, 982–984. [CrossRef]
34. Kankkunen, E.; Penttilä, P.; Peltola, K.; Bono, P. C-Reactive Protein and Immune-Related Adverse Events as Prognostic Biomarkers in Immune Checkpoint Inhibitor Treated Metastatic Renal Cell Carcinoma Patients. *Acta Oncol.* **2022**, *61*, 1240–1247. [CrossRef]
35. Roussel, E.; Kinget, L.; Verbiest, A.; Debruyne, P.R.; Baldewijns, M.; Van Poppel, H.; Albersen, M.; Beuselinck, B. C-Reactive Protein and Neutrophil-Lymphocyte Ratio Are Prognostic in Metastatic Clear-Cell Renal Cell Carcinoma Patients Treated with Nivolumab. *Urol. Oncol.* **2021**, *39*, 239.e17–239.e25. [CrossRef] [PubMed]
36. Ito, K.; Masunaga, A.; Tanaka, N.; Mizuno, R.; Shirotake, S.; Yasumizu, Y.; Ito, Y.; Miyazaki, Y.; Hagiwara, M.; Kanao, K.; et al. Impact of Inflammatory Marker Levels One Month after the First-Line Targeted Therapy Initiation on Progression-Free Survival Prediction in Patients with Metastatic Clear Cell Renal Cell Carcinoma. *Jpn. J. Clin. Oncol.* **2019**, *49*, 69–76. [CrossRef] [PubMed]
37. Abuhelwa, A.Y.; Bellmunt, J.; Kichenadasse, G.; McKinnon, R.A.; Rowland, A.; Sorich, M.J.; Hopkins, A.M. C-Reactive Protein Provides Superior Prognostic Accuracy than the IMDC Risk Model in Renal Cell Carcinoma Treated with Atezolizumab/Bevacizumab. *Front. Oncol.* **2022**, *12*, 918993. [CrossRef]
38. Fukuda, S.; Saito, K.; Yasuda, Y.; Kijima, T.; Yoshida, S.; Yokoyama, M.; Ishioka, J.; Matsuoka, Y.; Kageyama, Y.; Fujii, Y. Impact of C-Reactive Protein Flare-Response on Oncological Outcomes in Patients with Metastatic Renal Cell Carcinoma Treated with Nivolumab. *J. Immunother. Cancer* **2021**, *9*, e001564. [CrossRef] [PubMed]
39. Takamatsu, K.; Mizuno, R.; Tanaka, N.; Takeda, T.; Morita, S.; Matsumoto, K.; Kosaka, T.; Shinojima, T.; Kikuchi, E.; Asanuma, H.; et al. Prognostic Value of Serum C-Reactive Protein Level Prior to Second-Line Treatment in Intermediate Risk Metastatic Renal Cell Carcinoma Patients. *Int. J. Clin. Oncol.* **2019**, *24*, 1069–1074. [CrossRef]
40. Noguchi, G.; Nakaigawa, N.; Umemoto, S.; Kobayashi, K.; Shibata, Y.; Tsutsumi, S.; Yasui, M.; Ohtake, S.; Suzuki, T.; Osaka, K.; et al. C-Reactive Protein at 1 Month after Treatment of Nivolumab as a Predictive Marker of Efficacy in Advanced Renal Cell Carcinoma. *Cancer Chemother. Pharmacol.* **2020**, *86*, 75–85. [CrossRef]
41. Klümper, N.; Schmucker, P.; Hahn, O.; Höh, B.; Mattigk, A.; Banek, S.; Ellinger, J.; Heinzelbecker, J.; Sikic, D.; Eckstein, M.; et al. C-reactive Protein Flare-response Predicts Long-term Efficacy to First-line anti-PD-1-based Combination Therapy in Metastatic Renal Cell Carcinoma. *Clin. Transl. Immunol.* **2021**, *10*, e1358. [CrossRef] [PubMed]
42. Song, M.; Graubard, B.I.; Rabkin, C.S.; Engels, E.A. Neutrophil-to-Lymphocyte Ratio and Mortality in the United States General Population. *Sci. Rep.* **2021**, *11*, 464. [CrossRef] [PubMed]
43. Lalani, A.-K.A.; Xie, W.; Martini, D.J.; Steinharter, J.A.; Norton, C.K.; Krajewski, K.M.; Duquette, A.; Bossé, D.; Bellmunt, J.; Van Allen, E.M.; et al. Change in Neutrophil-to-Lymphocyte Ratio (NLR) in Response to Immune Checkpoint Blockade for Metastatic Renal Cell Carcinoma. *J. Immunother. Cancer* **2018**, *6*, 5. [CrossRef]

44. Jeyakumar, G.; Kim, S.; Bumma, N.; Landry, C.; Silski, C.; Suisham, S.; Dickow, B.; Heath, E.; Fontana, J.; Vaishampayan, U. Neutrophil Lymphocyte Ratio and Duration of Prior Anti-Angiogenic Therapy as Biomarkers in Metastatic RCC Receiving Immune Checkpoint Inhibitor Therapy. *J. Immunother. Cancer* **2017**, *5*, 82. [CrossRef]
45. Nishiyama, N.; Hirobe, M.; Kikushima, T.; Matsuki, M.; Takahashi, A.; Yanase, M.; Ichimatsu, K.; Egawa, M.; Hayashi, N.; Negishi, T.; et al. The Neutrophil-Lymphocyte Ratio Has a Role in Predicting the Effectiveness of Nivolumab in Japanese Patients with Metastatic Renal Cell Carcinoma: A Multi-Institutional Retrospective Study. *BMC Urol.* **2020**, *20*, 110. [CrossRef] [PubMed]
46. Ikarashi, D.; Kato, Y.; Kato, R.; Kanehira, M.; Takata, R.; Obara, W. Inflammatory Markers for Predicting Responses to Nivolumab in Patients with Metastatic Renal Cell Carcinoma. *Int. J. Urol.* **2020**, *27*, 350–351. [CrossRef] [PubMed]
47. Zahoor, H.; Barata, P.C.; Jia, X.; Martin, A.; Allman, K.D.; Wood, L.S.; Gilligan, T.D.; Grivas, P.; Ornstein, M.C.; Garcia, J.A.; et al. Patterns, Predictors and Subsequent Outcomes of Disease Progression in Metastatic Renal Cell Carcinoma Patients Treated with Nivolumab. *J. Immunother. Cancer* **2018**, *6*, 107. [CrossRef] [PubMed]
48. Conroy, M.; Naidoo, J. Immune-Related Adverse Events and the Balancing Act of Immunotherapy. *Nat. Commun.* **2022**, *13*, 392. [CrossRef]
49. Martini, D.J.; Goyal, S.; Liu, Y.; Evans, S.T.; Olsen, T.A.; Case, K.; Magod, B.L.; Brown, J.T.; Yantorni, L.; Russler, G.A.; et al. Immune-Related Adverse Events as Clinical Biomarkers in Patients with Metastatic Renal Cell Carcinoma Treated with Immune Checkpoint Inhibitors. *Oncologist* **2021**, *26*, e1742–e1750. [CrossRef]
50. Ikeda, T.; Ishihara, H.; Nemoto, Y.; Tachibana, H.; Fukuda, H.; Yoshida, K.; Takagi, T.; Iizuka, J.; Hashimoto, Y.; Ishida, H.; et al. Prognostic Impact of Immune-Related Adverse Events in Metastatic Renal Cell Carcinoma Treated with Nivolumab plus Ipilimumab. *Urol. Oncol.* **2021**, *39*, 735.e9–735.e16. [CrossRef]
51. Ishihara, H.; Takagi, T.; Kondo, T.; Homma, C.; Tachibana, H.; Fukuda, H.; Yoshida, K.; Iizuka, J.; Kobayashi, H.; Okumi, M.; et al. Association between Immune-Related Adverse Events and Prognosis in Patients with Metastatic Renal Cell Carcinoma Treated with Nivolumab. *Urol. Oncol.* **2019**, *37*, 355.e21–355.e29. [CrossRef]
52. Thomson, R.J.; Moshirfar, M.; Ronquillo, Y. Tyrosine Kinase Inhibitors. In *StatPearls*; StatPearls Publishing: Treasure Island, FL, USA, 2023.
53. Lai, Y.; Zhao, Z.; Zeng, T.; Liang, X.; Chen, D.; Duan, X.; Zeng, G.; Wu, W. Crosstalk between VEGFR and Other Receptor Tyrosine Kinases for TKI Therapy of Metastatic Renal Cell Carcinoma. *Cancer Cell Int.* **2018**, *18*, 31. [CrossRef]
54. Mauge, L.; Mejean, A.; Fournier, L.; Pereira, H.; Etienne-Grimaldi, M.-C.; Levionnois, E.; Caty, A.; Abadie-Lacourtoisie, S.; Culine, S.; Le Moulec, S.; et al. Sunitinib Prior to Planned Nephrectomy in Metastatic Renal Cell Carcinoma: Angiogenesis Biomarkers Predict Clinical Outcome in the Prospective Phase II PREINSUT Trial. *Clin. Cancer Res.* **2018**, *24*, 5534–5542. [CrossRef]
55. Jilaveanu, L.B.; Puligandla, M.; Weiss, S.A.; Wang, X.V.; Zito, C.; Flaherty, K.T.; Boeke, M.; Neumeister, V.; Camp, R.L.; Adeniran, A.; et al. Tumor Microvessel Density as a Prognostic Marker in High-Risk Renal Cell Carcinoma Patients Treated on ECOG-ACRIN E2805. *Clin. Cancer Res.* **2018**, *24*, 217–223. [CrossRef]
56. Oudard, S.; Benhamouda, N.; Escudier, B.; Ravel, P.; Tran, T.; Levionnois, E.; Negrier, S.; Barthelemy, P.; Berdah, J.; Gross-Goupil, M.; et al. Decrease of Pro-Angiogenic Monocytes Predicts Clinical Response to Anti-Angiogenic Treatment in Patients with Metastatic Renal Cell Carcinoma. *Cells* **2021**, *11*, 17. [CrossRef] [PubMed]
57. Xu, W.; Puligandla, M.; Manola, J.; Bullock, A.J.; Tamasauskas, D.; McDermott, D.F.; Atkins, M.B.; Haas, N.B.; Flaherty, K.; Uzzo, R.G.; et al. Angiogenic Factor and Cytokine Analysis among Patients Treated with Adjuvant VEGFR TKIs in Resected Renal Cell Carcinoma. *Clin. Cancer Res.* **2019**, *25*, 6098–6106. [CrossRef] [PubMed]
58. Hakimi, A.A.; Voss, M.H.; Kuo, F.; Sanchez, A.; Liu, M.; Nixon, B.G.; Vuong, L.; Ostrovnaya, I.; Chen, Y.-B.; Reuter, V.; et al. Transcriptomic Profiling of the Tumor Microenvironment Reveals Distinct Subgroups of Clear Cell Renal Cell Cancer: Data from a Randomized Phase III Trial. *Cancer Discov.* **2019**, *9*, 510–525. [CrossRef] [PubMed]
59. Tanaka, T.; Narazaki, M.; Kishimoto, T. IL-6 in Inflammation, Immunity, and Disease. *Cold Spring Harb. Perspect. Biol.* **2014**, *6*, a016295. [CrossRef]
60. Ishibashi, K.; Koguchi, T.; Matsuoka, K.; Onagi, A.; Tanji, R.; Takinami-Honda, R.; Hoshi, S.; Onoda, M.; Kurimura, Y.; Hata, J.; et al. Interleukin-6 Induces Drug Resistance in Renal Cell Carcinoma. *Fukushima J. Med. Sci.* **2018**, *64*, 103–110. [CrossRef]
61. Pilskog, M.; Bostad, L.; Edelmann, R.J.; Akslen, L.A.; Beisland, C.; Straume, O. Tumour Cell Expression of Interleukin 6 Receptor α Is Associated with Response Rates in Patients Treated with Sunitinib for Metastatic Clear Cell Renal Cell Carcinoma: Interleukin 6 Receptor α in Renal Cancer. *J. Pathol. Clin. Res.* **2018**, *4*, 114–123. [CrossRef]
62. Pilskog, M.; Nilsen, G.H.; Beisland, C.; Straume, O. Elevated Plasma Interleukin 6 Predicts Poor Response in Patients Treated with Sunitinib for Metastatic Clear Cell Renal Cell Carcinoma. *Cancer Treat. Res. Commun.* **2019**, *19*, 100127. [CrossRef]
63. Bellmunt, J.; Esteban, E.; Del Muro, X.G.; Sepúlveda, J.M.; Maroto, P.; Gallardo, E.; Del Alba, A.G.; Etxaniz, O.; Guix, M.; Larriba, J.L.G.; et al. Pazopanib as Second-Line Antiangiogenic Treatment in Metastatic Renal Cell Carcinoma After Tyrosine Kinase Inhibitor (TKI) Failure: A Phase 2 Trial Exploring Immune-Related Biomarkers for Testing in the Post-Immunotherapy/TKI Era. *Eur. Urol. Oncol.* **2021**, *4*, 502–505. [CrossRef]
64. Zizzari, I.G.; Napoletano, C.; Di Filippo, A.; Botticelli, A.; Gelibter, A.; Calabrò, F.; Rossi, E.; Schinzari, G.; Urbano, F.; Pomati, G.; et al. Exploratory Pilot Study of Circulating Biomarkers in Metastatic Renal Cell Carcinoma. *Cancers* **2020**, *12*, 2620. [CrossRef]
65. Montemagno, C.; Hagege, A.; Borchiellini, D.; Thamphya, B.; Rastoin, O.; Ambrosetti, D.; Iovanna, J.; Rioux-Leclercq, N.; Porta, C.; Negrier, S.; et al. Soluble Forms of PD-L1 and PD-1 as Prognostic and Predictive Markers of Sunitinib Efficacy in Patients with Metastatic Clear Cell Renal Cell Carcinoma. *OncoImmunology* **2020**, *9*, 1846901. [CrossRef]

66. Takamatsu, K.; Mizuno, R.; Omura, M.; Morita, S.; Matsumoto, K.; Shinoda, K.; Kosaka, T.; Takeda, T.; Shinojima, T.; Kikuchi, E.; et al. Prognostic Value of Baseline Serum C-Reactive Protein Level in Intermediate-Risk Group Patients with Metastatic Renal-Cell Carcinoma Treated by First-Line Vascular Endothelial Growth Factor-Targeted Therapy. *Clin. Genitourin. Cancer* **2018**, *16*, e927–e933. [CrossRef]
67. Wang, B.; Gu, W.; Wan, F.; Shi, G.; Ye, D. Prognostic Significance of the Dynamic Changes of Systemic Inflammatory Response in Metastatic Renal Cell Carcinoma. *Int. Braz. J. Urol.* **2019**, *45*, 89–99. [CrossRef]
68. Takagi, T.; Fukuda, H.; Kondo, T.; Ishihara, H.; Yoshida, K.; Kobayashi, H.; Iizuka, J.; Okumi, M.; Ishida, H.; Tanabe, K. Prognostic Markers for Refined Stratification of IMDC Intermediate-Risk Metastatic Clear Cell Renal Cell Carcinoma Treated with First-Line Tyrosine Kinase Inhibitor Therapy. *Target. Oncol.* **2019**, *14*, 179–186. [CrossRef]
69. Teishima, J.; Ohara, S.; Shinmei, S.; Inoue, S.; Hayashi, T.; Mochizuki, H.; Mita, K.; Shigeta, M.; Matsubara, A. Normalization of C-Reactive Protein Levels Following Cytoreductive Nephrectomy in Patients with Metastatic Renal Cell Carcinoma Treated with Tyrosine Kinase Inhibitors Is Associated with Improved Overall Survival. *Urol. Oncol.* **2018**, *36*, 339.e9–339.e15. [CrossRef] [PubMed]
70. Yasuda, Y.; Saito, K.; Yuasa, T.; Uehara, S.; Kawamura, N.; Yokoyama, M.; Ishioka, J.; Matsuoka, Y.; Yamamoto, S.; Okuno, T.; et al. Early Response of C-Reactive Protein as a Predictor of Survival in Patients with Metastatic Renal Cell Carcinoma Treated with Tyrosine Kinase Inhibitors. *Int. J. Clin. Oncol.* **2017**, *22*, 1081–1086. [CrossRef] [PubMed]
71. Erdogan, B.; Kostek, O.; Bekir Hacioglu, M.; Gokyer, A.; Kucukarda, A.; Ozcan, E.; Gokmen, I.; Uzunoglu, S.; Cicin, I. Is Early Change in Systemic Inflammatory Markers Associated with Treatment Response in Patients Who Received Pazopanib? *J. BUON* **2021**, *26*, 2196–2201. [PubMed]
72. Bayat Mokhtari, R.; Homayouni, T.S.; Baluch, N.; Morgatskaya, E.; Kumar, S.; Das, B.; Yeger, H. Combination Therapy in Combating Cancer. *Oncotarget* **2017**, *8*, 38022–38043. [CrossRef]
73. Rossi, E.; Bersanelli, M.; Gelibter, A.J.; Borsellino, N.; Caserta, C.; Doni, L.; Maruzzo, M.; Mosca, A.; Pisano, C.; Verzoni, E.; et al. Combination Therapy in Renal Cell Carcinoma: The Best Choice for Every Patient? *Curr. Oncol. Rep.* **2021**, *23*, 147. [CrossRef]
74. McDermott, D.F.; Huseni, M.A.; Atkins, M.B.; Motzer, R.J.; Rini, B.I.; Escudier, B.; Fong, L.; Joseph, R.W.; Pal, S.K.; Reeves, J.A.; et al. Clinical Activity and Molecular Correlates of Response to Atezolizumab Alone or in Combination with Bevacizumab versus Sunitinib in Renal Cell Carcinoma. *Nat. Med.* **2018**, *24*, 749–757. [CrossRef]
75. Martini, J.-F.; Plimack, E.R.; Choueiri, T.K.; McDermott, D.F.; Puzanov, I.; Fishman, M.N.; Cho, D.C.; Vaishampayan, U.; Rosbrook, B.; Fernandez, K.C.; et al. Angiogenic and Immune-Related Biomarkers and Outcomes Following Axitinib/Pembrolizumab Treatment in Patients with Advanced Renal Cell Carcinoma. *Clin. Cancer Res.* **2020**, *26*, 5598–5608. [CrossRef]
76. Msaouel, P.; Goswami, S.; Thall, P.F.; Wang, X.; Yuan, Y.; Jonasch, E.; Gao, J.; Campbell, M.T.; Shah, A.Y.; Corn, P.G.; et al. A Phase 1–2 Trial of Sitravatinib and Nivolumab in Clear Cell Renal Cell Carcinoma Following Progression on Antiangiogenic Therapy. *Sci. Transl. Med.* **2022**, *14*, eabm6420. [CrossRef]
77. Kamai, T.; Kijima, T.; Tsuzuki, T.; Nukui, A.; Abe, H.; Arai, K.; Yoshida, K.-I. Increased Expression of Adenosine 2A Receptors in Metastatic Renal Cell Carcinoma Is Associated with Poorer Response to Anti-Vascular Endothelial Growth Factor Agents and Anti-PD-1/Anti-CTLA4 Antibodies and Shorter Survival. *Cancer Immunol. Immunother.* **2021**, *70*, 2009–2021. [CrossRef]
78. Novelli, G.; Ciccacci, C.; Borgiani, P.; Papaluca Amati, M.; Abadie, E. Genetic Tests and Genomic Biomarkers: Regulation, Qualification and Validation. *Clin. Cases Miner. Bone* **2008**, *5*, 149–154.
79. Incorvaia, L.; Fanale, D.; Badalamenti, G.; Brando, C.; Bono, M.; De Luca, I.; Algeri, L.; Bonasera, A.; Corsini, L.R.; Scurria, S.; et al. A "Lymphocyte MicroRNA Signature" as Predictive Biomarker of Immunotherapy Response and Plasma PD-1/PD-L1 Expression Levels in Patients with Metastatic Renal Cell Carcinoma: Pointing towards Epigenetic Reprogramming. *Cancers* **2020**, *12*, 3396. [CrossRef] [PubMed]
80. Epaillard, N.; Simonaggio, A.; Elaidi, R.; Azzouz, F.; Braychenko, E.; Thibault, C.; Sun, C.-M.; Moreira, M.; Oudard, S.; Vano, Y.-A. BIONIKK: A Phase 2 Biomarker Driven Trial with Nivolumab and Ipilimumab or VEGFR Tyrosine Kinase Inhibitor (TKI) in Naïve Metastatic Kidney Cancer. *Bull. Cancer* **2020**, *107*, eS22–eS27. [CrossRef] [PubMed]
81. Miao, D.; Margolis, C.A.; Gao, W.; Voss, M.H.; Li, W.; Martini, D.J.; Norton, C.; Bossé, D.; Wankowicz, S.M.; Cullen, D.; et al. Genomic Correlates of Response to Immune Checkpoint Therapies in Clear Cell Renal Cell Carcinoma. *Science* **2018**, *359*, 801–806. [CrossRef]
82. Kim, Y.J.; Kang, Y.; Kim, J.S.; Sung, H.H.; Jeon, H.G.; Jeong, B.C.; Seo, S.I.; Jeon, S.S.; Lee, H.M.; Park, D.; et al. Potential of Circulating Tumor DNA as a Predictor of Therapeutic Responses to Immune Checkpoint Blockades in Metastatic Renal Cell Carcinoma. *Sci. Rep.* **2021**, *11*, 5600. [CrossRef]
83. Dietz, S.; Sültmann, H.; Du, Y.; Reisinger, E.; Riediger, A.L.; Volckmar, A.-L.; Stenzinger, A.; Schlesner, M.; Jäger, D.; Hohenfellner, M.; et al. Patient-Specific Molecular Alterations Are Associated with Metastatic Clear Cell Renal Cell Cancer Progressing under Tyrosine Kinase Inhibitor Therapy. *Oncotarget* **2017**, *8*, 74049–74057. [CrossRef]
84. Maroto, P.; Esteban, E.; Fernández Parra, E.; Mendez-Vidal, M.; Domenech, M.; Pérez-Valderrama, B.; Calderero, V.; Perez-Gracia, J.; Grande, E.; Algaba, F. HIF Pathway and C-Myc as Biomarkers for Response to Sunitinib in Metastatic Clear-Cell Renal Cell Carcinoma. *OncoTargets Ther.* **2017**, *10*, 4635–4643. [CrossRef]
85. Wierzbicki, P.; Klacz, J.; Kotulak-Chrzaszcz, A.; Wronska, A.; Stanislawowski, M.; Rybarczyk, A.; Ludziejewska, A.; Kmiec, Z.; Matuszewski, M. Prognostic Significance of VHL, HIF1A, HIF2A, VEGFA and P53 Expression in Patients with Clear-cell Renal Cell Carcinoma Treated with Sunitinib as First-line Treatment. *Int. J. Oncol.* **2019**, *55*, 371–390. [CrossRef]

86. Nayak, B.; Panaiyadiyan, S.; Singh, P.; Karmakar, S.; Kaushal, S.; Seth, A. Role of Circulating Tumor Cells in Patients with Metastatic Clear-Cell Renal Cell Carcinoma. *Urol. Oncol. Semin. Orig. Investig.* **2021**, *39*, 135.e9–135.e15. [CrossRef]
87. Gudkov, A.; Shirokorad, V.; Kashintsev, K.; Sokov, D.; Nikitin, D.; Anisenko, A.; Borisov, N.; Sekacheva, M.; Gaifullin, N.; Garazha, A.; et al. Gene Expression-Based Signature Can Predict Sorafenib Response in Kidney Cancer. *Front. Mol. Biosci.* **2022**, *9*, 753318. [CrossRef] [PubMed]
88. Crona, D.J.; Skol, A.D.; Leppänen, V.-M.; Glubb, D.M.; Etheridge, A.S.; Hilliard, E.; Peña, C.E.; Peterson, Y.K.; Klauber-DeMore, N.; Alitalo, K.K.; et al. Genetic Variants of VEGFA and FLT4 Are Determinants of Survival in Renal Cell Carcinoma Patients Treated with Sorafenib. *Cancer Res.* **2019**, *79*, 231–241. [CrossRef] [PubMed]
89. Dorff, T.B.; Longmate, J.A.; Pal, S.K.; Stadler, W.M.; Fishman, M.N.; Vaishampayan, U.N.; Rao, A.; Pinksi, J.K.; Hu, J.S.; Quinn, D.I.; et al. Bevacizumab Alone or in Combination with TRC105 for Patients with Refractory Metastatic Renal Cell Cancer: Bevacizumab and TRC105 in Renal Cancer. *Cancer* **2017**, *123*, 4566–4573. [CrossRef]
90. Bamias, A.; Karavasilis, V.; Gavalas, N.; Tzannis, K.; Samantas, E.; Aravantinos, G.; Koutras, A.; Gkerzelis, I.; Kostouros, E.; Koutsoukos, K.; et al. The Combination of Bevacizumab/Temsirolimus after First-Line Anti-VEGF Therapy in Advanced Renal-Cell Carcinoma: A Clinical and Biomarker Study. *Int. J. Clin. Oncol.* **2019**, *24*, 411–419. [CrossRef]
91. Palomero, L.; Bodnar, L.; Mateo, F.; Herranz-Ors, C.; Espín, R.; García-Varelo, M.; Jesiotr, M.; Ruiz De Garibay, G.; Casanovas, O.; López, J.I.; et al. EVI1 as a Prognostic and Predictive Biomarker of Clear Cell Renal Cell Carcinoma. *Cancers* **2020**, *12*, 300. [CrossRef] [PubMed]
92. Voss, M.H.; Chen, D.; Reising, A.; Marker, M.; Shi, J.; Xu, J.; Ostrovnaya, I.; Seshan, V.E.; Redzematovic, A.; Chen, Y.-B.; et al. PTEN Expression, Not Mutation Status in TSC1, TSC2, or mTOR, Correlates with the Outcome on Everolimus in Patients with Renal Cell Carcinoma Treated on the Randomized RECORD-3 Trial. *Clin. Cancer Res.* **2019**, *25*, 506–514. [CrossRef] [PubMed]
93. Flaifel, A.; Xie, W.; Braun, D.A.; Ficial, M.; Bakouny, Z.; Nassar, A.H.; Jennings, R.B.; Escudier, B.; George, D.J.; Motzer, R.J.; et al. PD-L1 Expression and Clinical Outcomes to Cabozantinib, Everolimus, and Sunitinib in Patients with Metastatic Renal Cell Carcinoma: Analysis of the Randomized Clinical Trials METEOR and CABOSUN. *Clin. Cancer Res.* **2019**, *25*, 6080–6088. [CrossRef] [PubMed]
94. Motzer, R.J.; Powles, T.; Atkins, M.B.; Escudier, B.; McDermott, D.F.; Alekseev, B.Y.; Lee, J.-L.; Suarez, C.; Stroyakovskiy, D.; De Giorgi, U.; et al. Final Overall Survival and Molecular Analysis in IMmotion151, a Phase 3 Trial Comparing Atezolizumab Plus Bevacizumab vs Sunitinib in Patients with Previously Untreated Metastatic Renal Cell Carcinoma. *JAMA Oncol.* **2022**, *8*, 275–280. [CrossRef]
95. Shaikh, F.; Dupont-Roettger, D.; Dehmeshki, J.; Awan, O.; Kubassova, O.; Bisdas, S. The Role of Imaging Biomarkers Derived from Advanced Imaging and Radiomics in the Management of Brain Tumors. *Front. Oncol.* **2020**, *10*, 559946. [CrossRef]
96. Tabei, T.; Nakaigawa, N.; Kaneta, T.; Ikeda, I.; Kondo, K.; Makiyama, K.; Hasumi, H.; Hayashi, N.; Kawahara, T.; Izumi, K.; et al. Early Assessment with 18F-2-Fluoro-2-Deoxyglucose Positron Emission Tomography/Computed Tomography to Predict Short-Term Outcome in Clear Cell Renal Carcinoma Treated with Nivolumab. *BMC Cancer* **2019**, *19*, 298. [CrossRef] [PubMed]
97. Drljevic-Nielsen, A.; Donskov, F.; Mains, J.R.; Andersen, M.B.; Thorup, K.; Thygesen, J.; Rasmussen, F. Prognostic Utility of Parameters Derived from Pretreatment Dual-Layer Spectral-Detector CT in Patients with Metastatic Renal Cell Carcinoma. *Am. J. Roentgenol.* **2022**, *218*, 867–876. [CrossRef]
98. Drljevic-Nielsen, A.; Mains, J.R.; Thorup, K.; Andersen, M.B.; Rasmussen, F.; Donskov, F. Early Reduction in Spectral Dual-Layer Detector CT Parameters as Favorable Imaging Biomarkers in Patients with Metastatic Renal Cell Carcinoma. *Eur. Radiol.* **2022**, *32*, 7323–7334. [CrossRef]
99. Martini, D.J.; Olsen, T.A.; Goyal, S.; Liu, Y.; Evans, S.T.; Magod, B.; Brown, J.T.; Yantorni, L.; Russler, G.A.; Caulfield, S.; et al. Body Composition Variables as Radiographic Biomarkers of Clinical Outcomes in Metastatic Renal Cell Carcinoma Patients Receiving Immune Checkpoint Inhibitors. *Front. Oncol.* **2021**, *11*, 707050. [CrossRef] [PubMed]
100. Malone, E.R.; Sim, H.-W.; Stundzia, A.; Pierre, S.; Metser, U.; O'Malley, M.; Sacher, A.G.; Sridhar, S.S.; Hansen, A.R. Predictive Radiomics Signature for Treatment Response to Nivolumab in Patients with Advanced Renal Cell Carcinoma. *Can. Urol. Assoc. J.* **2021**, *16*, E94–E101. [CrossRef] [PubMed]
101. Mittlmeier, L.M.; Unterrainer, M.; Rodler, S.; Todica, A.; Albert, N.L.; Burgard, C.; Cyran, C.C.; Kunz, W.G.; Ricke, J.; Bartenstein, P.; et al. 18F-PSMA-1007 PET/CT for Response Assessment in Patients with Metastatic Renal Cell Carcinoma Undergoing Tyrosine Kinase or Checkpoint Inhibitor Therapy: Preliminary Results. *Eur. J. Nucl. Med. Mol. Imaging* **2021**, *48*, 2031–2037. [CrossRef] [PubMed]
102. Zheng, B.; Shin, J.H.; Li, H.; Chen, Y.; Guo, Y.; Wang, M. Comparison of Radiological Tumor Response Based on iRECIST and RECIST 1.1 in Metastatic Clear-Cell Renal Cell Carcinoma Patients Treated with Programmed Cell Death-1 Inhibitor Therapy. *Korean J. Radiol.* **2021**, *22*, 366–375. [CrossRef]
103. Park, H.J.; Qin, L.; Bakouny, Z.; Krajewski, K.M.; Van Allen, E.M.; Choueiri, T.K.; Shinagare, A.B. Computed Tomography Texture Analysis for Predicting Clinical Outcomes in Patients with Metastatic Renal Cell Carcinoma Treated with Immune Checkpoint Inhibitors. *Oncologist* **2022**, *27*, 389–397. [CrossRef]
104. Udayakumar, D.; Zhang, Z.; Xi, Y.; Dwivedi, D.K.; Fulkerson, M.; Haldeman, S.; McKenzie, T.; Yousuf, Q.; Joyce, A.; Hajibeigi, A.; et al. Deciphering Intratumoral Molecular Heterogeneity in Clear Cell Renal Cell Carcinoma with a Radiogenomics Platform. *Clin. Cancer Res.* **2021**, *27*, 4794–4806. [CrossRef]

105. Nakaigawa, N.; Kondo, K.; Kaneta, T.; Tateishi, U.; Minamimoto, R.; Namura, K.; Ueno, D.; Kobayashi, K.; Kishida, T.; Ikeda, I.; et al. FDG PET/CT after First Molecular Targeted Therapy Predicts Survival of Patients with Renal Cell Carcinoma. *Cancer Chemother. Pharmacol.* **2018**, *81*, 739–744. [CrossRef]
106. Hall, P.E.; Shepherd, S.T.C.; Brown, J.; Larkin, J.; Jones, R.; Ralph, C.; Hawkins, R.; Chowdhury, S.; Boleti, E.; Bahl, A.; et al. Radiological Response Heterogeneity Is of Prognostic Significance in Metastatic Renal Cell Carcinoma Treated with Vascular Endothelial Growth Factor-Targeted Therapy. *Eur. Urol. Focus* **2020**, *6*, 999–1005. [CrossRef] [PubMed]
107. Go, H.; Kang, M.J.; Kim, P.-J.; Lee, J.-L.; Park, J.Y.; Park, J.-M.; Ro, J.Y.; Cho, Y.M. Development of Response Classifier for Vascular Endothelial Growth Factor Receptor (VEGFR)-Tyrosine Kinase Inhibitor (TKI) in Metastatic Renal Cell Carcinoma. *Pathol. Oncol. Res.* **2019**, *25*, 51–58. [CrossRef] [PubMed]
108. Mytsyk, Y.; Pasichnyk, S.; Dutka, I.; Dats, I.; Vorobets, D.; Skrzypczyk, M.; Uteuliyev, Y.; Botikova, A.; Gazdikova, K.; Kubatka, P.; et al. Systemic Treatment of the Metastatic Renal Cell Carcinoma: Usefulness of the Apparent Diffusion Coefficient of Diffusion-Weighted MRI in Prediction of Early Therapeutic Response. *Clin. Exp. Med.* **2020**, *20*, 277–287. [CrossRef] [PubMed]
109. Wu, G.; Liu, G.; Kong, W.; Qu, J.; Suo, S.; Liu, X.; Xu, J.; Zhang, J. Assessment of Response to Anti-Angiogenic Targeted Therapy in Pulmonary Metastatic Renal Cell Carcinoma: R2* Value as a Predictive Biomarker. *Eur. Radiol.* **2017**, *27*, 3574–3582. [CrossRef]
110. Navani, V.; Ernst, M.; Wells, J.C.; Yuasa, T.; Takemura, K.; Donskov, F.; Basappa, N.S.; Schmidt, A.; Pal, S.K.; Meza, L.; et al. Imaging Response to Contemporary Immuno-Oncology Combination Therapies in Patients with Metastatic Renal Cell Carcinoma. *JAMA Netw. Open* **2022**, *5*, e2216379. [CrossRef]
111. Labadie, B.W.; Liu, P.; Bao, R.; Crist, M.; Fernandes, R.; Ferreira, L.; Graupner, S.; Poklepovic, A.S.; Duran, I.; Maleki Vareki, S.; et al. BMI, irAE, and Gene Expression Signatures Associate with Resistance to Immune-Checkpoint Inhibition and Outcomes in Renal Cell Carcinoma. *J. Transl. Med.* **2019**, *17*, 386. [CrossRef]
112. Herrmann, T.; Mione, C.; Montoriol, P.-F.; Molnar, I.; Ginzac, A.; Durando, X.; Mahammedi, H. Body Mass Index, Sarcopenia, and Their Variations in Predicting Outcomes for Patients Treated with Nivolumab for Metastatic Renal Cell Carcinoma. *Oncology* **2022**, *100*, 114–123. [CrossRef]
113. Ueki, H.; Hara, T.; Okamura, Y.; Bando, Y.; Terakawa, T.; Furukawa, J.; Harada, K.; Nakano, Y.; Fujisawa, M. Association between Sarcopenia Based on Psoas Muscle Index and the Response to Nivolumab in Metastatic Renal Cell Carcinoma: A Retrospective Study. *Investig. Clin. Urol.* **2022**, *63*, 415–424. [CrossRef]
114. McKay, R.R.; Vu, P.; Albiges, L.K.; Lin, X.; Simantov, R.; Temel, J.S.; Choueiri, T.K. The Effect of Weight Change During Treatment with Targeted Therapy in Patients with Metastatic Renal Cell Carcinoma. *Clin. Genitourin. Cancer* **2019**, *17*, 443–450.e1. [CrossRef]
115. Ishihara, H.; Takagi, T.; Kondo, T.; Fukuda, H.; Yoshida, K.; Iizuka, J.; Tanabe, K. Effect of Changes in Skeletal Muscle Mass on Oncological Outcomes During First-Line Sunitinib Therapy for Metastatic Renal Cell Carcinoma. *Target. Oncol.* **2018**, *13*, 745–755. [CrossRef]
116. Janisch, F.; Klotzbücher, T.; Marks, P.; Kienapfel, C.; Meyer, C.P.; Yu, H.; Fühner, C.; Hillemacher, T.; Mori, K.; Mostafei, H.; et al. Predictive Value of De Ritis Ratio in Metastatic Renal Cell Carcinoma Treated with Tyrosine-Kinase Inhibitors. *World J. Urol.* **2021**, *39*, 2977–2985. [CrossRef] [PubMed]
117. Kang, M.; Yu, J.; Sung, H.H.; Jeon, H.G.; Jeong, B.C.; Park, S.H.; Jeon, S.S.; Lee, H.M.; Choi, H.Y.; Seo, S.I. Prognostic Impact of the Pretreatment Aspartate Transaminase/Alanine Transaminase Ratio in Patients Treated with First-Line Systemic Tyrosine Kinase Inhibitor Therapy for Metastatic Renal Cell Carcinoma. *Int. J. Urol.* **2018**, *25*, 596–603. [CrossRef] [PubMed]
118. Zhang, H.; Zhang, X.; Zhu, X.; Ni, Y.; Dai, J.; Zhu, S.; Sun, G.; Wang, Z.; Chen, J.; Zhao, J.; et al. The Impact of Renal Impairment on Survival Outcomes in Patients with Metastatic Renal Cell Carcinoma Treated with Tyrosine Kinase Inhibitors. *Cancer Control* **2020**, *27*, 107327482097714. [CrossRef] [PubMed]
119. Aktepe, O.H.; Guner, G.; Guven, D.C.; Taban, H.; Yildirim, H.C.; Sahin, T.K.; Ardic, F.S.; Yeter, H.H.; Yuce, D.; Erman, M. Impact of Albumin to Globulin Ratio on Survival Outcomes of Patients with Metastatic Renal Cell Carcinoma. *Turk. J. Urol.* **2021**, *47*, 113–119. [CrossRef]
120. Roussel, E.; Kinget, L.; Verbiest, A.; Boeckx, B.; Zucman-Rossi, J.; Couchy, G.; Caruso, S.; Baldewijns, M.; Joniau, S.; Van Poppel, H.; et al. Molecular Underpinnings of Glandular Tropism in Metastatic Clear Cell Renal Cell Carcinoma: Therapeutic Implications. *Acta Oncol.* **2021**, *60*, 1499–1506. [CrossRef]
121. Martini, D.J.; Kline, M.R.; Liu, Y.; Shabto, J.M.; Carthon, B.C.; Russler, G.A.; Yantorni, L.; Hitron, E.E.; Caulfield, S.; Goldman, J.M.; et al. Novel Risk Scoring System for Metastatic Renal Cell Carcinoma Patients Treated with Cabozantinib. *Cancer Treat. Res. Commun.* **2021**, *28*, 100393. [CrossRef]
122. Shirotake, S.; Kondo, H.; Okabe, T.; Makino, S.; Araki, R.; Komatsuda, A.; Kaneko, G.; Nishimoto, K.; Oyama, M. Early Tumor Shrinkage as a Predictive Factor of Metastatic Renal Cell Carcinoma in Molecular Targeted Therapy: A Single Institutional Study. *Mol. Clin. Oncol.* **2018**, *10*, 125–131. [CrossRef]
123. Kammerer-Jacquet, S.-F.; Brunot, A.; Bensalah, K.; Campillo-Gimenez, B.; Lefort, M.; Bayat, S.; Ravaud, A.; Dupuis, F.; Yacoub, M.; Verhoest, G.; et al. Hilar Fat Infiltration: A New Prognostic Factor in Metastatic Clear Cell Renal Cell Carcinoma with First-Line Sunitinib Treatment. *Urol. Oncol. Semin. Orig. Investig.* **2017**, *35*, 603.e7–603.e14. [CrossRef]
124. Pieretti, A.C.; Shapiro, D.D.; Westerman, M.E.; Hwang, H.; Wang, X.; Segarra, L.A.; Campbell, M.T.; Tannir, N.M.; Jonasch, E.; Matin, S.F.; et al. Tumor Diameter Response in Patients with Metastatic Clear Cell Renal Cell Carcinoma Is Associated with Overall Survival. *Urol. Oncol. Semin. Orig. Investig.* **2021**, *39*, 837.e9–837.e17. [CrossRef]

125. Shi, H.; Cao, C.; Wen, L.; Zhang, L.; Zhang, J.; Ma, J.; Shou, J.; Li, C. Prognostic Value of the Ratio of Maximum to Minimum Diameter of Primary Tumor in Metastatic Clear Cell Renal Cell Carcinoma. *BMC Urol.* **2022**, *22*, 95. [CrossRef]
126. Yildiz, I.; Bilici, A.; Karadurmuş, N.; Ozer, L.; Tural, D.; Kaplan, M.A.; Akman, T.; Bayoglu, I.V.; Uysal, M.; Yildiz, Y.; et al. Prognostic Factors for Survival in Metastatic Renal Cell Carcinoma Patients with Brain Metastases Receiving Targeted Therapy. *Tumori J.* **2018**, *104*, 444–450. [CrossRef] [PubMed]
127. Mi, H.; Bivalacqua, T.J.; Kates, M.; Seiler, R.; Black, P.C.; Popel, A.S.; Baras, A.S. Predictive Models of Response to Neoadjuvant Chemotherapy in Muscle-Invasive Bladder Cancer Using Nuclear Morphology and Tissue Architecture. *Cell Rep. Med.* **2021**, *2*, 100382. [CrossRef]
128. Mi, H.; Sivagnanam, S.; Betts, C.B.; Liudahl, S.M.; Jaffee, E.M.; Coussens, L.M.; Popel, A.S. Quantitative Spatial Profiling of Immune Populations in Pancreatic Ductal Adenocarcinoma Reveals Tumor Microenvironment Heterogeneity and Prognostic Biomarkers. *Cancer Res.* **2022**, *82*, 4359–4372. [CrossRef] [PubMed]
129. Zhang, D.; Ni, Y.; Wang, Y.; Feng, J.; Zhuang, N.; Li, J.; Liu, L.; Shen, W.; Zheng, J.; Zheng, W.; et al. Spatial Heterogeneity of Tumor Microenvironment Influences the Prognosis of Clear Cell Renal Cell Carcinoma. *J. Transl. Med.* **2023**, *21*, 489. [CrossRef] [PubMed]
130. Wessels, F.; Schmitt, M.; Krieghoff-Henning, E.; Kather, J.N.; Nientiedt, M.; Kriegmair, M.C.; Worst, T.S.; Neuberger, M.; Steeg, M.; Popovic, Z.V.; et al. Deep Learning Can Predict Survival Directly from Histology in Clear Cell Renal Cell Carcinoma. *PLoS ONE* **2022**, *17*, e0272656. [CrossRef]
131. Cheng, J.; Han, Z.; Mehra, R.; Shao, W.; Cheng, M.; Feng, Q.; Ni, D.; Huang, K.; Cheng, L.; Zhang, J. Computational Analysis of Pathological Images Enables a Better Diagnosis of TFE3 Xp11.2 Translocation Renal Cell Carcinoma. *Nat. Commun.* **2020**, *11*, 1778. [CrossRef]
132. Cheng, J.; Zhang, J.; Han, Y.; Wang, X.; Ye, X.; Meng, Y.; Parwani, A.; Han, Z.; Feng, Q.; Huang, K. Integrative Analysis of Histopathological Images and Genomic Data Predicts Clear Cell Renal Cell Carcinoma Prognosis. *Cancer Res.* **2017**, *77*, e91–e100. [CrossRef]

cancers

MDPI

Review

Disturbances in Nitric Oxide Cycle and Related Molecular Pathways in Clear Cell Renal Cell Carcinoma

Corina Daniela Ene [1,2], Mircea Tampa [3,4,*], Simona Roxana Georgescu [3,4,*], Clara Matei [3], Iulia Maria Teodora Leulescu [4], Claudia Ioana Dogaru [4], Mircea Nicolae Penescu [1,2] and Ilinca Nicolae [4]

1 Department of Nephrology, Carol Davila Clinical Hospital of Nephrology, 010731 Bucharest, Romania; koranik85@yahoo.com (C.D.E.); mirceapenescu4@gmail.com (M.N.P.)
2 Department of Nephrology, "Carol Davila" University of Medicine and Pharmacy, 020021 Bucharest, Romania
3 Department of Dermatology, "Carol Davila" University of Medicine and Pharmacy, 020021 Bucharest, Romania; matei_clara@yahoo.com
4 Department of Dermatology, "Victor Babes" Clinical Hospital for Infectious Diseases, 030303 Bucharest, Romania; iulialeulescu@gmail.com (I.M.T.L.); claudia_ioana97@yahoo.com (C.I.D.); drnicolaei@yahoo.ro (I.N.)
* Correspondence: dermatology.mt@gmail.com (M.T.); srg.dermatology@gmail.com (S.R.G.)

Simple Summary: In this article, we analyze the current state of research on nitric oxide biosynthesis metabolites in clear cell renal cell carcinoma and the metabolic pathways that amplify nitric oxide production in the tumour microenvironment. Special attention is given to nitric oxide homeostasis disruption, mechanisms of nitric oxide biosynthesis and signalling, analysis of nitric oxide bimodal effects, quantitative analysis of nitric oxide, metabolites ureagenic cycle and glutamine metabolism, arginine metabolism and depletion, hyperammonemia, branched-chain amino acids catabolism and nitric oxide-based therapy for cancer. Clarifying these issues will contribute to the development of personalized medicine for patients with clear cell renal carcinoma.

Abstract: It is important to note that maintaining adequate levels of nitric oxide (NO), the turnover, and the oxidation level of nitrogen are essential for the optimal progression of cellular processes, and alterations in the NO cycle indicate a crucial step in the onset and progression of multiple diseases. Cellular accumulation of NO and reactive nitrogen species in many types of tumour cells is expressed by an increased susceptibility to oxidative stress in the tumour microenvironment. Clear cell renal cell carcinoma (ccRCC) is a progressive metabolic disease in which tumour cells can adapt to metabolic reprogramming to enhance NO production in the tumour space. Understanding the factors governing NO biosynthesis metabolites in ccRCC represents a relevant, valuable approach to studying NO-based anticancer therapy. Exploring the molecular processes mediated by NO, related disturbances in molecular pathways, and NO-mediated signalling pathways in ccRCC could have significant therapeutic implications in managing and treating this condition.

Keywords: nitric oxide; signalling; ccRCC

Citation: Ene, C.D.; Tampa, M.; Georgescu, S.R.; Matei, C.; Leulescu, I.M.T.; Dogaru, C.I.; Penescu, M.N.; Nicolae, I. Disturbances in Nitric Oxide Cycle and Related Molecular Pathways in Clear Cell Renal Cell Carcinoma. *Cancers* **2023**, *15*, 5797. https://doi.org/10.3390/cancers15245797

Academic Editor: José I. López

Received: 13 November 2023
Revised: 3 December 2023
Accepted: 9 December 2023
Published: 11 December 2023

1. Introduction

It has recently been noted that the evolution of renal tumours is regulated at the molecular level, and the impact of metabolic remodelling on tumour biology is largely unknown [1–4].

Renal cell carcinoma (RCC) affects 400,000 patients annually worldwide, causing 2.4/100,000 deaths [5]. Clear cell renal carcinoma (ccRCC) is the most common type of kidney cancer, accounting for 70–75% of RCC cases [5,6]. ccRCC originates from tubular epithelial cells and is orchestrated by epigenetic alterations. Genomic loss of the Von Hippel-Lindau (VHL) tumour suppressor, arginase 2, argininosuccinate synthetase (enzymes involved in the urea cycle), the progression of arginine, glutamine, tryptophan, glutathione, cysteine/methionine metabolism, polyamine, and activation of HIF2a-mediated pro-oncogenic signalling are the most well-known metabolic disturbances in ccRCC [1,3,7].

ccRCC is a progressive metabolic disease in which tumour cells adapt to interconnected metabolic and epigenetic reprogramming, along with alterations in nitrogen homeostasis [5,8–10]. Tumours actively modulate the host's metabolism to increase their nitrogen supply and promote their growth and progression [4,10–13].

The regulation of nitrogen supply in renal tumours is a complex process and is influenced by disruptions in NO homeostasis, ureagenic cycle dysregulation, low levels of arginine, loss of VHL, mitochondrial dysfunction, HIF expression, disruption of branched-chain amino acid metabolism, nucleotide synthesis, the presence of endogenous competitive NOS inhibitors [1,3–6,12–14].

In this article, we analyze the current state of research on NO biosynthesis metabolites in ccRCC and the metabolic pathways that amplify NO production in the tumour microenvironment. Understanding the role of NO biosynthesis in the pathogenesis of ccRCC helps in identifying effective therapeutic means for the prevention and treatment of diseases associated with the alteration of the L-arginine–NO molecular pathway.

2. The Disruption of NO Homeostasis in ccRCC

NO has emerged as a molecule of interest in carcinogenesis and tumour progression due to its bimodal role in various cellular processes [4]. Tumour cells require distinct NO concentrations, promoting either a pro-tumoural phenotype or an anti-tumoural phenotype. Low to moderate levels may promote tumourigenesis, whilst higher levels would exert anti-tumour effects [8]. The effects of NO in cancer cells appear to depend on the origin of NO, the type, activity, and localization of NOS isoforms, concentration, and duration of NO exposure, temporal and spatial regulation at various levels of the NO cycle, NO-mediated signalling pathways, the metabolic phenotype of cells in the tumour microenvironment, and cellular sensitivity to NO. In the tumour microenvironment, low to moderate levels of NO (NO) derived from tumour and endothelial cells can activate angiogenesis, promoting an aggressive phenotype. Conversely, high levels of NO derived from M1 macrophages and Th1 lymphocytes can exert an anti-tumoural effect, providing protection against cancer [1,8,15,16]. Therefore, identifying the NO equilibrium state is particularly important for ccRCC biology.

2.1. The Disruption of NO Biosynthesis in ccRCC

Currently, there are two major pathways for NO generation in vivo: the L-arginine–NO pathway and the Nitrate–Nitrite–NO pathway (Figure 1) [2,4,17,18].

In the oxidative pathway, NOSs convert plasma L-arginine into NO in equimolar amounts. L-arginine comes from endogenous sources (de novo synthesis, protein turnover) and exogenous sources (diet). Microorganisms carry out the reductive nitrate–nitrite–NO processes at the oral and intestinal levels. Nitrates come from exogenous or endogenous sources (diet, water, environment). Under normoxia, the total daily production of NO in the entire body was estimated at 1100 micromoles. The relative contribution of the classical L-arginine–NO pathway represents 90% of the total NO. In comparison, the average rate of NO synthesis through the alternative nitrate–nitrite–NO pathway is approximately 10% [4,17]. Recent approaches in nephrology attempt to analyze ccRCC and autosomal dominant polycystic kidney disease (ADPKD) as arginine-auxotrophic metabolic diseases. These conditions are associated with impaired capacity for recycling or synthesizing intracellular arginine through the urea cycle pathway [19–21]. In malignancy, L-arginine functions as both an onconutrient and an immunonutrient [22]. A recent study has documented the prospective relationship between initial serum arginine concentrations and the risk of cancer in hypertensive participants. Higher serum levels of arginine demonstrate a significantly increased risk of cancer [23].

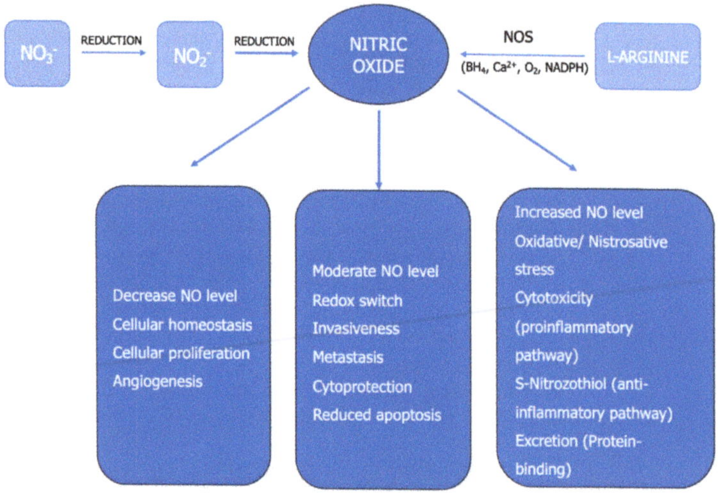

Figure 1. NO biochemistry. Concentration-dependent effects of NO in cancer. NO—nitric oxide, NOS—NO synthase, $NO_3{}^-$—nitrate, $NO_2{}^-$—nitrite, NADPH—nicotinamide adenine dinucleotide phosphate, BH4—6R-5,6,7,8-tetrahydrobiopterin.

In ccRCC, NO biosynthesis and the mechanisms that regulate the NO cycle are disrupted. At the renal level, NO levels are regulated through several mechanisms. These include the availability of L-arginine (affecting renal L-arginine biosynthesis, endothelial transport), the competition between NOS and other metabolic pathways for NO utilization, elevated circulating levels of ADMA (asymmetric dimethylarginine), an endogenous competitive NOS inhibitor (increased protein methylation, protein catabolism rate to supply free ADMA, reduced ADMA catabolism via DDAH-dimethylarginine dimethylaminohydrolase), and the distribution of NOS at the renal level [24,25].

Three isoforms of NO synthetases (NOSs) are known to be involved in NO synthesis: endothelial (NOS1 or eNOS), inducible (NOS2 or iNOS), and neuronal (NOS3 or nNOS) [16,26].

A variant of NOS in red blood cells (eNOS-RBC) has recently been studied for its activity in ischemia-reperfusion injury [15,16].

NOSs require nicotinamide adenine dinucleotide phosphate (NADPH), flavin mononucleotide (FMN), flavin adenine dinucleotide (FAD), and 6R-5,6,7,8-tetrahydrobiopterin (BH4) as cofactors. In the kidney, eNOS was identified in vascular endothelium and certain epithelia (thick ascending loop of Henle, collecting duct). At the same time, nNOS and nNOS mRNA were detected in the macula densa, efferent arterioles, collecting duct, Bowman's capsule, and iNOS was observed in the inner medullary collecting duct, immune cells, and tumour cells [27].

2.2. *The Disruption of NO Signaling in ccRCC*

The molecular mechanisms governing the ubiquitous nature of NO transduction and its impact on biological processes involve, on the one hand, the direct interaction between NO and potential cellular targets and, on the other hand, the formation of other reactive nitrogen species that possess signalling properties themselves [18].

The processes mediated by NO in the tumour environment involve either cyclic guanosine monophosphate (cGMP)-dependent or cGMP-independent reactions. In the cGMP-dependent pathway, NO binds to the heme component of soluble guanylate cyclase (sGC), promoting the conversion of guanosine triphosphate (GTP) to cGMP, activating protein kinases and phosphodiesterases and cyclic nucleotide channels. In the NO/sGC/cGMP pathway, low NO levels produced by constitutive NOSs regulate renal physiology. The phenomenon concludes rapidly after removing the initial stimulus from the tumour microenvironment, explained by the short half-life of NO, estimated at 0.1–2 s [12,15,28]. Disruption of NO/sGC/cGMP signalling results in reduced cGMP levels and the emergence of severe kidney pathologies [29].

In the cGMP-independent pathway, high levels of NO produced by iNOS induce oxidative and nitrosative stress, damage macromolecules, generate mutagenic and carcinogenic compounds (S-nitrosothiols, N-nitrosamines), halt the cell cycle, trigger antiapoptotic responses, senescence, and promote tumour progression. Recent quantified post-translational modifications in ccRCC (glycosylation, TDH, protein carbonylation, proteolytic cleavage, peroxidation, methylation, phosphorylation) accompany tumour development and metastasis promotion [3]. The disruption of the NO cycle in tumours is associated with the alteration of NF-kB, AP-1, and CREB pathways, as well as MAPK/PI3K, Ras, G-proteins, mitochondrial cytochrome c oxidase [12,15,28]. NO acts as a pleiotropic messenger, thus influencing the ccRCC progression.

Among the numerous NO-mediated signalling mechanisms, the interaction of NOSs–NO, the coupling of NOSs-$O_2(-)$, post-translational protein changes, and epigenetic regulation play a significant role in the ccRCC progression.

The NOSs-NO signalling, which has recently been revealed, affects NO bioavailability through cGMP-dependent mechanisms. Constitutive NOSs modulate the release of mediators from the cells for a short period of time and in small quantities. NOS3/cGMP facilitates tumour angiogenesis and signal transduction. Low levels of NO produced by constitutive NOSs prevent leukocyte adhesion, inhibit platelet aggregation, induce immune and inflammatory responses, and regulate cell growth, apoptosis, transduction, and survival. iNOS, expressed in epithelial and immune cells after induction by cytokines or bacterial products, generates large amounts of NO over the long term and cytotoxic products [12,15,28,30]. iNOS mRNA and protein expressed in A498 and A704 RCC cells under hypoxic conditions ensure the hypoxic adaptation of RCC and the ability to form tumour thrombi [31]. NOSi mRNA expressed by mouse RCC macrophages, RENCA, modulates tumour growth and angiogenesis through NO production via microRNA miR-146a [32].

The mechanism of NOS–NO uncoupling is associated with superoxide or hydroperoxide synthesis, leading to reduced NO generation. Disruption of NOS-NO signalling is caused by arginine and BH4 deficiency, NOS dimerisation failure, the presence of methylated and nitrated forms of arginine, glutathionylation, overexpression of ROS and RNS [18].

Post-translational changes mediated by NO (nitrosylation, nitrosation, nitration, transnitrosylation) can give rise to biomolecules with signalling properties. Nitrosylation (direct interaction between NO and transition metals) inhibits mitochondrial cytochrome C oxidase and inactivates Fe-SH and Fe-nonheme sites. Nitrosation (covalent binding of NO(+) to -SH and the formation of S-nitrosothiols) affects redox homeostasis.

Nitration (binding $-NO_2$ to organic substrates) modifies the activity of tyrosine in proteins, unsaturated fatty acids, and nucleotides [3,18]. S-nitrosylation plays various regulatory roles in metabolism, apoptosis, protein phosphorylation, and transcription factor activation [30]. S-nitroso groups (SNO) can be generated through transnitrosylation, the transfer of an NO fragment between thiols and S-nitrosothiols (thiol/nitrosothiol exchange) [33]. This mechanism is also the basis for denitrosylation of proteins, which is the necessary process to remove NO from S-nitrosothiols.

Transnitrosylation (transfer of an NO fragment between thiols and S-nitrosothiols) is a process that removes NO from S-nitrosothiols and constitutes a mechanism for denitrosylation of proteins and regeneration of protein thiol groups.

Denitrosylation is catalyzed by S-nitroso glutathione reductase (GSNOR), S-nitroso-CoA reductase (ScoR), and thioredoxin reductase (TrxR). S-nitrosylation is an inhibitory mechanism for crucial enzymes that are involved in cell metabolism [34].

Epigenetic regulation mediated by NO has gained considerable interest in recent years. Under neoplastic conditions, NO influences histone post-translational modifications, DNA methylation, and microRNA status [8,18]. Long non-coding RNAs affect the epigenetic landscape by regulating genes or post-translational modification of proteins, leading to abnormal signal transduction and the induction of malignant transformation in ccRCC [5].

2.3. Bimodal Effects of NO in ccRCC

NO levels can serve as an indicator of cellular homeostasis (Figure 1). NO is a primary regulator of tumour progression, with the ability to modulate multiple cellular processes dynamically. Lower concentrations (picomolar to nanomolar range) of NO are present under normal physiological conditions within a cell, but a sudden increase in concentration (micromolar range) results in the development of pathological processes. NO exhibits dichotomous effects on various crucial processes in cancer biology. It is shown to have both pro-tumourigenic and anti-tumourigenic effects, depending on the concentration, source of NO, type, activity, and localization of NOS isoforms, as well as the composition of the tumour microenvironment.

Low NO flux, similar to that generated by constitutive NOSs, maintains cellular homeostasis, drives cell cycle progression, promotes proliferation, neovascularization, angiogenesis, tumour growth and spread, processes governed by the activation of the HIF1α and VEGF signalling pathways [18,30]. Moderate NO levels maintain redox balance in the tumour microenvironment, promote increased invasiveness, metastasis, and cytoprotection, and inhibit apoptosis [18,30].

High NO levels, similar to those generated by iNOS (inducible NOS), induce oxidative and nitrosative stress [18,30]. Elevated NO can lead to DNA damage (single and double-strand breaks, impaired repair), mitochondrial dysfunction, lipid interactions, and the activation of major oncogenic pathways that enhance survival, proliferation, and metastasis in aggressive tumours [18,30].

Additionally, following oxidative reactions, NO can give rise to reactive metabolites, causing the disruption of intracellular and intercellular signalling and the inactivation of transcription factors and effectors within various transduction pathways [5,30].

These bimodal effects of NO emphasize the importance of careful regulation of NO levels, which can be a critical therapeutic target in managing ccRCC.

Similarly, the effects of NO in carcinogenesis are influenced by the metabolic phenotypes of the cells that constitute the tumour microenvironment. The tumour microenvironment consists of cancer cells, stromal cells, immune cells, and endothelial cells, each with distinct metabolic characteristics [9]. NO produced by cancer cells promotes cancer progression, increasing the aggressiveness of these cells in vivo. NOS2 in immune mediators has anti-tumour and pro-inflammatory effects. NO derived from endothelial cells mediates the elimination of disseminated tumour cells, and NO from stromal cells inhibits tumour growth [30].

The subcellular localization and activation of NOSs influence the selectivity of NO targets. NOS1 and NOS3, constitutive isoforms, generate nanomolar concentrations of NO over very short periods, while NOS2, the inducible isoform, produces micromolar concentrations of NO over more extended periods. Most evidence suggests an increase in total NO production in ccRCC patients. This could reflect the activation of iNOS in neoplastic and inflammatory conditions [3,4] The regulation of NOS2 synthesis and production in the tumour microenvironment affects tumour progression.

The level of NO can act as a redox switch. Small amounts of NO regulate anti-inflammatory and antioxidant responses under normal physiological conditions. In contrast, high concentrations of NO in tissues regulate the function of iNOS and promote the oxidation of macromolecules [18].

The NO level is considered to be an indicator of cellular redox status and has an approximate value of 300 nM under physiological conditions [18,30]. Different NOS isoforms in the tumour microenvironment have distinct cellular targets. NO derived from NOSc supports carcinogenesis, while NO derived from NOSi becomes cytotoxic to cancer cells (Figure 1). NO plays a cytostatic/cytotoxic role in cancer cells through multiple mechanisms, including metabolic reprogramming, DNA synthesis inhibition, apoptosis regulation, and necrosis.

NO mediates tumour promotion through genotoxic reactions (DNA damage), anti-apoptotic activity (regulation of caspases and death-associated proteins), regulation of the NO/NOS2–p53 axis, which is critical in defining the apoptotic mechanisms of tumours, upregulation of angiogenesis (production of proangiogenic factors), metastatic actions (overexpression of MMPs and VEGF), and suppression of the immune response (reduction in leukocyte infiltration) [30,35].

NOSs–NO plays a significant role in tumour suppression [9,35]. iNOS is essential for the development, maturation, and differentiation of T cells, B cells, monocytes, and dendritic cells [35]. In response to tumour progression, NO and RNS stimulate the passage of immune cells from the bloodstream around cancer cells, the polarization of tumour-associated macrophages (TAMs), the transactivation of eNOS/iNOS, the stabilization of HIF1α, the transcription of proangiogenic factors, the activation of S-nitrosylation, the inactivation of Janus kinase 3, early response kinase, and protein kinase B, which prevent the IL-2 response.

Immune-activating M1 macrophages and immune-suppressing M2 macrophages differentiate based on the presence of arginase and iNOS. Both arginase and iNOS catalyze arginine in different ways (Figure 1. M1 macrophages, with high iNOS expression, produce large amounts of NO and use NO/RNS for cytotoxic clearance. The host's NO/NOSi-dependent defence mechanism associated with M1 macrophages provides survival benefits to cancer cells and supports malignancy [9,35].

NO adapts the metabolism to the needs of tumour cells by inhibiting metabolic enzymes and regulating mitochondrial modulators involved in cancer-associated growth, antioxidant responses, and metabolic rewiring. NO regulates the electron transport chain complexes (cytochrome c oxidase, complex IV), enzymes in the tricarboxylic acid cycle (aconitase, α-ketoglutarate dehydrogenase, succinate dehydrogenase), enzymes involved in fatty acid oxidation, branched-chain amino acid metabolism, pyruvate kinase M2, and the stability of the TRAP1/SIRT3 complex (Tumour Necrosis Factor Receptor-Associated Protein 1/sirtuin 3). At low concentrations, NO competes for the catalytic site, whereas at high concentrations, inhibition occurs through oxidative post-translational modifications. NO generated by NOS1 activates SIRT3, while NO generated by NOSi inactivates SIRT3. These mitochondrial events simultaneously provide the capacity to cope with oxidative stress and adapt to metabolic changes [34–39].

2.4. Quantitative Determination of NO Metabolites in Biological Samples

The quantitative determination of NO in vitro and in vivo is particularly challenging due to its rapid diffusion through membranes, biological interferences, a pronounced tendency to auto oxidize, short lifespan (seconds), low equilibrium concentration (nanomoles), and the heightened reactivity of NO and its natural derivatives (NO, NO_2, N_2O_3, NO_2^-, which rapidly react with haemoglobin, glutathione, sulfhydryls, and unsaturated fatty acids).

Various techniques have been developed for measuring NO and its metabolites in biological systems. The primary methods for quantifying NO and its derivatives include [40,41]:

1. Spectrophotometry: Utilizing azoic dyes and the Griess test.
2. Fluorescence: Employing reagents such as Diaminofluorescein (DAF-2).
3. Luminescence: Using luciferin–luciferase assays.
4. Electrochemical: Employing amperometric NO microelectrodes.
5. Tandem Mass Spectrometry: Including mass spectrometry in tandem (MS/MS) and electrospray ionization mass spectrometry (ESI-MS/MS).
6. Liquid Chromatography-Mass Spectrometry (LC-MS): Combining liquid chromatography with mass spectrometry
7. Electron Paramagnetic Resonance (EPR): Employed for NO detection.
8. HPLC: High-Performance Liquid Chromatography is another technique.

9. Antibody-Based Methods: These encompass immunohistochemical, immunoblotting, and enzyme-linked immunosorbent assays (ELISAs).
10. Chemiluminescence: A method based on the detection of light emission.
11. UV-Visible Absorption Spectrum: Measuring the absorption of UV-visible light.

Accurate identification and quantification of the rate of NO formation or degradation, precursors, derived species, molecular targets, and the consequences of disturbances in the NO cycle and related molecular pathways, alongside pharmacological and omics methods, will help solve the role of NO in ccRCC (Table 1).

Table 1. No Synthesis Metabolites in ccRCC.

NO Parameters (References)	Biological Systems	Results	Conclusions
Calcium-dependent and calcium-independent NO synthetase [42]	Human kidney/RCC, proximal tubular cell lines HN4, HN51.	Calcium-dependent NOS activity, identified in all the samples studied, was downregulated in RCC compared to non-malignant renal tissues studied; calcium-independent NOS activity was inconsistently expressed in the renal tissue.	NO exerted cytostatic effects on cultured renal cells.
NOS1, NOS2, NOS3 [43]	Non-neoplastic renal tissues and RCC	In non-neoplastic tissues, NOS3 immunoreactivity was increased and NOS2 was reduced compared to RCC. The NOS expression was correlated with tumour size and a poor prognosis.	NOS3 as a predictive factor in RCC
NOS, sGC, nitrotirosine [44]	Normal and tumoural renal tissue (benign and malignant tumours).	NOS1 is downregulated in malignant tissues and associated with the tumour grade; sGC is present in all renal tumours; nitrotyrosine is present in normal renal parenchyma and tumour tissues.	Autocrine signalling of NO is similar in normal and non-malignant renal tissues and altered in malignant tissues
Nitrites [45]	Serum (apparently healthy women diagnosed with RCC)	Elevated serum level in patients with RCC	Elevated serum nitrite levels are associated with a low risk of renal cancer.
ASS1, ASL, Arg2 [46,47]	RCC tissue samples and control	mRNA and ASS1, ASL, Arg2 activity are reduced in RCC vs. normal kidney Altered urea cycle-metabolic pathway in RCC.	Attenuation of the cytotoxic effects of NO. ASS1, ASL, Arg2 -metabolic suppressors in RCC.
NOSi-ARN [31]	RCC and control tissue samples	mRNA and iNOS protein present in tumour thrombi in patients with RCC and in A498 and A704 cells under hypoxic conditions.	It mediates the formation of tumour thrombi and hypoxic adaptation.
Arg2, ASS1 [46,48]	Normal and malignant renal cell lines	The expression of enzymes in the urea cycle is downregulated in RCC compared to the control. Deficiency of enzymes in the urea cycle disrupts polyamine synthesis, conservation of pyridoxal phosphate, arginine auxotrophy, infiltration of cytotoxic T cells in the tumour tissue, and immunosuppression in the tumour microenvironment.	Arg2 and ASS1 are potential metabolic suppressors of renal tumourigenesis.
ASS1, ADI (E.C.3.5.3.6) [35]	Biopsy samples, animal models, cell lines.	Low or undetectable ASS1 in RCC, present in normal proximal tubule epithelium. Exogenous ADI (arginine deiminase) determines antiproliferative and antiangiogenic effects in vivo on RENCA tumour cells and extends the survival of tumour-bearing mice.	Arginine deprivation via ADI—an antitumour strategy in RCC
Spermine, spermidine [49]	Normal and malignant human renal tissue	Spermidine levels and spermidine/spermine ratio increase; normal tissue < differentiated RCC < poorly differentiated RCC. The other polyamines do not show differences between normal tissues, tumours, and metastases.	Polyamines—biochemical markers for the malignancy of RCC
Diamine, spermidine, spermine [50]	Tissue, urine, blood	Elevated levels are correlated with the progression of RCC	Polyamines—tumour markers in RCC

Table 1. *Cont.*

NO Parameters (References)	Biological Systems	Results	Conclusions
Agmatinase (E. C.) [51]	Normal and malignant renal tissue	The expression and mRNA of agmatinase are decreased in RCC compared to benign renal tumours. Accumulated agmatine stimulates NOS3 and NOS2, leading to NO synthesis.	Reduced agmatinase increases the cytotoxic activity of NO in RCC
RNS, NO_2^- [52]	Cell cultures	JS-K, a NO donor, stimulates the increase in ROS (Reactive Oxygen Species) and RNS (Reactive Nitrogen Species), the reduction in GSH/GSSG (glutathione redox status), the increase in pro-apoptotic proteins (Bak, Bax), and the reduction in anti-apoptotic proteins (Bcl-2) in RCC (Renal Cell Carcinoma). JS-K induces apoptosis in cancer cells by modulating the production and signalling of NO, MAPK (Mitogen-Activated Protein Kinase), the ubiquitin–proteasome pathway, and the β-catenin/T-cell factor (TCF) signalling pathway	NO released by JS-K induces apoptosis in renal carcinoma cells by increasing the levels of ROS/RNS and induces chemosensitivity of tumour cells to doxorubicin
NOS, cGMP [53]	Experimental and human tumours.	The NOS enzymes are overexpressed in tumour tissues.	NO metabolites correlate with angiogenesis and tumour aggressiveness.
Dietary nitrite, nitrate intake [54]	Evaluation of nitrates and nitrites in food sources (41 articles, 13 types of cancer).	Nitrates from plant sources and nitrates in general do not affect the development of renal cancer. Nitrites from processed meat are associated with an increased risk of renal cancer < pancreatic cancer < thyroid cancer, stomach cancer, glioma, and others.	The nitrites/nitrates intake has specific effects on the type and site-specific risk of cancer
Phytonutrients [55–60]	Epidemiological studies on the role of dietary factors.	Phytonutrients play an essential role in cancer prevention. Dietary sources of nitrites and nitrates have a role in immunity and vascular function. Dietary sources of nitrites include vegetables, fruits, and processed meat.	Vitamin C inhibits endogenous nitrosation.
Red and processed meat [61]	Meta-analysis (12 case-control studies, 16 cohorts)	No statistically significant data were obtained between the consumption of red and processed meat, individual variables (BMI, smoking, total energy intake), and the development of renal cancer.	An independent relationship between meat consumption and the risk of renal cancer was not evident.
Red and processed meat [62]	Meta−analysis (23 eligible publications) on the association and impact of red meat consumption on RCC	Positive relationship between consumption of beef, salami, ham, bacon, sausages, hamburgers, and renal cancer.	Statistically significant positive association between red meat consumption and RCC
Dietary factors [63,64]	Report on 22 meta−analyses (566 publications).	No suggestive or convincing evidence between the consumption of foods, beverages, alcohol, macronutrients, micronutrients, and the incidence of RCC	The intake of vegetables and vitamin C is associated with the risk of RCC
Serum NO_2^-, NO_3^- [65]	RCC patients and control patients	No significant differences between patients and controls. Variations depending on the tumour grade.	NO exerts immunoregulatory effects in RCC
Arginase 2 [66]	Murin renal cell lines, normal and neoplastic.	Arginase 2 rapidly metabolizes L-arginine, suppresses tumour growth, and reduces the expression of CD3zeta.	Arginase 2 modulates the function of T cells, depleting arginine.
Arginase 2 [67]	Peripheral blood of metastatic RCC and control patients	Myeloid suppressor cells producing arginase present in patients with metastatic ccRCC	Arginase 2 regulates the availability of arginine.

Table 1. *Cont.*

NO Parameters (References)	Biological Systems	Results	Conclusions
Arginase 2 [7]			Arginase 2 supports the growth of ccRCC
BCAA, BCAT, ASS1 [1,21,68–70]	Cultured primary and metastatic renal cancer cells (omic study)	Transcriptionally suppressed BCAA catabolism, overexpressed BCAT, urea cycle, glutathione, cysteine/methionine, arginine, glutamine, tryptophan, reactivated polyamines in ccRCC show metabolic flexibility during tumour progression and offer invasive potential.	Altered metabolic advantage for cancer cell survival
SIRTs [34,36,71–73]	ccRCC cells, cell lines, genetic models, pharmacological models, omic studies, computational	SIRTs-1,3,6,7 maintain renal homeostasis. SIRT1 regulates NOSe in glomerular cells. NO regulates the stability of the SIRT3/TRAP1 complex. SIRT3 has dual effects on tumour growth. SIRT3 has antioxidant and anti-inflammatory effects in kidney disease. SIRT4 inhibits glutamine metabolism	Activating SIRTs before tumour initiation is a preventive strategy. NO/SIRT3 regulates mitochondrial biogenesis in ccRCC and cell sensitivity to antitumour therapy.

We observed systemic alterations in the levels of NO metabolites (NO_3^-, NO_2^-, $NO_x = NO_3^- + NO_2^-$) and overproduction of nitrotyrosine in cancer patients. The aberrant activation of NO signalling induced by the hypoxic microenvironment is associated with tumour progression [3,4].

The profile of NO (NO) in tumour cells and the microenvironment influences the rate of cancer progression, therapy effectiveness, and patient prognosis. In a recent study, it was noted that markers of nitrosative stress (3-nitrotyrosine, nitrite/nitrate) were elevated in the ccRCC group, correlating with the increased pathological stage of the tumour (TNM, histological grade, angioinvasion) [74].

3. Dysregulated Ureagenic Cycle—A Distinctive Sign in ccRCC

The fact that the dysregulation of the ureagenic cycle represents a distinctive feature of ccRCC is supported by the low levels of mRNA and the altered expression and function of representative enzymes (ASS1, ASL, ARG2) in tumour samples compared to normal kidney tissue [46]. The urea cycle is a metabolic pathway responsible for converting excess nitrogen from ammonia and aspartate into urea (Figure 2).

Disruptions in the urea cycle arise from enzymatic deficiencies and can additionally be induced by the dysfunction of transporters responsible for transferring mitochondrial aspartate into the cytosol. The aforementioned changes in enzyme expression and metabolites within the urea cycle at different stages of cancer development demonstrate the dynamic alterations in this metabolic cycle [75–78].

Some enzymes in the urea cycle are repressed in primary renal cancer cells, while ASS1 is epigenetically reactivated in metastatic populations [1]. Sensitive to the arginine levels in the microenvironment, the selective expression of ASS1 provides metastatic renal cancer cells with the ability to utilize nitrogen from BCAA catabolism to produce arginine [1]. Reduced ARG2 activity promotes ccRCC tumour growth through at least two distinct mechanisms: preserving pyridoxal phosphate and preventing the accumulation of toxic polyamines [7].

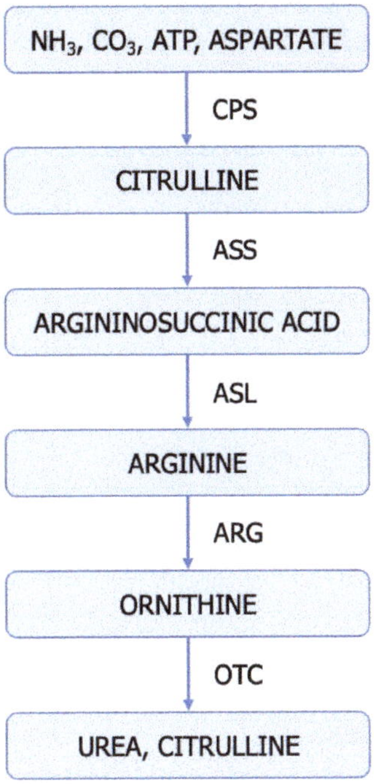

Figure 2. The ureagenic cycle: CPS-1—carbamoyl-phosphate synthetase I (E.C. 6.3.4.16); OTC, ornithine transcarbamylase (E.C. 2.1.3.3); ASS—argininosuccinate synthetase (E.C. 6.3.4.5); ASL—argininosuccinate lyase (E.C. 4.3.2.1); ARG—arginase (E.C. 3.5.3.1).

In conclusion, the dysregulation of the urea cycle in cancer is accompanied by the inactivation of ASS1, activation of CAD (carbamoyl-phosphate synthetase 2, aspartate transcarbamylase, and dihydroorotase), overexpression of NO, enhanced pyrimidine synthesis at the expense of purine, mutagenesis, increased cell proliferation, invasiveness, and survival, poor prognosis, and increased immunotherapy efficacy [1,10,76,78]. Disruption of urea cycle enzymes in ccRCC contributes to the rerouting of carbon and nitrogen towards the generation of anabolic biomass [46]. Dysregulation of the urea cycle is a metabolically advantageous phenomenon for cancer cell proliferation.

4. The Upregulation of Glutamine: An Alternative Source of Nitrogen for ccRCC

Amino acids such as glutamine, arginine, aspartate, alanine, glycine, and serine serve as significant nitrogen sources for cancer cells, immune cells, endothelial cells, and stromal cells within the tumour microenvironment (Figure 3) [9]. Cancer cells metabolize glutamine (Gln) differently than normal cells, requiring an enzyme known as glutaminase (E.C. 3.5.1.2). ccRCC cells exhibit higher levels of glutamine, glutamate, and the SLC1A transporter compared to normal renal tissue [79]. Hypoxic cells, particularly those with a VHL deficiency, employ glutaminase to obtain glutamate, which indirectly plays a pivotal role in pyrimidine and lipid synthesis as well as antioxidative processes (Figure 2). Human ccRCC renal cells utilize HIF2α-mediated reductive carboxylation to sustain de novo pyrimidine biosynthesis [80–82]. Recent studies suggest a reciprocal regulation between Gln and NO [83]. Gln, an essential amino acid, competitively inhibits Cit availability for interacting with L-Arg, thereby regulating macrophage NO production capacity.

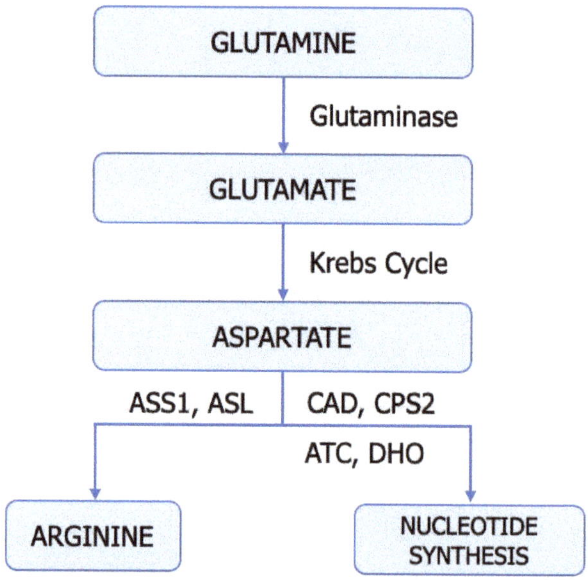

Figure 3. Utilization of glutamine in tumours. ASL—argininosuccinate lyase (E.C. 4.3.2.1); ASS1—argininosuccinate synthase (E.C. 6.3.4.5), CAD—trifunctional protein, CPS 2—carbamoyl-phosphate synthase 2 (E.C. 6.3.5.5), ATC—aspartate transcarbamylase (E.C. 2.1.3.2) DHO—dihydroorotase (E.C. 3.5.2.3).

Moreover, Gln modulates NO synthase to catalyze the conversion of Arg to NO. Glutamine appears to be a significant modulator interfering with citrulline-mediated NO production [80,81,84]. On the other hand, activated glutamate receptors and NO reduce glutamine synthetase (E.C. 6.3.1.2) activity, while Gln inhibits ASS activity and NO synthesis. Through this mechanism, glutamine decreases intracellular arginine concentration and NO release. The importance of citrulline and glutamine concentrations in NO synthesis must be interpreted with caution, as both metabolites can be modulated through independent NOS mechanisms [83]. As a newly discovered molecule, mitochondrial protein Sirtuin 4 (SIRT4) has been linked to alternative glutamine metabolism and regulation of the tumour microenvironment. In ccRCC cells, SIRT4 promotes apoptosis by increasing ROS. Down-regulation of SIRT4 in ccRCC promoted HO-1 expression in hypoxic cells, counteracting the pro-apoptotic effect of SIRT4. Moreover, SIRT4 regulates ROS and HO-1 expression through Akt and P38MAPK phosphorylation in ccRCC [5].

The nucleotide imbalance in tumours is associated with multiple transversions that propagate from DNA to RNA and proteins, leading to the production of immunogenic neoantigens [10]. Pyrimidine synthesis plays a significant role in carcinogenesis, affecting the prognosis of patients and their response to immunotherapy.

5. Cellular Arginine Depletion—A Proliferation Strategy in ccRCC

A significant metabolic defect in tumour biology is represented by the altered intrinsic ability of cancer cells to synthesize arginine. In tumour metabolism, arginine plays a role in signalling, epigenetic regulation, and immunomodulation [14,85]. Arginine is a conditionally essential amino acid synthesized under conditions of rapid cell growth through two pathways: the citrulline–NO cycle and the intestinal–renal axis, utilizing ASS and ASL. Arginine is metabolized into NO and citrulline by NOS, into agmatine and CO_2 by ADC, into ornithine and urea by arginase, into creatine by AGAT, and citrulline and NH_3 by ADI (Figure 4) [22].

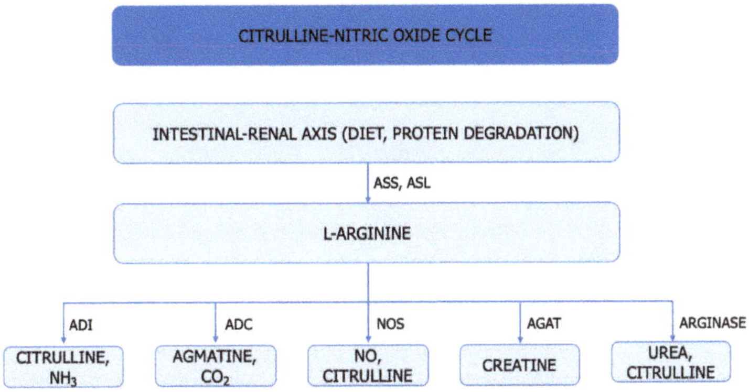

Figure 4. Arginine Metabolism. ASS (E.C. 6.3.4.5: argininosuccinate synthetase, ASL (E.C. 4.3.2.1): argininosuccinate lyase, ADI (E.C. 3.5.3.6.): arginine deiminase, ADC (E.C. 4.1.1.19) arginine decarboxylase, NOS (E.C. 1.14.13.39): NO synthase, AGAT (E.C. 2.1.4.1.) arginine:glycine amidinotransferase.

Arginase 1 is preferentially expressed in the liver, while arginase 2 is primarily identified in the kidney. In renal carcinogenesis, arginine supports rapid tumour growth and induces T-cell dysfunction [66,75]. Arginine regulates NO production in cancer cells [21].

Arginine regulates NO production in cancer cells [21]. In renal carcinogenesis, arginine facilitates rapid tumour growth and induces T-cell dysfunction [66,75].

ASS and ASL, which are cytosolic enzymes involved in nitrogen metabolism, impact ccRCC growth by depleting cellular aspartate reserves and disrupting pyrimidine production. Loss of ASS1 promotes cancer proliferation by diverting its substrate, aspartate, toward pyrimidine synthesis (carbamoyl-phosphate synthetase 2 (CPS2), aspartate transcarbamylase (ATC), and dihydroorotase [10]. Metastatic renal cancer cells reactivate ASS1, an enzyme suppressed in primary ccRCC, in order to maintain the invasive potential of metastatic renal cancer cells in vitro and in vivo and to modulate the sensitivity of metastatic cancer cells to arginine depletion [1].

Arginase 2, preferentially expressed in the kidneys, plays a significant role in the proliferation of tumour cells and the reduction in L-arginine availability [66,67]. Reduced ARG2 activity has been found to promote ccRCC tumour growth through multiple mechanisms. Arginase 2 suppresses renal carcinoma progression by depleting the biosynthetic cofactor pyridoxal phosphate and increasing polyamine toxicity.[7] Cancer cells rely on glutamine for survival and proliferation. VHL loss-dependent reprogramming of Arg is necessary to maintain the NO reservoir in cancer cells [7].

6. Hyperammonemia in ccRCC

The catabolism of glutamine and other amino acids is accompanied by the secretion of ammonia, leading to the loss of amino groups from the cell and the accumulation of ammonia in the tumour microenvironment [47,48,78,86]. Extracellular ammonia is viewed as either a toxic cellular by-product of amino acid metabolism that needs to be metabolized into a non-toxic form, such as urea, for excretion from the body or recycled into the central amino acid metabolism to maximize nitrogen utilization and support tumour biomass [47,78,87].

The accumulation of ammonia has allowed glutamate dehydrogenase to function in reductive amination, accelerating the incorporation of nitrogen from ammonia back into amino acids (Figure 4).

The progression of ccRCC depends on changes in ammonia metabolism [7,48]. Ammonia accumulates in the tumour microenvironment and induces metabolic reprogramming of T cells. Increasing ammonia clearance reduces tumour size, enhances survival, reactivates

T cells, and improves the effectiveness of anti-PD-L1 therapy [7,48]. Recently, it has been reported that during axitinib treatment for metastatic renal carcinoma, hyperammonemia developed due to associated thyroid disorders [88].

Ammonia activates NO production by stimulating arginine uptake and regulating the distribution of ADMA and SDMA. The intracellular L-Arg/ADMA ratio modulates iNOS activity and NO levels [89]. These findings warrant further investigations into the role of ammonia as a modulator of NO production.

7. The Reduction in BCAA Catabolism in ccRCC

Branched-chain amino acids (BCAA), namely valine, leucine, and isoleucine, belong to a group of essential amino acids. BCAAs can be directly incorporated into proteins or participate in other essential metabolic pathways for tumourigenesis [14,90]. In VHL-deficient renal cancer cells, BCAA catabolism represents an indirect source of nitrogen for nucleotide and non-essential amino acid biosynthesis (Figure 5) at all stages of tumour evolution [1,14]. BCAA catabolism in renal cancer cells is linked to the glutamate–glutamine axis, transcriptional sensitivity to VHL restoration, epigenetic modification, enhanced tumour immunity, and ferroptosis [14]. Hypoxia suppresses BCAA catabolism in specific tissues but regulates the expression of SLC7A5 and BCAT1 in tumours [40,62]. BCAT1 and BCAT2, which are responsible for BCAA degradation, can be considered immunosuppressive factors affecting the cancer cell's survival ability via the NO cycle (Figure 6) [91].

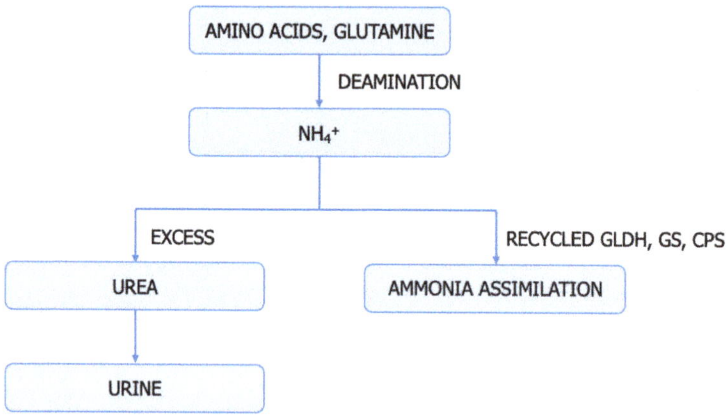

Figure 5. Ammonia metabolism. GLDH—Glutamate dehydrogenase (E.C. 1.4.1.3), GS—Glutamine synthetase (E.C. 6.3.1.2), CPS—Carbamoyl phosphate synthetase (E.C. 6.3.4.16).

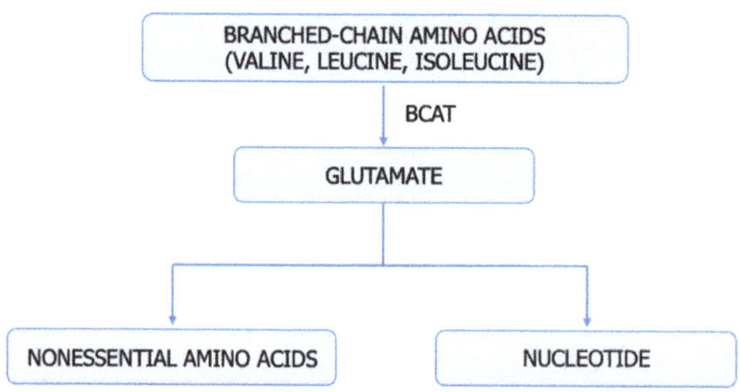

Figure 6. BCAA metabolism in cancer. BCATs—branched-chain amino acid transaminases (E.C. 2.6.1.42).

The decrease in glutamate derived from BCAAs can be compensated by increasing glutaminase activity, which generates glutamate from glutamine. The inhibitory effect of BCAAs on lipid peroxidation and NO scavenging activity allows the development of antioxidant and anti-inflammatory products in the pharmaceutical industry [92].

8. Endogenous Inhibitors of NO Synthesis

Methylarginines and DDAH enzymes are endogenous modulators of NO production. Methylarginines are endogenous metabolites obtained through the post-translational N-methylation of arginine residues incorporated into proteins, catalyzed by protein methyltransferases (PRMT), and released into the cytosol following proteolysis. Cationic amino acid transporters (CAT1, CAT2A, and CAT2B) facilitate the transport of methylarginines across cell membranes. The primary methylarginines include monomethylarginine (NMMA), asymmetric dimethylarginine (ADMA), and symmetric dimethylarginine (SDMA). ADMA and NMMA compete with L-arginine for binding to the active site of NOS, acting as competitive inhibitors for all isoforms. SDMA competes with CAT transporters for L-arginine, indirectly exerting inhibitory effects on NO synthesis [3,4,24,26,89].

Enzymes dimethylarginine dimethylaminohydrolases (DDAHs) metabolize ADMA and NMMA into L-citrulline, dimethylamine, or monomethylamine. They are vital components in maintaining the homeostatic control of NO [12,26,93]. The DDAH/ADMA/NO pathway is summarized in Figure 7.

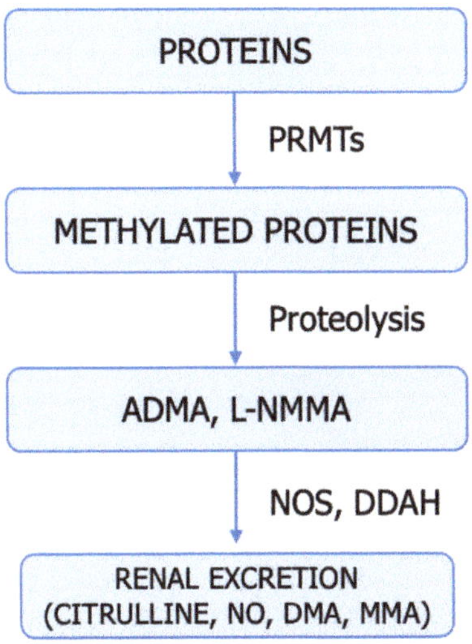

Figure 7. PRMT–DDAH–ADMA–NO Axis. PRMT1 (E.C. 2.1.1.319)—protein–arginine–methyltransferase, DDAH (E.C. 3.5.3.18)—dimethylarginine dimethylaminohydrolase, NOS (E.C. 1.14.1,39.)—NO synthase, ADMA—asymmetric dimethylarginine, NMMA—monomethylarginine, DMA—dimethylamine.

The pathophysiology of increased ADMA and SDMA in malignancies can have a complex origin. The rate of synthesis and degradation, inflammation, oxidative stress, antioxidants, redox status, proteinuria, endothelial function, functional activity of PRMTs, and DDAHs influence the circulating levels of methylarginine [3,4,24,89].

The dysregulation of the DDAH/ADMA/NO pathway, resulting in locally increased NO availability, is often associated with promoting tumour angiogenesis, growth, invasion, and metastasis. These findings warrant further investigation into the role of the PRMT–DDAH–ADMA axis in NO production [12,89].

9. The Inactivation of VHL and the Accumulation of HIFs—Essential Characteristics of ccRCC

ccRCC, whether sporadic or genetic, is characterized by frequent mutations in genes located on chromosome 3p, including VHL, PBRM1, BAP1, SETD2 H3K36, KDM6A, KDM5C, PTEN, mTOR, PIK3CA, TP53 [6]. The inactivation of VHL is accompanied by the accumulation of hypoxia-inducible factors (HIFs) in the tumour microenvironment, their interaction with NO synthases, the regulation of intratumoural NO production, and the development of renal tumours (Figure 8) [94].

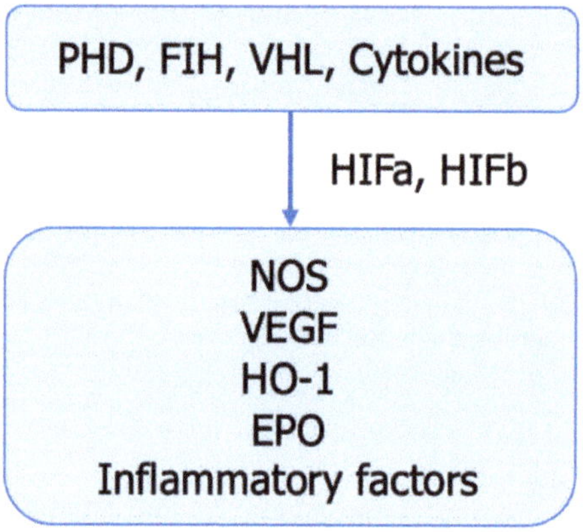

Figure 8. VHL—HIFs-NOS interaction. HIF—hypoxia-inducible factors; PHD—prolyl-4-hydroxylases (E.C. 1.14.11.2); FIH—factor-inhibiting HIF, VHL—von Hippel—Lindau protein, EPO—erythropoietin, VEGF—vascular endothelial growth factor, HO-1—heme oxygenase-1 (E.C. 1.14.14.18), NOS—NO synthase (E.C. 1.14.1,39), GLUTs—glucose transporters.

HIFs (HIF1, HIF2, HIF3) are heterodimers composed of alpha and beta subunits with distinct spatial distributions, specific inductions, and well-defined biological roles during tumour hypoxia (Figure 8). HIF-1α is primarily located in the renal tubular epithelium, while HIF-2α is mainly found in stromal and endothelial cells. HIFs play a dual role in the hypoxic microenvironment of tumour cells. HIF-1α is involved in the acute phase, while HIF-2α and HIF-3α are more frequently involved in chronic hypoxia. HIF-2α is considered an oncogene in ccRCC, whereas HIF1α likely has a tumour-suppressive function. However, a finer balance between HIF1α and HIF2α has also been proposed. HIF1α is essential in tumour initiation and glucose metabolism reprogramming, while HIF2α regulates biosynthetic pathways such as lipid metabolism, ribosome biogenesis, and the transcriptional activity of numerous factors. HIF-α combines with HIF-β in the nucleus to mediate the transcription of target genes. Moreover, both HIFs are crucial for the immune reprogramming of ccRCC tumours [6,94].

Many studies have suggested that HIF-1α can inhibit tumour growth while HIF-2α promotes tumour growth and metastasis [94]. HIF activation suppresses ASS1 and redirects aspartate toward nucleotide biosynthesis. In the metastatic population [1]. NOSs (endothelial, neuronal, inducible), upregulated by HIFs, promote NO release into the tumour microenvironment.

10. NO-Based Therapy for ccRCC

The interactions between cancer cells and the tumour microenvironment play crucial roles in the progression of ccRCC. By producing growth factors or cytokines and accentuating hypoxia and necrosis, RCC cells can promote the attraction and activation of non-tumour cells. Recent studies have identified a subpopulation of CD133+/CD24+ cells in ccRCC specimens that exhibited self-renewal capacity and clonogenic multipotency.

These cells are referred to as "cancer stem cells" (CSC), utilizing signalling pathways similar to those that control cell fate during early embryogenesis [95,96]. As discussed in this article, the NO/NOS system is interconnected with various tumour-promoting processes in the tumour microenvironment. In recent years, numerous studies have reported the design and development of NO-based nanomedicines, with the intention of being used in cancer treatment. In this regard, current and future objectives of pharmacotherapy are focused on the development of NO donors encapsulated in liposomes, antibody-NO conjugates, NO donors encapsulated in exosomes, iNOS overexpression (gene therapy), and tissue-selective donors that release NO in a controlled manner. Delivered at high concentrations, NO inhibits tumour growth and enhances apoptosis/ferroptosis of tumour cells by generating both reactive oxygen species and reactive nitrogen species [97,98].

11. Conclusions and Future Perspectives

In recent years, the study of the tumour microenvironment has gained significant attention [2,4,13]. Studies on tumour microenvironment metabolism have revealed that altering cells' metabolic profiles provides a survival advantage for growth and survival under challenging conditions.

Excessive NO production is primarily attributed to inducible NO synthase (iNOS) induction during oxidative stress and inflammatory states, which play a crucial role in ccRCC development. This article observed that the reprogramming of NO metabolism and the failure of immune surveillance in preventing malignant conditions are crucial factors in ccRCC progression.

Patients with ccRCC develop disturbances in most of the NO synthesis metabolites, worsening with ccRCC progression. NO metabolites may serve as promising markers for ccRCC stage and severity. Further research is needed to establish their clinical potential. NO affects both tumour and immune metabolism and exerts a significant regulatory effect on progression and immunotherapy. Therefore, the study aims to determine the metabolic characteristics of NO metabolites, which could help identify molecular features that provide opportunities for targeted metabolism. Clarifying these issues will contribute to developing personalized medicine for ccRCC patients. In recent years, numerous studies have reported the design and development of NO-based nanomedicines. These include NO donors encapsulated in liposomes, antibody–NO conjugates, NO donors encapsulated in exosomes, ligands selective for NOS isoforms, and tissue-selective donors that release NO in a controlled manner.

Author Contributions: Conceptualization, C.D.E. and I.N.; writing—original draft preparation, C.D.E., I.N., I.M.T.L. and C.I.D.; writing—review and editing, M.T., C.M., M.N.P. and S.R.G.; supervision, I.N., C.D.E., M.T. and S.R.G. All authors have read and agreed to the published version of the manuscript.

Funding: This research received no external funding.

Conflicts of Interest: The authors declare no conflict of interest.

References

1. Sciacovelli, M.; Dugourd, A.; Jimenez, L.V.; Yang, M.; Nikitopoulou, E.; Costa, A.S.H.; Tronci, L.; Caraffini, V.; Rodrigues, P.; Schmidt, C.; et al. Nitrogen partitioning between branched-chain amino acids and urea cycle enzymes sustains renal cancer progression. *bioRxiv* **2021**. [CrossRef]
2. Ene, C.D.; Nicolae, I. The Inflammatory Profile Orchestrated by Inducible Nitric Oxide Synthase in Systemic Lupus Erythematosus. *J. Pers. Med.* **2023**, *13*, 934. [CrossRef] [PubMed]
3. Ene, C.D.; Penescu, M.N.; Georgescu, S.R.; Tampa, M.; Nicolae, I. Posttranslational modifications pattern in clear cell renal cell carcinoma. *Metabolites* **2021**, *11*, 10. [CrossRef]
4. Ene, C.D.; Nicolae, I. Hypoxia-Nitric Oxide Axis and the Associated Damage Molecular Pattern in Cutaneous Melanoma. *J. Pers. Med.* **2022**, *12*, 1646. [CrossRef] [PubMed]
5. Zhang, H.; Yu, L.; Chen, J.; Liu, L.; Yang, X.; Cui, H.; Yue, G. Role of Metabolic Reprogramming of Long non-coding RNA in Clear Cell Renal Cell Carcinoma. *J. Cancer* **2022**, *13*, 691–705. [CrossRef] [PubMed]

6. Czyzyk-Krzeska, M.F.; Landero Figueroa, J.A.; Gulati, S.; Cunningham, J.T.; Meller, J.; Shamsaei, B.; Vemuri, B.; Plas, D.R. Molecular and metabolic subtypes in sporadic and inherited clear cell renal cell carcinoma. *Genes* **2021**, *12*, 388. [CrossRef] [PubMed]
7. Ochocki, J.D.; Khare, S.; Hess, M.; Ackerman, D.; Qiu, B.; Daisak, J.I.; Worth, A.J.; Lin, N.; Lee, P.; Xie, H.; et al. Arginase 2 Suppresses Renal Carcinoma Progression via Biosynthetic Cofactor Pyridoxal Phosphate Depletion and Increased Polyamine Toxicity. *Cell Metab.* **2018**, *27*, 1263–1280.e6. [CrossRef]
8. Hu, Y.; Xiang, J.; Su, L.; Tang, X. The regulation of nitric oxide in tumor progression and therapy. *J. Int. Med. Res.* **2020**, *48*, 0300060520905985. [CrossRef]
9. Kurmi, K.; Haigis, M.C. Nitrogen Metabolism in Cancer and Immunity. *Trends Cell Biol.* **2020**, *30*, 408–424. [CrossRef]
10. Lee, J.S.; Adler, L.; Karathia, H.; Carmel, N.; Rabinovich, S.; Auslander, N.; Keshet, R.; Stettner, N.; Silberman, A.; Agemy, L.; et al. Urea Cycle Dysregulation Generates Clinically Relevant Genomic and Biochemical Signatures. *Cell* **2018**, *174*, 1559–1570.e22. [CrossRef]
11. Fukumura, D.; Kashiwagi, S.; Jain, R.K. The role of nitric oxide in tumour progression. *Nat. Rev. Cancer* **2006**, *6*, 521–534. [CrossRef]
12. Hulin, J.A.; Gubareva, E.A.; Jarzebska, N.; Rodionov, R.N.; Mangoni, A.A.; Tommasi, S. Inhibition of Dimethylarginine Dimethylaminohydrolase (DDAH) Enzymes as an Emerging Therapeutic Strategy to Target Angiogenesis and Vasculogenic Mimicry in Cancer. *Front. Oncol.* **2020**, *9*, 1455. [CrossRef]
13. Ene, C.V.; Nicolae, I.; Geavlete, B.; Geavlete, P.; Ene, C.D. IL-6 Signaling Link between Inflammatory Tumor Microenvironment and Prostatic Tumorigenesis. *Anal. Cell. Pathol.* **2022**, *2022*, 1–10. [CrossRef]
14. Wei, Z.; Liu, X.; Cheng, C.; Yu, W.; Yi, P. Metabolism of Amino Acids in Cancer. *Front. Cell Dev. Biol.* **2021**, *8*, 603837. [CrossRef]
15. Somasundaram, V.; Basudhar, D.; Bharadwaj, G.; No, J.H.; Ridnour, L.A.; Cheng, R.Y.S.; Fujita, M.; Thomas, D.D.; Anderson, S.K.; McVicar, D.W.; et al. Molecular mechanisms of nitric oxide in cancer progression, signal transduction, and metabolism. *Antioxidants Redox Signal.* **2019**, *30*, 1124–1143. [CrossRef] [PubMed]
16. Wang, Y.; Chen, W.; Zhou, J.; Wang, Y.; Wang, H.; Wang, Y. Nitrate Metabolism and Ischemic Cerebrovascular Disease: A Narrative Review. *Front. Neurol.* **2022**, *13*, 735181. [CrossRef] [PubMed]
17. Ghasemi, A. Quantitative aspects of nitric oxide production from nitrate and nitrite. *Excli J.* **2022**, *21*, 470–486. [CrossRef] [PubMed]
18. Lundberg, J.O.; Weitzberg, E.; Gladwin, M.T. The nitrate-nitrite-nitric oxide pathway in physiology and therapeutics. *Nat. Rev. Drug Discov.* **2008**, *7*, 156–167. [CrossRef]
19. Di Meo, N.A.; Lasorsa, F.; Rutigliano, M.; Loizzo, D.; Ferro, M.; Stella, A.; Bizzoca, C.; Vincenti, L.; Pandolfo, S.D.; Autorino, R.; et al. Renal Cell Carcinoma as a Metabolic Disease: An Update on Main Pathways, Potential Biomarkers, and Therapeutic Targets. *Int. J. Mol. Sci.* **2022**, *23*, 14360. [CrossRef] [PubMed]
20. Trott, J.F.; Hwang, V.J.; Ishimaru, T.; Chmiel, K.J.; Zhou, J.X.; Shim, K.; Stewart, B.J.; Mahjoub, M.R.; Jen, K.Y.; Barupal, D.K.; et al. Arginine reprogramming in ADPKD results in arginine-dependent cystogenesis. *Am. J. Physiol. Ren. Physiol.* **2018**, *315*, F1855–F1868. [CrossRef]
21. Keshet, R.; Erez, A. Arginine and the metabolic regulation of nitric oxide synthesis in cancer. *DMM Dis. Model. Mech.* **2018**, *11*, dmm033332. [CrossRef] [PubMed]
22. Albaugh, V.L.; Pinzon-Guzman, C.; Barbul, A. Arginine—Dual roles as an onconutrient and immunonutrient. *J. Surg. Oncol.* **2017**, *115*, 273–280. [CrossRef] [PubMed]
23. Liu, T.; Wang, X.; Jia, P.; Liu, C.; Wei, Y.; Song, Y.; Li, S.; Liu, L.; Wang, B.; Shi, H. Association between serum arginine levels and cancer risk: A community-based nested case-control study. *Front. Nutr.* **2022**, *9*, 1069113. [CrossRef] [PubMed]
24. Baylis, C. Nitric oxide deficiency in chronic kidney disease. *Am. J. Physiol. Ren. Physiol.* **2008**, *294*, F1–F9. [CrossRef] [PubMed]
25. Morou-Bermúdez, E.; Torres-Colón, J.E.; Bermúdez, N.S.; Patel, R.P.; Joshipura, K.J. Pathways Linking Oral Bacteria, Nitric Oxide Metabolism, and Health. *J. Dent. Res.* **2022**, *101*, 623–631. [CrossRef] [PubMed]
26. Piechowicz, J.; Gamian, A.; Chukwu, O.; Polak-Jonkisz, D. Nitric Oxide Synthesis Metabolites—As Potential Markers in Chronic Kidney Disease in Children. *Curr. Issues Mol. Biol.* **2022**, *44*, 3518–3532. [CrossRef] [PubMed]
27. Carlström, M. Nitric oxide signalling in kidney regulation and cardiometabolic health. *Nat. Rev. Nephrol.* **2021**, *17*, 575–590. [CrossRef]
28. Khan, F.H.; Dervan, E.; Bhattacharyya, D.D.; McAuliffe, J.D.; Miranda, K.M.; Glynn, S.A. The role of nitric oxide in cancer: Master regulator or not? *Int. J. Mol. Sci.* **2020**, *21*, 9393. [CrossRef]
29. Krishnan, S.M.; Kraehling, J.R.; Eitner, F.; Bénardeau, A.; Sandner, P. The impact of the nitric oxide (no)/soluble guanylyl cyclase (sGC) signaling cascade on kidney health and disease: A preclinical perspective. *Int. J. Mol. Sci.* **2018**, *19*, 1712. [CrossRef]
30. Mishra, D.; Patel, V.; Banerjee, D. Nitric Oxide and S-Nitrosylation in Cancers: Emphasis on Breast Cancer. *Breast Cancer Basic Clin. Res.* **2020**, *14*, 1178223419882688. [CrossRef]
31. Hara, N.; Bilim, V.; Kasahara, T.; Obara, K.; Saito, K.; Takahashi, K.; Tomita, Y. Inducible Nitric Oxide Synthase in Renal Cell Carcinoma: Expression in Tumor Thrombi and Induction under Hypoxic Conditions. *Anticancer. Res.* **2003**, *23*, 4641–4649.
32. Perske, C.; Lahat, N.; Levin, S.S.; Bitterman, H.; Hemmerlein, B.; Rahat, M.A. Loss of inducible nitric oxide synthase expression in the mouse renal cell carcinoma cell line RENCA is mediated by MicroRNA miR-146a. *Am. J. Pathol.* **2010**, *177*, 2046–2054. [CrossRef] [PubMed]

33. Nakamura, T.; Oh, C.K.; Zhang, X.; Tannenbaum, S.R.; Lipton, S.A. Protein transnitrosylation signaling networks contribute to inflammaging and neurodegenerative disorders. *Antioxidants Redox Signal.* **2021**, *35*, 531–550. [CrossRef] [PubMed]
34. Faienza, F.; Rasola, A.; Filomeni, G. Nitric oxide-based regulation of metabolism: Hints from TRAP1 and SIRT3 crosstalk. *Front. Mol. Biosci.* **2022**, *9*, 942729. [CrossRef]
35. Yoon, S.; Eom, G.H.; Kang, G. Nitrosative stress and human disease: Therapeutic potential of denitrosylation. *Int. J. Mol. Sci.* **2021**, *22*, 9794. [CrossRef]
36. Gu, Y.R.; Kim, J.; Na, J.C.; Han, W.K. Mitochondrial metabolic reprogramming by SIRT3 regulation ameliorates drug resistance in renal cell carcinoma. *PLoS ONE* **2022**, *17*, e0269432. [CrossRef] [PubMed]
37. Sanchez-Martin, C.; Serapian, S.A.; Colombo, G.; Rasola, A. Dynamically Shaping Chaperones. Allosteric Modulators of HSP90 Family as Regulatory Tools of Cell Metabolism in Neoplastic Progression. *Front. Oncol.* **2020**, *10*, 1177. [CrossRef] [PubMed]
38. Poderoso, J.J.; Helfenberger, K.; Poderoso, C. The effect of nitric oxide on mitochondrial respiration. *Nitric Oxide Biol. Chem.* **2019**, *88*, 61–72. [CrossRef]
39. Lowenstein, C.J. Metabolism reprogrammed by the nitric oxide signalling molecule. *Nature* **2019**, *565*, 33–34. [CrossRef]
40. Zhang, S.X.; Marzluff, E.M.; Lindgren, C.A. Quantitative determination of nitric oxide from tissue samples using liquid chromatography—Mass spectrometry. *MethodsX* **2021**, *8*, 101412. [CrossRef]
41. Möller, M.N.; Rios, N.; Trujillo, M.; Radi, R.; Denicola, A.; Alvarez, B. Detection and quantification of nitric oxide-derived oxidants in biological systems. *J. Biol. Chem.* **2019**, *294*, 14776–14802. [CrossRef] [PubMed]
42. Jansson, O.T.; Morcos, E.; Brundin, L.; Beroerheim, U.S.R.; Adolfsson, J.; Wiklund, N.P. Nitric oxide synthase activity in human renal cell carcinoma. *J. Urol.* **1998**, *160*, 556–560. [CrossRef]
43. Zequi, C.; de, S.; Fregnani, J.H.G.T.; Favaretto, R.L.; Costa, W.H.; Madeira Campos, R.S.; Fonseca, F.P.; Guimaraes, G.C.; Soares, F.A.; da Cunha, I.W.; et al. The impact of immunohistochemical expression of nitric oxide synthases on clinical and pathological features of renal cell carcinoma. *World J. Urol.* **2013**, *31*, 1197–1203. [CrossRef] [PubMed]
44. Renaudin, K.; Denis, M.G.; Karam, G.; Vallette, G.; Buzelin, F.; Laboisse, C.L.; Jarry, A. Loss of NOSI expression in high-grade renal cell carcinoma associated with a shift of NO signalling. *Br. J. Cancer* **2004**, *90*, 2364–2369. [CrossRef] [PubMed]
45. Ali, M.A.; Akhmedkhanov, A.; Zeleniuch-Jaquotte, A.; Toniolo, P.; Frenkel, K.; Huang, X. Reliability of serum iron, ferritin, nitrite, and association with risk of renal cancer in women. *Cancer Detect. Prev.* **2003**, *27*, 116–121. [CrossRef] [PubMed]
46. Khare, S.; Kim, L.C.; Lobel, G.; Doulias, P.-T.; Ischiropoulos, H.; Nissim, I.; Keith, B.; Simon, M.C. ASS1 and ASL suppress growth in clear cell renal cell carcinoma via altered nitrogen metabolism. *Cancer Metab.* **2021**, *9*, 1–16. [CrossRef] [PubMed]
47. Lucarelli, G.; Ferro, M.; Ditonno, P.; Battaglia, M. The urea cycle enzymes act as metabolic suppressors in clear cell renal cell carcinoma. *Transl. Cancer Res.* **2018**, *7*, S766–S769. [CrossRef]
48. Carney, E.F. Altered ammonia metabolism in ccRCC. *Nat. Rev. Nephrol.* **2018**, *14*, 476. [CrossRef]
49. Minoru, M.; Masao, O.; Toshihiko, K.; Takao, S.; Kenji, S.; Teruo, N. Concentrations of polyamines in renal cell carcinoma. *Clin. Chim. Acta* **1978**, *87*, 93–99. [CrossRef]
50. Koida, T. The clinical significance of tissue, blood and urine polyamine in renal cell carcinoma. *Japanese J. Urol.* **1992**, *83*, 1228–1237. [CrossRef]
51. Dallmann, K.; Junker, H.; Balabanov, S.; Zimmermann, U.; Giebel, J.; Walther, R. Human agmatinase is diminished in the clear cell type of renal cell carcinoma. *Int. J. Cancer* **2004**, *108*, 342–347. [CrossRef] [PubMed]
52. Xie, J.; Chen, L.; Huang, D.; Yue, W.; Chen, J.; Liu, C. A nitric oxide-releasing prodrug promotes apoptosis in human renal carcinoma cells: Involvement of reactive oxygen species. *Open Chem.* **2021**, *19*, 635–645. [CrossRef]
53. Morbidelli, L.; Donnini, S.; Ziche, M. Role of Nitric Oxide in the Modulation of Angiogenesis. *Curr. Pharm. Des.* **2005**, *9*, 521–530. [CrossRef] [PubMed]
54. Said Abasse, K.; Essien, E.E.; Abbas, M.; Yu, X.; Xie, W.; Jinfang, S.; Akter, L.; Cote, A. Association between Dietary Nitrate, Nitrite Intake, and Site-Specific Cancer Risk: A Systematic Review and Meta-Analysis. *Nutrients* **2022**, *14*, 666. [CrossRef]
55. Kamal, N.; Ilowefah, M.A.; Hilles, A.R.; Anua, N.A.; Awin, T.; Alshwyeh, H.A.; Aldosary, S.K.; Jambocus, N.G.S.; Alosaimi, A.A.; Rahman, A.; et al. Genesis and Mechanism of Some Cancer Types and an Overview on the Role of Diet and Nutrition in Cancer Prevention. *Molecules* **2022**, *27*, 1794. [CrossRef]
56. Hord, N.G.; Tang, Y.; Bryan, N.S. Food sources of nitrates and nitrites: The physiologic context for potential health benefits. *Am. J. Clin. Nutr.* **2009**, *90*, 1–10. [CrossRef]
57. Mirvish, S.S. Experimental evidence for inhibition of N-nitroso compound formation as a factor in the negative correlation between vitamin C consumption and the incidence of certain cancers. *Cancer Res.* **1994**, *54*, 1948s–1951s.
58. Bartsch, H.; Ohshima, H.; Pignatelli, B. Inhibitors of endogenous nitrosation mechanisms and implications in human cancer prevention. *Mutat. Res. Fundam. Mol. Mech. Mutagen.* **1988**, *202*, 307–324. [CrossRef]
59. Bogovski, P.; Bogovski, S. Special report animal species in which n-nitroso compounds induce cancer. *Int. J. Cancer* **1981**, *27*, 471–474. [CrossRef]
60. Brock, K.E.; Gridley, G.; Chiu, B.C.H.; Ershow, A.G.; Lynch, C.F.; Cantor, K.P. Dietary fat and risk of renal cell carcinoma in the USA: A case-control study. *Br. J. Nutr.* **2009**, *101*, 1228–1238. [CrossRef]
61. Alexander, D.D.; Cushing, C.A. Quantitative assessment of red meat or processed meat consumption and kidney cancer. *Cancer Epidemiol.* **2009**, *32*, 340–351. [CrossRef] [PubMed]

62. Zhang, S.; Wang, Q.; He, J. Correction: Intake of red and processed meat and risk of renal cell carcinoma: A meta-analysis of observational studies. *Oncotarget* **2018**, *9*, 29018. [CrossRef]
63. Liao, Z.; Fang, Z.; Gou, S.; Luo, Y.; Liu, Y.; He, Z.; Li, X.; Peng, Y.; Fu, Z.; Li, D.; et al. The role of diet in renal cell carcinoma incidence: An umbrella review of meta-analyses of observational studies. *BMC Med.* **2022**, *20*, 39. [CrossRef] [PubMed]
64. Ma, L.; Hu, L.; Feng, X.; Wang, S. Nitrate and nitrite in health and disease. *Aging Dis.* **2018**, *9*, 938–945. [CrossRef] [PubMed]
65. Sözen, S.; Coskun, U.; Sancak, B.; Bukan, N.; Günel, N.; Tunc, L.; Bozkirli, I. Serum levels of interleukin-18 and nitrite+nitrate in renal cell carcinoma patients with different tumor stage and grade. *Neoplasma* **2004**, *51*, 25–29.
66. Tate, D.J.; Vonderhaar, D.J.; Caldas, Y.A.; Metoyer, T.; Patterson IV, J.R.; Aviles, D.H.; Zea, A.H. Effect of arginase II on L-arginine depletion and cell growth in murine cell lines of renal cell carcinoma. *J. Hematol. Oncol.* **2008**, *1*, 14. [CrossRef] [PubMed]
67. Zea, A.H.; Rodriguez, P.C.; Atkins, M.B.; Hernandez, C.; Signoretti, S.; Zabaleta, J.; McDermott, D.; Quiceno, D.; Youmans, A.; O'Neill, A.; et al. Arginase-producing myeloid suppressor cells in renal cell carcinoma patients: A mechanism of tumor evasion. *Cancer Res.* **2005**, *65*, 3044–3048. [CrossRef]
68. Clark, D.J.; Dhanasekaran, S.M.; Petralia, F.; Pan, J.; Song, X.; Hu, Y.; da Veiga Leprevost, F.; Reva, B.; Lih, T.S.M.; Chang, H.Y.; et al. Integrated Proteogenomic Characterization of Clear Cell Renal Cell Carcinoma. *Cell* **2019**, *179*, 964–983.e31. [CrossRef]
69. Hakimi, A.A.; Reznik, E.; Lee, C.H.; Creighton, C.J.; Brannon, A.R.; Luna, A.; Aksoy, B.A.; Liu, E.M.; Shen, R.; Lee, W.; et al. An Integrated Metabolic Atlas of Clear Cell Renal Cell Carcinoma. *Cancer Cell* **2016**, *29*, 104–116. [CrossRef]
70. Pandey, N.; Lanke, V.; Vinod, P.K. Network-based metabolic characterization of renal cell carcinoma. *Sci. Rep.* **2020**, *10*, 5955. [CrossRef]
71. Liu, H.; Li, S.; Liu, X.; Chen, Y.; Deng, H. SIRT3 Overexpression Inhibits Growth of Kidney Tumor Cells and Enhances Mitochondrial Biogenesis. *J. Proteome Res.* **2018**, *17*, 3143–3152. [CrossRef]
72. Morigi, M.; Perico, L.; Benigni, A. Sirtuins in Renal Health and Disease. *J. Am. Soc. Nephrol.* **2018**, *29*, 1799–1809. [CrossRef]
73. Costa-Machado, L.F.; Fernandez-Marcos, P.J. The sirtuin family in cancer. *Cell Cycle* **2019**, *18*, 2164–2196. [CrossRef]
74. Galiniak, S.; Biesiadecki, M.; Mołoń, M.; Olech, P.; Balawender, K. Serum Oxidative and Nitrosative Stress Markers in Clear Cell Renal Cell Carcinoma. *Cancers* **2023**, *15*, 3995. [CrossRef] [PubMed]
75. Poillet-Perez, L.; Xie, X.; Zhan, L.; Yang, Y.; Sharp, D.W.; Hu, Z.S.; Su, X.; Maganti, A.; Jiang, C.; Lu, W.; et al. Autophagy maintains tumour growth through circulating arginine. *Nature* **2018**, *563*, 569–573. [CrossRef] [PubMed]
76. Hajaj, E.; Sciacovelli, M.; Frezza, C.; Erez, A. The context-specific roles of urea cycle enzymes in tumorigenesis. *Mol. Cell* **2021**, *81*, 3749–3759. [CrossRef]
77. Nagamani, S.C.S.; Erez, A. A metabolic link between the urea cycle and cancer cell proliferation. *Mol. Cell. Oncol.* **2016**, *3*, e1127314. [CrossRef] [PubMed]
78. Bai, C.; Wang, H.; Dong, D.; Li, T.; Yu, Z.; Guo, J.; Zhou, W.; Li, D.; Yan, R.; Wang, L.; et al. Urea as a By-Product of Ammonia Metabolism Can Be a Potential Serum Biomarker of Hepatocellular Carcinoma. *Front. Cell Dev. Biol.* **2021**, *9*, 677. [CrossRef]
79. Hoerner, C.R.; Chen, V.J.; Fan, A.C. The "achilles heel" of metabolism in renal cell carcinoma: Glutaminase inhibition as a rational treatment strategy. *Kidney Cancer* **2019**, *3*, 15–29. [CrossRef]
80. Okazaki, A.; Gameiro, P.; Stephanopoulos, G.; Iliopoulos, O. Abstract 1123: Glutaminase inhibitors suppress pyrimidine synthesis and promote DNA replication stress in VHL -deficient human renal cancer cells. *Cancer Res.* **2015**, *75*, 1123. [CrossRef]
81. Kaushik, A.K.; Tarangelo, A.; Boroughs, L.K.; Ragavan, M.; Zhang, Y.; Wu, C.Y.; Li, X.; Ahumada, K.; Chiang, J.C.; Tcheuyap, V.T.; et al. In vivo characterization of glutamine metabolism identifies therapeutic targets in clear cell renal cell carcinoma. *Sci. Adv.* **2022**, *8*, eabp8293. [CrossRef] [PubMed]
82. Nabi, S.; Kessler, E.R.; Bernard, B.; Flaig, T.W.; Lam, E.T. Renal cell carcinoma: A review of biology and pathophysiology. *F1000Research* **2018**, *7*, 307. [CrossRef] [PubMed]
83. Pérez-Neri, I.; Ramírez-Bermúdez, J.; Ojeda-López, C.; Montes, S.; Soto-Hernández, J.L.; Ríos, C. Glutamine and citrulline concentrations reflect nitric oxide synthesis in the human nervous system. *Neurologia* **2020**, *35*, 96–104. [CrossRef] [PubMed]
84. Bryk, J.; Ochoa, J.B.; Correia, M.I.T.D.; Munera-Seeley, V.; Popovic, P.J. Effect of citrulline and glutamine on nitric oxide production in RAW 264.7 cells in an arginine-depleted environment. *J. Parenter. Enter. Nutr.* **2008**, *32*, 377–383. [CrossRef]
85. Chen, C.L.; Hsu, S.C.; Ann, D.K.; Yen, Y.; Kung, H.J. Arginine signaling and cancer metabolism. *Cancers* **2021**, *13*, 3541. [CrossRef] [PubMed]
86. Ochocki, J.; Lin, N.; Qiu, B.; Simon, M.C. Abstract C32: Metabolic advantages of urea cycle misregulation in renal cell carcinoma. *Cancer Res.* **2013**, *73*, C32. [CrossRef]
87. Spinelli, J.B.; Yoon, H.; Ringel, A.E.; Jeanfavre, S.; Clish, C.B.; Haigis, M.C. Metabolic recycling of ammonia via glutamate dehydrogenase supports breast cancer biomass. *Science* **2017**, *358*, 941–946. [CrossRef]
88. Kimura, S.; Fujisaki, Y.; Onizuka, C.; Hasuike, S.; Sato, Y.; Mukai, S.; Kamoto, T. A case of hyperammonemia occurring during treatment of metastatic renal cell carcinoma with axitinib. *IJU Case Reports* **2023**, *6*, 206–210. [CrossRef]
89. Milewski, K.; Bogacińska-Karaś, M.; Fręśko, I.; Hilgier, W.; Jaźwiec, R.; Albrecht, J.; Zielińska, M. Ammonia reduces intracellular asymmetric dimethylarginine in cultured astrocytes stimulating its y+LAT2 carrier-mediated loss. *Int. J. Mol. Sci.* **2017**, *18*, 2308. [CrossRef]
90. Peng, H.; Wang, Y.; Luo, W. Multifaceted role of branched-chain amino acid metabolism in cancer. *Oncogene* **2020**, *39*, 6747–6756. [CrossRef]

91. Ananieva, E.A.; Wilkinson, A.C. Branched-chain amino acid metabolism in cancer. *Curr. Opin. Clin. Nutr. Metab. Care* **2018**, *21*, 64–70. [CrossRef] [PubMed]
92. Jin, H.J.; Lee, J.H.; Kim, D.H.; Kim, K.T.; Lee, G.W.; Choi, S.J.; Chang, P.S.; Paik, H.D. Antioxidative and nitric oxide scavenging activity of branched-chain amino acids. *Food Sci. Biotechnol.* **2015**, *24*, 1555–1558. [CrossRef]
93. Chachaj, A.; Wiśniewski, J.; Rybka, J.; Butrym, A.; Biedroń, M.; Krzystek-Korpacka, M.; Fleszar, M.G.; Karczewski, M.; Wróbel, T.; Mazur, G.; et al. Asymmetric and symmetric dimethylarginines and mortality in patients with hematological malignancies—A prospective study. *PLoS ONE* **2018**, *13*, e0197148. [CrossRef]
94. Li, Z.L.; Wang, B.; Wen, Y.; Wu, Q.L.; Lv, L.L.; Liu, B.C. Disturbance of Hypoxia Response and Its Implications in Kidney Diseases. *Antioxidants Redox Signal.* **2022**, *37*, 936–955. [CrossRef]
95. Lasorsa, F.; Rutigliano, M.; Milella, M.; Ferro, M.; Pandolfo, S.D.; Crocetto, F.; Autorino, R.; Battaglia, M.; Ditonno, P.; Lucarelli, G. Cancer Stem Cells in Renal Cell Carcinoma: Origins and Biomarkers. *Int. J. Mol. Sci.* **2023**, *24*, 13179. [CrossRef]
96. Lasorsa, F.; Rutigliano, M.; Milella, M.; Ferro, M.; Pandolfo, S.D.; Crocetto, F.; Tataru, O.S.; Autorino, R.; Battaglia, M.; Ditonno, P.; et al. Cellular and Molecular Players in the Tumor Microenvironment of Renal Cell Carcinoma. *J. Clin. Med.* **2023**, *12*, 3888. [CrossRef]
97. Sinha, B.K. Can Nitric Oxide-Based Therapy Be Improved for the Treatment of Cancers? A Perspective. *Int. J. Mol. Sci.* **2023**, *24*, 13611. [CrossRef]
98. Zhao, Y.; Ouyang, X.; Peng, Y.; Peng, S. Stimuli responsive nitric oxide-based nanomedicine for synergistic therapy. *Pharmaceutics* **2021**, *13*, 1917. [CrossRef]

cancers

MDPI

Article

Surgical Trends and Complications in Partial and Radical Nephrectomy: Results from the GRAND Study

Nikolaos Pyrgidis, Gerald Bastian Schulz, Christian Stief, Iulia Blajan, Troya Ivanova, Annabel Graser and Michael Staehler *

Department of Urology, LMU University Hospital, LMU Munich, 81377 Munich, Germany; nikolaos.pyrgidis@med.uni-muenchen.de (N.P.); gerald.schulz@med.uni-muenchen.de (G.B.S.); christian.stief@med.uni-muenchen.de (C.S.); iulia.blajan@med.uni-muenchen.de (I.B.); troya.ivanova@med.uni-muenchen.de (T.I.); annabel.graser@med.uni-muenchen.de (A.G.)

* Correspondence: michael.staehler@med.uni-muenchen.de; Tel.: +49-152-5484-7591; Fax: +49-89-4400-78890

Simple Summary: Studies about the current trends in renal cancer surgery and its perioperative outcomes are lacking. Using the nationwide data of Germany from 2005 to 2021, we found that the utilization of partial nephrectomy substantially increased, while the utilization of radical nephrectomy substantially decreased in the last years. Patients selected for radical nephrectomy had more comorbidities and risk factors compared to patients selected for partial nephrectomy. Our analyses suggest that patients undergoing radical nephrectomy present worse perioperative morbidity and mortality, as well as prolonged hospitalization, compared to patients undergoing partial nephrectomy.

Abstract: Background: We aimed to evaluate the current trends in renal cancer surgery, as well as to compare the perioperative outcomes of partial versus radical nephrectomy. Methods: We used the GeRmAn Nationwide inpatient Data (GRAND), provided by the Research Data Center of the Federal Bureau of Statistics (2005–2021). We report the largest study in the field, with 317,843 patients and multiple patient-level analyses. Results: Overall, 123,924 (39%) patients underwent partial and 193,919 (61%) underwent radical nephrectomy in Germany from 2005 to 2021. Of them, 57,308 (18%) were operated on in low-, 142,702 (45%) in intermediate-, and 117,833 (37%) in high-volume centers. A total of 249,333 (78%) patients underwent open, 44,994 (14%) laparoscopic, and 23,516 (8%) robotic nephrectomy. The number of patients undergoing renal surgery remained relatively stable from 2005 to 2021. Over the study period, the utilization of partial nephrectomy increased threefold, while radical nephrectomy decreased by about 40%. After adjusting for major risk factors in the multivariate regression analysis, radical nephrectomy was associated with 3.2-fold higher odds (95% CI: 3.2 to 3.9, $p < 0.001$) of 30-day mortality, longer hospitalization by 1.9 days (95% CI: 1.9 to 2, $p < 0.001$), and higher inpatient costs by EUR 1778 (95% CI: 1694 to 1862, $p < 0.001$) compared to partial nephrectomy. Furthermore, radical nephrectomy had a higher risk of in-hospital transfusion ($p < 0.001$), sepsis ($p < 0.001$), acute respiratory failure ($p < 0.001$), acute kidney disease ($p < 0.001$), acute thromboembolism ($p < 0.001$), surgical wound infection ($p < 0.001$), ileus ($p < 0.001$), intensive care unit admission ($p < 0.001$), and pancreatitis ($p < 0.001$). Conclusions: More patients are offered partial nephrectomy in Germany. Patients undergoing radical nephrectomy present with a higher rate of concomitant risk factors and have increased perioperative morbidity and mortality, prolonged hospitalization, and increased in-hospital costs.

Keywords: cohort study; partial nephrectomy; radical nephrectomy; perioperative outcomes; mortality

Citation: Pyrgidis, N.; Schulz, G.B.; Stief, C.; Blajan, I.; Ivanova, T.; Graser, A.; Staehler, M. Surgical Trends and Complications in Partial and Radical Nephrectomy: Results from the GRAND Study. *Cancers* **2024**, *16*, 97. https://doi.org/10.3390/cancers16010097

Academic Editors: José I. López and Claudia Manini

Received: 7 December 2023
Revised: 20 December 2023
Accepted: 22 December 2023
Published: 24 December 2023

1. Introduction

Renal cancer accounts for over 3% of all new cancer cases worldwide, affecting more than 430,000 individuals every year [1]. The increasing prevalence of risk factors for renal cancer, such as obesity, hypertension, and chronic renal disease, has contributed to a rising incidence of renal cancer globally [2]. Accordingly, the recent technological improvements, in combination with the wide implementation of cross-sectional imaging, have led to earlier diagnosis of renal cancer in many patients [3]. Therefore, 75% of all newly diagnosed renal masses are asymptomatic, incidental findings and smaller than 7 cm in diameter. About 80% of all surgically resected renal tumors are malignant in the final histology [4].

Despite the recent advancements in other local or systemic therapies, partial nephrectomy is considered the standard treatment for patients with suspected renal cancer, if technically feasible [5,6]. Partial nephrectomy has supplanted radical nephrectomy as the preferred treatment modality, given that it is associated with superior functional and equivalent oncological outcomes [7]. Radical nephrectomy remains the treatment of choice for tumors in which partial nephrectomy is not possible [8]. Both partial and radical nephrectomy can be performed with either an open or a minimally invasive approach [9]. Nevertheless, nephrectomy is associated with perioperative mortality and morbidity, as well as with prolonged hospital stay, intensive care unit admission, and significant treatment-related costs [10,11].

Currently, there is a global trend toward the centralization of complex urological operations in healthcare systems [12]. This shift is supported by the accumulating evidence indicating that increased annual hospital volume leads to improved perioperative outcomes for major urological operations. [13]. Therefore, the EAU Guideline Panel on Renal Cell Carcinoma recommends that hospitals should annually perform at least forty partial nephrectomies [14]. Nevertheless, this recommendation is based on a low level of evidence derived from retrospective studies with relatively low numbers of included patients. Indeed, there is a paucity of existing studies attempting to identify a hospital volume threshold for annual kidney cancer surgery cases that may improve perioperative outcomes. Similarly, studies assessing the perioperative complications in patients undergoing kidney cancer surgery are lacking. In this scope, we aimed to evaluate the current trends in partial and radical nephrectomy, and to compare the perioperative outcomes of partial versus radical nephrectomy through the largest study in the field.

2. Methods

2.1. *GeRmAn Nationwide Inpatient Data (GRAND)*

For the present analysis, we used the GeRmAn Nationwide inpatient Data (GRAND). The GRAND study contains all German inpatient data from 2005 to 2021 apart from military, psychiatric, and forensic cases. These data are stored in an anonymized format at the Research Data Center of the German Bureau of Statistics, and they were retrieved for further analysis upon agreement (LMU—4710-2022). To ensure anonymity, the Research Data Center excludes patient groups with fewer than three baseline characteristics or inpatient complications and does not allow hospital-level comparisons for any outcome. After the 2004 implementation of a diagnosis- and procedure-related remuneration system in Germany (German diagnosis-related groups (DRGs)), all hospitals need to transfer the in-hospital patient data (e.g., coexisting conditions, surgical procedures, perioperative outcomes) to the Institute for the Hospital Remuneration System to receive their remuneration. These patient data are coded based on the International Statistical Classification of Diseases and Related Health Problems, 10th revision, German modification (ICD-10-GM) and the German Procedure Classification (OPS).

2.2. *Selection Criteria*

We included all patients undergoing radical (OPS code: 5-554.4, 5-554.a) and partial nephrectomy (OPS code: 5-553) for suspected renal tumors. To obtain patient data on procedures, concurrent conditions, and inpatient complications, we used the available diagnostic and procedural codes (ICD-10-GM and OPS). The primary outcome of the present analysis was to assess surgical trends in patients undergoing radical or partial nephrectomy. Secondary outcomes included the effects of radical versus partial nephrectomy on 30-day mortality and perioperative complications (i.e., transfusion, sepsis, acute respiratory failure, acute kidney disease, acute thromboembolism, surgical wound infection, ileus, intensive care unit admission, and pancreatitis). We also analyzed hospital revenues and length of hospital stay. Moreover, we compared 30-day mortality and perioperative complications in patients undergoing radical versus partial nephrectomy separately in high-, intermediate-, and low-volume centers. Given that there is no consensus on the definition of high-volume centers, we defined high-volume centers as those that perform at least 100 nephrectomies (partial and radical) per year. Similarly, intermediate-volume centers were defined as those that perform between 40 and 99 nephrectomies per year, and low-volume centers were defined as those that perform less than 40 nephrectomies per year.

2.3. Data Synthesis and Statistical Analysis

Our research team did not have direct access to patient-level data. Thus, all statistical analyses were performed on our behalf by the Research Data Center of the German Bureau of Statistics, based on R codes developed by our research team (source: Research Data Center of the Federal Bureau of Statistics, DRG Statistics 2005–2021; own calculations). Subsequently, the summary results were provided to our research group for further evaluation. Approval by an ethics committee or informed patient consent was not required based on the German legislation.

All hospitals performing renal surgeries were identified through their postal codes and were further subclassified based on their annual caseload as low-volume centers (<40 cases/year), intermediate-volume centers fulfilling the EAU recommendation (40–99 cases/year), and high-volume centers (≥100 cases/year). The corresponding comparisons among low- (<40 cases/year), intermediate- (40–99 cases/year), and high-volume centers (≥100 cases/year) were performed with the chi-squared test and the Kruskal–Wallis test. Accordingly, all comparisons between patients undergoing partial versus radical nephrectomy were performed with the chi-squared test and the Mann–Whitney U test. All continuous variables were calculated as medians with interquartile ranges (IQRs), and all categorical variables were calculated as frequencies with proportions.

We conducted multiple multivariable logistic and linear regression analyses to evaluate the effect of the type of surgery and the effect of the annual hospital caseload on inpatient outcomes (i.e., 30-day mortality, perioperative complications, length of hospital stay, and hospital revenues). All regression models were adjusted for sex, age, obesity, history of chronic obstructive pulmonary disease, chronic heart failure, chronic kidney disease, cerebrovascular accident, hypertension, and diabetes, as well as for the surgical approach and the year of operation. Odds ratios (ORs) with 95% confidence intervals (CIs) were estimated for all logistic models, and two-sided p-values lower than 0.05 were considered statistically significant. The log-rank test with Kaplan–Meier analyses was used to assess the effects of radical versus partial nephrectomy on 30-day mortality.

3. Results

3.1. Baseline Characteristics

A total of 317,843 patients with a median age of 66 years (IQR: 56–74) underwent kidney cancer surgery, with 188,123 (59%) being male; 179,386 (56%) had hypertension, 65,605 (21%) has chronic kidney disease, and 57,155 (18%) had diabetes. Overall, 123,924 (39%) patients underwent partial and 193,919 (61%) underwent radical nephrectomy. A total of 249,333 (78%) patients underwent open, 44,994 (14%) laparoscopic, and 23,516 (7.4%) robotic nephrectomy. Based on the annual hospital caseload volume for kidney cancer surgery, 57,308 (18%) operations (17,108 (30%) partial and 40,200 (70%) radical nephrectomies) were performed in low-volume centers, 142,702 (45%) operations (52,966 (37%) partial and 89,736 (63%) radical nephrectomies) in intermediate-volume centers, and 117,833 (37%) operations (53,850 (46%) partial and 63,983 (54%) radical nephrectomies) in high-volume centers.

Patients undergoing radical nephrectomy were older ($p < 0.001$) and had higher proportions of diabetes ($p < 0.001$), chronic heart failure ($p < 0.001$), chronic obstructive pulmonary disease ($p < 0.001$), chronic kidney disease ($p < 0.001$), cerebrovascular disease

($p < 0.001$), and dementia ($p < 0.001$) compared to patients undergoing partial nephrectomy. Laparoscopic or robotic surgical approaches were preferred more often in patients undergoing partial nephrectomy ($p < 0.001$). The latter was also observed in the separate analyses for low-, intermediate-, and high-volume centers. The baseline characteristics of all patients undergoing renal surgery are presented in Table 1, and the corresponding baseline characteristics of patients operated on in low-, intermediate-, and high-volume centers are presented in the Supplementary Materials (Table S1).

Table 1. Baseline characteristics of the included patients based on the type of renal cancer surgery: Variables are presented as medians with interquartile ranges or as frequencies with proportions. The Mann–Whitney test was performed for comparisons between continuous variables, and the chi-squared test was used for categorical variables. The bold cells indicate statistically significant *p*-values.

Characteristic	Overall, n = 317,843	Partial Nephrectomy, n = 123,924	Radical Nephrectomy, n = 193,919	*p*-Value
Age (years)	66 (56–74)	65 (56–73)	67 (56–75)	**<0.001**
Males	188,123 (59%)	77,117 (62%)	111,006 (57%)	**<0.001**
Diabetes	57,155 (18%)	21,319 (17%)	35,836 (18%)	**<0.001**
Chronic heart failure	20,140 (6.3%)	5722 (4.6%)	14,418 (7.4%)	**<0.001**
Chronic obstructive pulmonary disease	23,142 (7.3%)	8694 (7%)	14,448 (7.5%)	**<0.001**
Chronic kidney disease	65,605 (21%)	16,380 (13%)	49,225 (25%)	**<0.001**
Cerebrovascular disease	7607 (2.4%)	2126 (1.7%)	5481 (2.8%)	**<0.001**
Dementia	3689 (1.2%)	693 (0.6%)	2996 (1.5%)	**<0.001**
Hypertension	179,386 (56%)	70,382 (57%)	109,004 (56%)	**0.001**
Obesity	32,008 (10%)	12,384 (10%)	19,624 (10%)	0.25
Operative technique				**<0.001**
Open	249,333 (78%)	89,227 (72%)	160,106 (83%)	
Laparoscopic	44,994 (14%)	15,528 (13%)	29,466 (15%)	
Robotic	23,516 (7.4%)	19,169 (15%)	4347 (2.2%)	

The number of patients undergoing renal surgery increased moderately from 17,360 cases in 2005 to 18,686 in 2021. Interestingly, the number of patients undergoing partial nephrectomy substantially increased throughout these years, from 3358 cases in 2005 to 10,153 in 2021, while the number of patients undergoing radical nephrectomy substantially decreased throughout the same period, from 14,002 cases in 2005 to 8533 in 2021. The latter was also observed in low-, intermediate-, and high-volume centers. Nevertheless, the increase in partial nephrectomies and the decrease in radical nephrectomies were steeper in high-volume centers. In particular, in high-volume centers, 1427 (2.6%) partial nephrectomies and 4634 (7.2%) radical nephrectomies were performed in 2005, compared to 4485 (8.3%) partial nephrectomies and 2857 (4.5%) radical nephrectomies in 2021. The number of patients undergoing renal surgery was not affected during the COVID-19 pandemic. The annual trends for radical and partial nephrectomy are depicted in Figure 1, whereas the corresponding annual trends for radical and partial nephrectomy in low-, intermediate-, and high-volume centers are depicted in the Supplementary Materials (Figure S1).

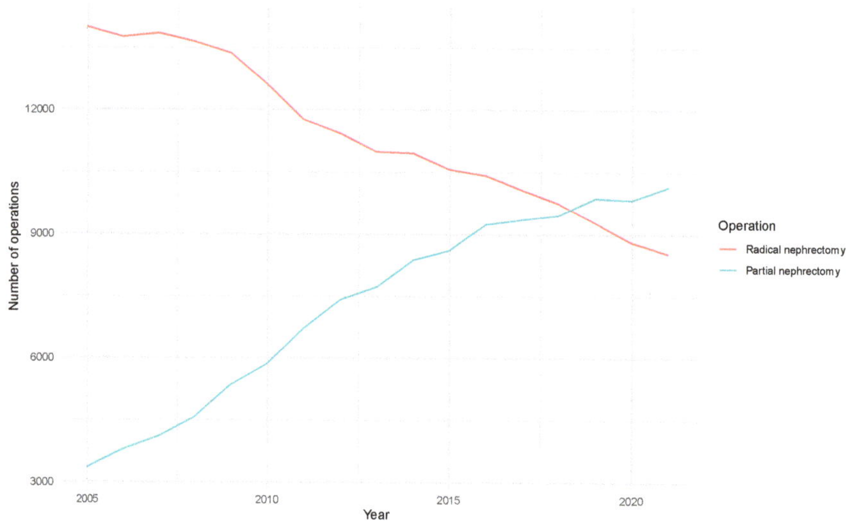

Figure 1. The annual trends for radical and partial nephrectomy.

3.2. *Effects of Renal Surgery on Perioperative Morbidity, Mortality, Hospital Stay, and Costs*

Overall, radical nephrectomy, compared to partial nephrectomy, was significantly associated with higher odds of transfusion (25% versus 12%; OR: 2, 95% CI: 1.9 to 2, $p < 0.001$), sepsis (3.1% versus 1%; OR: 2.6, 95% CI: 2.4 to 2.8, $p < 0.001$), acute respiratory failure (5.4% versus 3.6%; OR: 1.6, 95% CI: 1.5 to 1.7, $p < 0.001$), acute kidney disease (5.9% versus 4%; OR: 1.6, 95% CI: 1.5 to 1.7), acute thromboembolism (0.9% versus 0.5%; OR: 1.9, 95% CI: 1.7 to 2, $p < 0.001$), surgical wound infection (0.7% versus 0.3%; OR: 2, 95% CI: 1.8 to 2.2, $p < 0.001$), ileus (2% versus 1.3%; OR: 1.4, 95% CI: 1.3 to 1.5, $p < 0.001$), intensive care unit admission (21% versus 16%; OR: 1.2, 95% CI: 1.2 to 1.2, $p < 0.001$), and pancreatitis (0.3% versus 0.1%; OR: 2.1, 95% CI: 1.8 to 2.5, $p < 0.001$). Similarly, radical nephrectomy was associated with longer hospital stay by 1.9 days (95% CI: 1.9 to 2, $p < 0.001$) and higher inpatient costs by EUR 1778 (95% CI: 1694 to 1862, $p < 0.001$). The multivariable analysis is presented in Tables 2 and 3. In the separate multivariable analyses of patients undergoing renal surgery in low-, intermediate-, and high-volume centers, radical nephrectomy was associated with worse perioperative outcomes compared to partial nephrectomy. The corresponding findings are presented in the Supplementary Materials (Table S2).

A total of 4202 (1.3%) in-hospital deaths were observed within 30 days of surgery. Of these, 540 (0.4%) deaths occurred among patients undergoing partial nephrectomy and 3662 (1.9%) among patients undergoing radical nephrectomy. The numbers of 30-day in-hospital deaths after partial nephrectomy were 116 (0.7%) in low-volume centers, 234 (0.4%) in intermediate-volume centers, and 190 (0.4%) in high-volume centers. Meanwhile, the numbers of 30-day in-hospital deaths after radical nephrectomy were 917 (2.3%) in low-volume centers, 1704 (1.9%) in intermediate-volume centers, and 1041 (1.6%) in high-volume centers.

Radical nephrectomy was associated with 3.5-fold higher odds (95% CI: 3.2 to 3.9, $p < 0.001$) of 30-day mortality compared to partial nephrectomy in the whole study population. In low-volume centers, radical nephrectomy was associated with 2.8-fold higher odds (95% CI: 2.3 to 3.5, $p < 0.001$) of 30-day mortality compared to partial nephrectomy. In intermediate-volume centers, radical nephrectomy was associated with 3.6-fold higher odds (95% CI: 3.1 to 4.1, $p < 0.001$) of 30-day mortality compared to partial nephrectomy. In high-volume centers, radical nephrectomy was associated with 3.7-fold higher odds (95% CI: 3.2 to 4.4, $p < 0.001$) of 30-day mortality compared to partial nephrectomy. The latter was also observed in the time-to-death analysis for the whole study population, as well as for low-, intermediate-, and high-volume centers (log-rank test for all comparisons: $p < 0.001$). The corresponding Kaplan–Meier analyses are presented in Figure 2 and Figure S2.

Table 2. Multivariable logistic regression analysis for the effects of the type of surgery on transfusion, sepsis, acute respiratory failure, acute kidney disease, acute thromboembolism, and surgical wound infection. All models are adjusted for sex, age, obesity, history of chronic obstructive pulmonary disease, chronic heart failure, chronic kidney disease, cerebrovascular accident, hypertension, diabetes, surgical approach, and year of operation. The bold cells indicate statistically significant *p*-values. CI: confidence interval, OR: odds ratio.

Complications		Partial Nephrectomy	Radical Nephrectomy
Transfusion	Events	15,315 (12%)	49,169 (25%)
	OR (95% CI)	-	2 (1.9, 2)
	p-Value	-	**<0.001**
Sepsis	Events	1260 (1%)	6018 (3.1%)
	OR (95% CI)	-	2.6 (2.4, 2.8)
	p-Value	-	**<0.001**
Acute respiratory failure	Events	4448 (3.6%)	10,438 (5.4%)
	OR (95% CI)	-	1.6 (1.5, 1.7)
	p-Value	-	**<0.001**
Acute kidney disease	Events	4985 (4%)	11,408 (5.9%)
	OR (95% CI)	-	1.6 (1.5, 1.7)
	p-Value	-	**<0.001**
Acute thromboembolism	Events	630 (0.5%)	1802 (0.9%)
	OR (95% CI)	-	1.9 (1.7, 2)
	p-Value	-	**<0.001**
Surgical wound infection	Events	391 (0.3%)	1369 (0.7%)
	OR (95% CI)	-	2 (1.8, 2.2)
	p-Value	-	**<0.001**

Table 3. Multivariable linear and logistic regression analysis for the effects of the type of surgery on ileus, 30-day mortality, ICU admission, length of hospital stay, costs, and pancreatitis. All models are adjusted for sex, age, obesity, history of chronic obstructive pulmonary disease, chronic heart failure, chronic kidney disease, cerebrovascular accident, hypertension, diabetes, surgical approach, and year of operation. The bold cells indicate statistically significant *p*-values. CI: confidence interval, ICU: intensive care unit, OR: odds ratio.

Complications		Partial Nephrectomy	Radical Nephrectomy
Ileus	Events	1623 (1.3%)	3903 (2%)
	OR (95% CI)	-	1.4 (1.3, 1.5)
	p-Value	-	**<0.001**
30-Day mortality	Events	540 (0.4%)	3662 (1.9%)
	OR (95% CI)	-	3.5 (3.2, 3.9)
	p-Value	-	**<0.001**
ICU admission	Events	20,116 (16%)	41,355 (21%)
	OR (95% CI)	-	1.2 (1.2, 1.2)
	p-Value	-	**<0.001**

Table 3. *Cont.*

Complications		Partial Nephrectomy	Radical Nephrectomy
Length of hospital stay	Days	9 (7–11)	10 (8–15)
	Beta (95% CI)	-	1.9 (1.9, 2)
	p-Value	-	**<0.001**
Costs	EUR	7087 (6484–8122)	7400 (6484–10,492)
	Beta (95% CI)	-	1778 (1694, 1862)
	p-Value	-	**<0.001**
Pancreatitis	Events	166 (0.1%)	636 (0.3%)
	OR (95% CI)	-	2 (1.8, 2.5)
	p-Value	-	**<0.001**

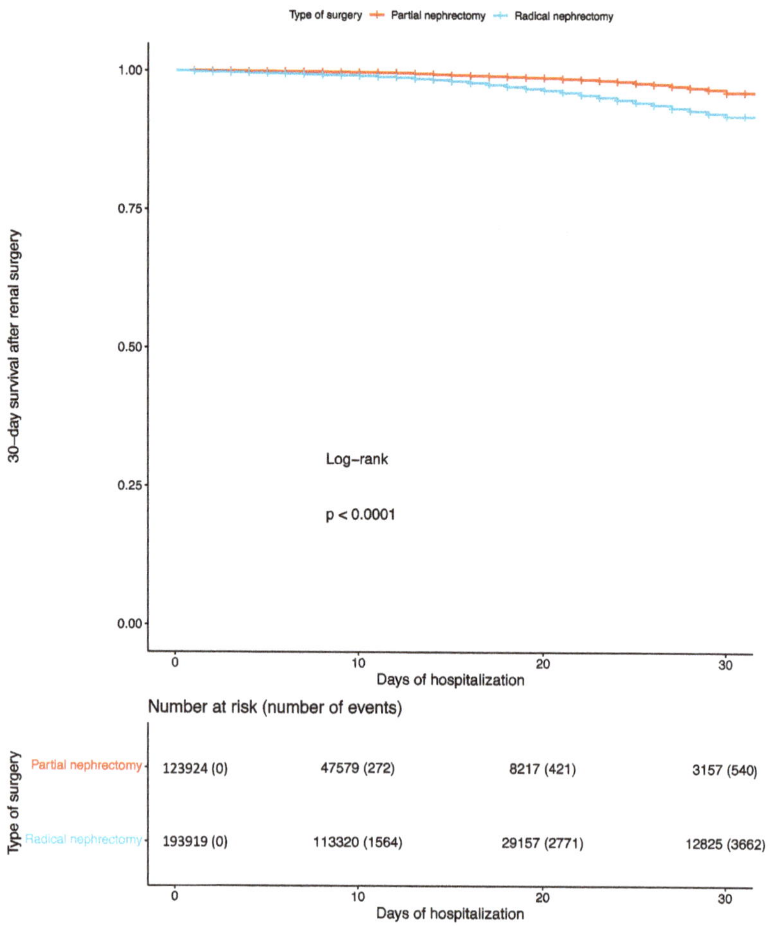

Figure 2. Kaplan–Maier curve for the 30-day survival in patients undergoing radical versus partial nephrectomy.

4. Discussion

The present high-volume study suggests that the annual cases of partial nephrectomy have undergone a threefold increase, while the annual cases of radical nephrectomy have undergone an important decrease of 40% in the last few years. Interestingly, the number of patients undergoing renal surgery has remained largely unchanged from 2005 to 2021. In high-volume centers, the proportion of partial to radical nephrectomies is higher compared to in intermediate- and low-volume centers. Furthermore, it seems that minimally invasive renal surgery is preferred in only a small proportion of patients requiring nephrectomy in Germany. It should be noted that patients undergoing radical nephrectomy displayed worse baseline characteristics compared to those undergoing partial nephrectomy. Nevertheless, after adjusting for these characteristics, we found that radical nephrectomy was associated with 195% higher odds of surgical wound infection, 160% higher odds of sepsis, 110% higher odds of pancreatitis, 100% higher odds of transfusion, 90% higher odds of acute thromboembolism, 60% higher odds of acute respiratory failure and acute kidney disease, 40% higher odds of ileus, and 20% higher odds of intensive care unit admission. Similarly, radical nephrectomy, compared to partial nephrectomy, was associated with 250% higher odds of 30-day in-hospital mortality, as well as with longer hospital stay and higher inpatient costs. The latter was also observed in the separate analysis between partial and radical nephrectomy in low-, intermediate-, and high-volume centers.

Our findings demonstrate that the utilization of partial nephrectomy has considerably increased in recent years. The latter might be predominantly attributed to the fact that current major guidelines recommend partial nephrectomy over radical nephrectomy whenever possible [14–16]. Studies from the US and the UK are in line with our findings demonstrating that the use of partial nephrectomy has increased significantly over time for both small and larger renal masses [17–19]. A large prospective study from Singapore suggested a shift towards nephron-sparing surgery for clinically localized tumors [20]. Accordingly, a nationwide study from the Netherlands demonstrated a clear increase in nephron-sparing management either with active surveillance, partial nephrectomy, or focal therapy over time for cT1a tumors. Conversely, for cT1b tumors, radical nephrectomy remained the most common treatment modality, although patients in high-volume centers more often underwent partial nephrectomy [21].

Still, despite the advancements in surgical techniques for nephrectomy, it seems that, in Germany, minimally invasive radical or partial nephrectomy is performed in a relatively small amount of patients compared to the US [22,23]. Even though the number of suspicious renal masses diagnosed every year in Germany has increased [24], the overall number of nephrectomy cases that are performed every year has remained stable in recent years [25,26]. The latter might be explained by the fact that an increasing number of patients with diagnosed renal masses undergo active surveillance or other ablative techniques [27–30]. Moreover, the fact that partial nephrectomy cases have increased by 300% while radical nephrectomy cases have decreased by 40% over the years might indicate that renal cancer is often diagnosed at less-advanced tumor stages. Indeed, large epidemiological studies suggest that the incidence of renal cancer continues to rise, mainly for early-stage tumors, whereas that of advanced stages has declined [31,32].

It should be highlighted that renal cancer surgery is associated with low in-hospital mortality and morbidity, which are higher in patients undergoing radical nephrectomy. Large prospective comparative studies indicate that in patients with similar renal masses in terms of diameter and location, partial and radical nephrectomy present comparable perioperative outcomes [8,33,34]. Still, in the present analysis, we could not perform an adjustment between the two groups in terms of their tumor characteristics. Based on the previous notion, patients undergoing radical nephrectomy presented worse baseline parameters and worse tumor characteristics, which might be the cause of the observed higher morbidity and mortality, as well as of the prolonged hospital stay and the increased costs compared to partial nephrectomy. It should be noted that patients with larger or more advanced tumors are more likely to require radical nephrectomy. The latter introduces a selection bias in the present analysis, since the inherent differences in tumor characteristics, the stage of the tumor, and patients' general condition may impact the estimated differences in terms of outcomes when comparing partial nephrectomy and radical nephrectomy.

In the absence of detailed tumor characteristics, it is difficult to determine how an advanced tumor stage might affect outcomes. In an attempt to overcome this selection bias, we adjusted for multiple important risk factors in the multivariate regression analysis. However, without specific tumor characteristics, there may be major residual confounding due to unmeasured variables related to tumor stage and extension. Nevertheless, it should be stressed that radical nephrectomy was associated with worse perioperative outcomes for all estimates in low-, intermediate-, and high-volume centers compared to partial nephrectomy.

As this report, to the best of our knowledge, is the largest study on trends and perioperative outcomes in renal cancer surgery, our data have limitations that need to be considered. First of all, our analyses were derived from retrospective administrative data and are prone to coding errors and misclassifications. Although these administrative data present a high degree of accuracy and are regularly evaluated by independent physician task forces from healthcare insurance companies, important information on renal cancer surgery is not collected. In particular, the tumor size and location, the patient's laboratory findings, the performance of a prior renal biopsy, the operative time, and the oncological status (i.e., histology findings, TNM classification, and surgical margins) are not available in the GRAND study. Similarly, data on mortality and morbidity after hospital discharge, readmission rates and causes of reoperation, functional outcomes, and follow-up data were not collected in the GRAND study. Moreover, the GRAND study does not provide information on the decision-making process between partial and radical nephrectomy, including patient preferences. Furthermore, the degree of baseline characteristics or perioperative complications such as chronic kidney disease, sepsis, or other perioperative complications cannot be retrieved. It was also beyond the scope of the present study to assess the number of operations performed in urological versus non-urological surgical departments (e.g., general surgery, pediatric surgery). Similarly, we did not consider exploring differences in the perioperative outcomes across different patient groups (such as younger patients, patients with obesity, or those with different levels of chronic kidney disease). Finally, our analyses are restricted to data for Germany and, thus, cannot be extrapolated to other healthcare systems. Accordingly, our findings may have limited generalizability, especially if the patient population undergoing radical nephrectomy differs systematically from those undergoing partial nephrectomy in terms of tumor characteristics and patients' general condition. Nevertheless, in an attempt to overcome these limitations, our holistic and critical approach, combined with the size and nature of the GRAND study, leads to solid conclusions.

5. Conclusions

The present high-volume, nationwide, real-world data from Germany demonstrate an increased utilization of partial nephrectomy in renal cancer surgery. In recent years, the annual cases of partial nephrectomy have exceeded those of radical nephrectomy. Only one-fourth of all patients undergoing renal surgery are treated with a minimally invasive surgical approach. Based on our findings, patients undergoing radical nephrectomy present with worse baseline characteristics and experience higher perioperative morbidity and mortality, prolonged length of hospital stay, and increased in-hospital costs compared to patients undergoing partial nephrectomy. Still, this study's conclusions about the superiority of partial nephrectomy in terms of perioperative outcomes should be interpreted with caution, as the decision to perform radical nephrectomy might be driven by clinical factors related to an advanced tumor stage or tumor location and/or anatomical complexity.

Supplementary Materials: The following supporting information can be downloaded at: https://www.mdpi.com/article/10.3390/cancers16010097/s1, Supplementary Material Table S1: Baseline characteristics in low-, intermediate- and high-volume centers performing renal surgeries; Supplementary Material Figure S1: Annual trends in low-, intermediate- and high-volume centers performing renal surgeries, Supplementary Material Table S2: Regression analysis in low-, intermediate- and high-volume centers performing renal surgeries; Supplementary Material Figure S2: 30-day survival in low-, intermediate- and high-volume centers performing renal surgeries.

Author Contributions: Conceptualization, N.P. and G.B.S.; Methodology, I.B.; Validation, T.I.; Formal Analysis, N.P. and M.S.; Resources, C.S. and M.S.; Data Curation, G.B.S.; Writing—Original Draft, N.P. and M.S.; Writing—Review and Editing, G.B.S., C.S., I.B., T.I. and A.G.; Supervision, A.G. and M.S.; Project Administration, C.S.; Funding Acquisition, C.S. All authors have read and agreed to the published version of the manuscript.

Funding: This research did not receive any specific grant from funding agencies in the public, commercial, or not-for-profit sectors.

Institutional Review Board Statement: Ethical approval, was not required for the present study, in accordance with the national legislation and institutional requirements.

Informed Consent Statement: Written informed consent from the participants was not required for the present study, in accordance with the national legislation and institutional requirements.

Data Availability Statement: All data used in this work are stored in an anonymized fashion at the German Federal Statistical Office.

Conflicts of Interest: The authors declare no conflict of interest.

References

1. Bukavina, L.; Bensalah, K.; Bray, F.; Carlo, M.; Challacombe, B.; Karam, J.A.; Kassouf, W.; Mitchell, T.; Montironi, R.; O'Brien, T.; et al. Epidemiology of Renal Cell Carcinoma: 2022 Update. *Eur. Urol.* **2022**, *82*, 529–542. [CrossRef] [PubMed]
2. Huang, J.; Leung, D.K.-W.; Chan, E.O.-T.; Lok, V.; Leung, S.; Wong, I.; Lao, X.-Q.; Zheng, Z.-J.; Chiu, P.K.-F.; Ng, C.-F.; et al. A Global Trend Analysis of Kidney Cancer Incidence and Mortality and Their Associations with Smoking, Alcohol Consumption, and Metabolic Syndrome. *Eur. Urol. Focus* **2022**, *8*, 200–209. [CrossRef] [PubMed]
3. Diana, P.; Klatte, T.; Amparore, D.; Bertolo, R.; Carbonara, U.; Erdem, S.; Ingels, A.; Kara, O.; Marandino, L.; Marchioni, M.; et al. Screening programs for renal cell carcinoma: A systematic review by the EAU young academic urologists renal cancer working group. *World J. Urol.* **2023**, *41*, 929–940. [CrossRef] [PubMed]
4. Kutikov, A.; Fossett, L.K.; Ramchandani, P.; Tomaszewski, J.E.; Siegelman, E.S.; Banner, M.P.; Van Arsdalen, K.N.; Wein, A.J.; Malkowicz, S.B. Incidence of benign pathologic findings at partial nephrectomy for solitary renal mass presumed to be renal cell carcinoma on preoperative imaging. *Urology* **2006**, *68*, 737–740. [CrossRef] [PubMed]
5. Siva, S.; Ali, M.; Correa, R.J.M.; Muacevic, A.; Ponsky, L.; Ellis, R.J.; Lo, S.S.; Onishi, H.; Swaminath, A.; McLaughlin, M.; et al. 5-year outcomes after stereotactic ablative body radiotherapy for primary renal cell carcinoma: An individual patient data meta-analysis from IROCK (the International Radiosurgery Consortium of the Kidney). *Lancet Oncol.* **2022**, *23*, 1508–1516. [CrossRef] [PubMed]
6. Van Poppel, H.; Da Pozzo, L.; Albrecht, W.; Matveev, V.; Bono, A.; Borkowski, A.; Marechal, J.-M.; Klotz, L.; Skinner, E.; Keane, T.; et al. A Prospective randomized EORTC intergroup phase 3 study comparing the complications of elective nephron-sparing surgery and radical nephrectomy for low-stage renal cell carcinoma. *Eur. Urol.* **2007**, *51*, 1606–1615. [CrossRef]
7. Kim, S.P.; Thompson, R.H.; Boorjian, S.A.; Weight, C.J.; Han, L.C.; Murad, M.H.; Shippee, N.D.; Erwin, P.J.; Costello, B.A.; Chow, G.K.; et al. Comparative effectiveness for survival and renal function of partial and radical nephrectomy for localized renal tumors: A systematic review and meta-analysis. *J. Urol.* **2012**, *188*, 51–57. [CrossRef]
8. Mir, M.C.; Derweesh, I.; Porpiglia, F.; Zargar, H.; Mottrie, A.; Autorino, R. Partial Nephrectomy Versus Radical Nephrectomy for Clinical T1b and T2 Renal Tumors: A Systematic Review and Meta-analysis of Comparative Studies. *Eur. Urol.* **2017**, *71*, 606–6177. [CrossRef]
9. Crocerossa, F.; Carbonara, U.; Cantiello, F.; Marchioni, M.; Ditonno, P.; Mir, M.C.; Porpiglia, F.; Derweesh, I.; Hampton, L.J.; Damiano, R.; et al. Robot-assisted Radical Nephrectomy: A Systematic Review and Meta-analysis of Comparative Studies. *Eur. Urol.* **2021**, *80*, 428–439. [CrossRef]
10. Sun, M.; Bianchi, M.; Trinh, Q.-D.; Abdollah, F.; Schmitges, J.; Jeldres, C.; Shariat, S.F.; Graefen, M.; Montorsi, F.; Perrotte, P.; et al. Hospital volume is a determinant of postoperative complications, blood transfusion and length of stay after radical or partial nephrectomy. *J. Urol.* **2012**, *187*, 405–410. [CrossRef]

11. Hsu, R.C.J.; Salika, T.; Maw, J.; Lyratzopoulos, G.; Gnanapragasam, V.J.; Armitage, J.N. Influence of hospital volume on nephrectomy mortality and complications: A systematic review and meta-analysis stratified by surgical type. *BMJ Open* **2017**, *7*, e016833. [CrossRef] [PubMed]
12. Xia, L.; Pulido, J.E.; Chelluri, R.R.; Strother, M.C.; Taylor, B.L.; Raman, J.D.; Guzzo, T.J. Hospital volume and outcomes of robot-assisted partial nephrectomy. *BJU Int.* **2018**, *121*, 900–907. [CrossRef] [PubMed]
13. Bruins, H.M.; Veskimäe, E.; Hernández, V.; Neuzillet, Y.; Cathomas, R.; Compérat, E.M.; Cowan, N.C.; Gakis, G.; Espinós, E.L.; Lorch, A.; et al. The Importance of Hospital and Surgeon Volume as Major Determinants of Morbidity and Mortality after Radical Cystectomy for Bladder Cancer: A Systematic Review and Recommendations by the European Association of Urology Muscle-invasive and Metastatic Bladder Cancer Guideline Panel. *Eur. Urol. Oncol.* **2020**, *3*, 131–144. [PubMed]
14. Ljungberg, B.; Albiges, L.; Abu-Ghanem, Y.; Bedke, J.; Capitanio, U.; Dabestani, S.; Fernández-Pello, S.; Giles, R.H.; Hofmann, F.; Hora, M.; et al. European Association of Urology Guidelines on Renal Cell Carcinoma: The 2022 Update. *Eur. Urol.* **2022**, *82*, 399–410. [CrossRef] [PubMed]
15. Bjurlin, M.A.; Walter, D.; Taksler, G.B.; Huang, W.C.; Wysock, J.S.; Sivarajan, G.; Loeb, S.; Taneja, S.S.; Makarov, D.V. National trends in the utilization of partial nephrectomy before and after the establishment of AUA guidelines for the management of renal masses. *Urology* **2013**, *82*, 1283–1290. [CrossRef]
16. Campbell, S.C.; Uzzo, R.G.; Karam, J.A.; Chang, S.S.; Clark, P.E.; Souter, L. Renal Mass and Localized Renal Cancer: Evaluation, Management, and Follow-up: AUA Guideline: Part II. *J. Urol.* **2021**, *206*, 209–218. [CrossRef]
17. Fero, K.; Hamilton, Z.A.; Bindayi, A.; Murphy, J.D.; Derweesh, I.H. Utilization and quality outcomes of cT1a, cT1b and cT2a partial nephrectomy: Analysis of the national cancer database. *BJU Int.* **2018**, *121*, 565–574. [CrossRef]
18. Plante, K.; Stewart, T.M.; Wang, D.; Bratslavsky, G.; Formica, M. Treatment trends, determinants, and survival of partial and radical nephrectomy for stage I renal cell carcinoma: Results from the National Cancer Data Base, 2004–2013. *Int. Urol. Nephrol.* **2017**, *49*, 1375–1381. [CrossRef]
19. Hsu, R.C.J.; Barclay, M.; Loughran, M.A.; Lyratzopoulos, G.; Gnanapragasam, V.J.; Armitage, J.N. Time trends in service provision and survival outcomes for patients with renal cancer treated by nephrectomy in England 2000–2010. *BJU Int.* **2018**, *122*, 599–609. [CrossRef]
20. Chen, K.; Lee, A.; Huang, H.H.; Tay, K.J.; Sim, A.; Lee, L.S.; Cheng, C.W.S.; Ng, L.G.; Ho, H.S.S.; Yuen, J.S.P. Evolving trends in the surgical management of renal masses over the past two decades: A contemporary picture from a large prospectively-maintained database. *Int. J. Urol.* **2019**, *26*, 465–474. [CrossRef]
21. Yildirim, H.; Schuurman, M.S.; Widdershoven, C.V.; Lagerveld, B.W.; Brink, L.v.D.; Ruiter, A.E.C.; Beerlage, H.P.; van Moorselaar, R.J.A.; Graafland, N.M.; Bex, A.; et al. Variation in the management of cT1 renal cancer by surgical hospital volume: A nationwide study. *BJUI Compass* **2023**, *4*, 455–463. [CrossRef] [PubMed]
22. Alameddine, M.; Koru-Sengul, T.; Moore, K.J.; Miao, F.; Sávio, L.F.; Nahar, B.; Prakash, N.S.; Venkatramani, V.; Jue, J.S.; Punnen, S.; et al. Trends in Utilization of Robotic and Open Partial Nephrectomy for Management of cT1 Renal Masses. *Eur. Urol. Focus* **2019**, *5*, 482–487. [CrossRef] [PubMed]
23. Xia, L.; Talwar, R.; Taylor, B.L.; Shin, M.H.; Berger, I.B.; Sperling, C.D.; Chelluri, R.R.; Zambrano, I.A.; Raman, J.D.; Guzzo, T.J. National trends and disparities of minimally invasive surgery for localized renal cancer, 2010 to 2015. *Urol. Oncol. Semin. Orig. Investig.* **2019**, *37*, 182.e17–182.e27. [CrossRef] [PubMed]
24. Volpe, A.; Cadeddu, J.A.; Cestari, A.; Gill, I.S.; Jewett, M.A.; Joniau, S.; Kirkali, Z.; Marberger, M.; Patard, J.J.; Staehler, M.; et al. Contemporary management of small renal masses. *Eur. Urol.* **2011**, *60*, 501–515. [CrossRef] [PubMed]
25. Ferlay, J.; Colombet, M.; Soerjomataram, I.; Parkin, D.M.; Piñeros, M.; Znaor, A.; Bray, F. Cancer statistics for the year 2020: An overview. *Int. J. Cancer* **2021**, *149*, 778–789. [CrossRef] [PubMed]
26. Ferlay, J.; Steliarova-Foucher, E.; Lortet-Tieulent, J.; Rosso, S.; Coebergh, J.W.W.; Comber, H.; Forman, D.; Bray, F. Cancer incidence and mortality patterns in Europe: Estimates for 40 countries in 2012. *Eur. J. Cancer* **2013**, *49*, 1374–1403. [CrossRef] [PubMed]
27. Palumbo, C.; A Mistretta, F.; Knipper, S.; Mazzone, E.; Pecoraro, A.; Tian, Z.; Perrotte, P.; Antonelli, A.; Montorsi, F.; Shariat, S.F.; et al. Assessment of local tumor ablation and non-interventional management versus partial nephrectomy in T1a renal cell carcinoma. *Minerva Urol. Nefrol.* **2020**, *72*, 350–359. [CrossRef]
28. Su, Z.T.; Patel, H.D.; Huang, M.M.; Alam, R.; Cheaib, J.G.; Pavlovich, C.P.; Allaf, M.E.; Pierorazio, P.M. Active Surveillance versus Immediate Intervention for Small Renal Masses: A Cost-Effectiveness and Clinical Decision Analysis. *J. Urol.* **2022**, *208*, 794–803. [CrossRef]
29. Correa, R.J.; Louie, A.V.; Zaorsky, N.G.; Lehrer, E.J.; Ellis, R.; Ponsky, L.; Kaplan, I.; Mahadevan, A.; Chu, W.; Swaminath, A.; et al. The Emerging Role of Stereotactic Ablative Radiotherapy for Primary Renal Cell Carcinoma: A Systematic Review and Meta-Analysis. *Eur. Urol. Focus* **2019**, *5*, 958–969. [CrossRef]
30. Flegar, L.; Thoduka, S.G.; Mahnken, A.H.; Figiel, J.; Heers, H.; Aksoy, C.; Eisenmenger, N.; Groeben, C.; Huber, J.; Zacharis, A. Focal Therapy for Renal Cancer: Comparative Trends in the USA and Germany from 2006 to 2020 and Analysis of the German Health Care Landscape. *Urol. Int.* **2023**, *107*, 396–405. [CrossRef]
31. Znaor, A.; Lortet-Tieulent, J.; Laversanne, M.; Jemal, A.; Bray, F. International variations and trends in renal cell carcinoma incidence and mortality. *Eur. Urol.* **2015**, *67*, 519–530. [CrossRef] [PubMed]

32. Chow, W.-H.; Dong, L.M.; Devesa, S.S. Epidemiology and risk factors for kidney cancer. *Nat. Rev. Urol.* **2010**, *7*, 245–257. [CrossRef] [PubMed]
33. Li, J.; Zhang, Y.; Teng, Z.; Han, Z. Partial nephrectomy versus radical nephrectomy for cT2 or greater renal tumors: A systematic review and meta-analysis. *Minerva Urol. Nefrol.* **2019**, *71*, 435–444. [CrossRef] [PubMed]
34. Pierorazio, P.M.; Johnson, M.H.; Patel, H.D.; Sozio, S.M.; Sharma, R.; Iyoha, E.; Bass, E.B.; Allaf, M.E. Management of Renal Masses and Localized Renal Cancer: Systematic Review and Meta-Analysis. *J. Urol.* **2016**, *196*, 989–999. [CrossRef]

cancers

MDPI

Article

Non-Metastatic Clear Cell Renal Cell Carcinoma Immune Cell Infiltration Heterogeneity and Prognostic Ability in Patients Following Surgery

Daniel D. Shapiro [1,2,*], Taja Lozar [3], Lingxin Cheng [4], Elliot Xie [4], Israa Laklouk [5], Moon Hee Lee [1], Wei Huang [6], David F. Jarrard [1], Glenn O. Allen [1], Rong Hu [6], Toshi Kinoshita [6], Karla Esbona [6], Paul F. Lambert [3], Christian M. Capitini [7], Christina Kendziorski [4] and Edwin Jason Abel [1]

1 Department of Urology, University of Wisconsin School of Medicine and Public Health, Madison, WI 53792, USA
2 William S. Middleton Memorial Veterans Hospital, Madison, WI 53705, USA
3 McArdle Laboratory for Cancer Research, University of Wisconsin, Madison, WI 53706, USA
4 Department of Biostatistics and Medical Informatics, University of Wisconsin School of Medicine and Public Health, Madison, WI 53792, USA; lcheng74@wisc.edu (L.C.)
5 Department of Pathology, University of California, Los Angeles, Los Angeles, CA 90024, USA; ilaklouk@mednet.ucla.edu
6 Department of Pathology, University of Wisconsin School of Medicine and Public Health, Madison, WI 53792, USA; rhu6@wisc.edu (R.H.); kesbona@wisc.edu (K.E.)
7 Department of Pediatrics, University of Wisconsin School of Medicine and Public Health, Madison, WI 53792, USA; ccapitini@pediatrics.wisc.edu
* Correspondence: ddshapiro@wisc.edu

Simple Summary: It is difficult to predict which patients with non-metastatic clear cell renal cell carcinoma (ccRCC) will develop metastatic disease after nephrectomy. Recent studies suggest that immune cell infiltration within ccRCC tumors may impact tumor progression. This study assessed the number and type of immune cells in non-metastatic ccRCC tumors that were surgically removed and their ability to predict which patients developed metastatic disease. We found that higher levels of a specific immune cell ($CD8^+$ T cells) were linked to a lower risk of progressive disease. Patients who did progress had more exhausted $CD8^+$ T cells in the tumor microenvironment. Additionally, our study design accounted for tumor heterogeneity by sampling tumors in multiple locations and showed differences in the spatial distribution of $CD8^+$ T cells in tumors that progressed to metastatic disease. With further validation, this study shows that $CD8^+$ T cell infiltration within ccRCC tumors could be used as a prognostic biomarker to predict progression to metastatic disease.

Abstract: Predicting which patients will progress to metastatic disease after surgery for non-metastatic clear cell renal cell carcinoma (ccRCC) is difficult; however, recent data suggest that tumor immune cell infiltration could be used as a biomarker. We evaluated the quantity and type of immune cells infiltrating ccRCC tumors for associations with metastatic progression following attempted curative surgery. We quantified immune cell densities in the tumor microenvironment and validated our findings in two independent patient cohorts with multi-region sampling to investigate the impact of heterogeneity on prognostic accuracy. For non-metastatic ccRCC, increased $CD8^+$ T cell infiltration was associated with a reduced likelihood of progression to metastatic disease. Interestingly, patients who progressed to metastatic disease also had increased percentages of exhausted $CD8^+$ T cells. Finally, we evaluated the spatial heterogeneity of the immune infiltration and demonstrated that patients without metastatic progression had $CD8^+$ T cells in closer proximity to ccRCC cells. These data strengthen the evidence for $CD8^+$ T cell infiltration as a prognostic biomarker in non-metastatic ccRCC and demonstrate that multi-region sampling may be necessary to fully characterize immune infiltration within heterogeneous tumors. Tumor $CD8^+$ T cell infiltration should be investigated as a biomarker in adjuvant systemic therapy clinical trials for high-risk non-metastatic RCC.

Citation: Shapiro, D.D.; Lozar, T.; Cheng, L.; Xie, E.; Laklouk, I.; Lee, M.H.; Huang, W.; Jarrard, D.F.; Allen, G.O.; Hu, R.; et al. Non-Metastatic Clear Cell Renal Cell Carcinoma Immune Cell Infiltration Heterogeneity and Prognostic Ability in Patients Following Surgery. *Cancers* **2024**, *16*, 478. https://doi.org/10.3390/cancers16030478

Academic Editors: José I. López and Claudia Manini

Received: 27 December 2023
Revised: 18 January 2024
Accepted: 21 January 2024
Published: 23 January 2024

Keywords: renal cell carcinoma; tumor heterogeneity; immune microenvironment; T cell

1. Introduction

Approximately two-thirds of renal cell cancer (RCC) patients have clinically localized disease at presentation, for which surgery is commonly recommended [1]. Metastatic progression is exceptionally uncommon for stage 1 tumors (<5%) following surgery. However, the risk of metastatic progression increases with higher stage tumors and varies significantly among patients. For example, in patients with pT2 (tumors of $\geq$7 cm confined to the kidney), ~30% of patients progress from non-metastatic to metastatic disease, and pT3 tumors (non-metastatic tumors invading the perinephric fat or venous system) have a progression rate of ~50–70% [2–5]. Current prognostic models use clinical and pathological variables but have limited prognostic ability to identify patients who are at high risk of developing metastatic disease after surgery [6]. Correa et al. demonstrated that commonly used prognostic models perform poorly, with limited predictive accuracy demonstrated by c-indices ranging between 0.56 and 0.69 [6].

Currently, biomarkers for RCC are not used clinically for localized, locally advanced, or metastatic RCCs. However, biomarkers are critically needed to guide treatment decisions for patients with all stages of disease. Useful biomarkers must be developed and validated using independent cohorts, and they should provide actionable information that informs decision-making. As such, high-risk non-metastatic RCC patients are an ideal cohort for biomarker-informed decision-making because patients have to choose whether or not to be treated with adjuvant immune checkpoint inhibitor therapy after surgery [7,8]. Potential benefits of treatment include improved disease-free survival compared to placebo. However, patients treated with adjuvant therapy also have a 32% overall risk of adverse events and a 20% risk of serious adverse events [8]. In this subpopulation, prognostic biomarkers to improve prognostic ability may facilitate treatment of the patients most likely to develop metastatic disease while avoiding adverse events in lower risk patients.

There is a strong biologic rationale to investigate the immune microenvironment as a prognostic biomarker for non-metastatic clear cell RCC because it is one of the most highly immune-infiltrated solid tumors [9,10]. Additionally, metastatic ccRCC is highly responsive to treatment with immune checkpoint inhibitor therapy (ICI), with multiple approved ICI therapies for metastatic disease and adjuvant therapy [8,11–15]. To date, a few studies have investigated the immune microenvironment of non-metastatic ccRCC as a prognostic biomarker, and studies that have investigated the immune microenvironment have demonstrated conflicting results regarding the prognostic capability of infiltrating immune cells [16–18]. The objective of this study was to evaluate the prognostic value of immune infiltration within non-metastatic ccRCC following surgery and investigate whether immune cell heterogeneity, cellular composition, and spatial distribution of immune cells are correlated with prognosis.

2. Materials and Methods

2.1. Discovery Cohort Patient Selection

Institutional review board approval at the University of Wisconsin—Madison was obtained for this study (#2018-0018). Patients from a single center were included if they underwent surgery for non-metastatic ccRCC. Patients were eligible if they had a primary tumor with a maximum pathologic tumor diameter $\geq$7 cm in size (AJCC T stage 2) with no clinical evidence of nodal or other metastatic disease (Figure 1A). Clinical and pathologic data were extracted from medical records. All pathologic slides from each tumor were evaluated by a trained genitourinary pathologist. A representative whole slide was selected that contained tumor tissue and the highest degree of lymphocyte infiltration for further quantitative analysis.

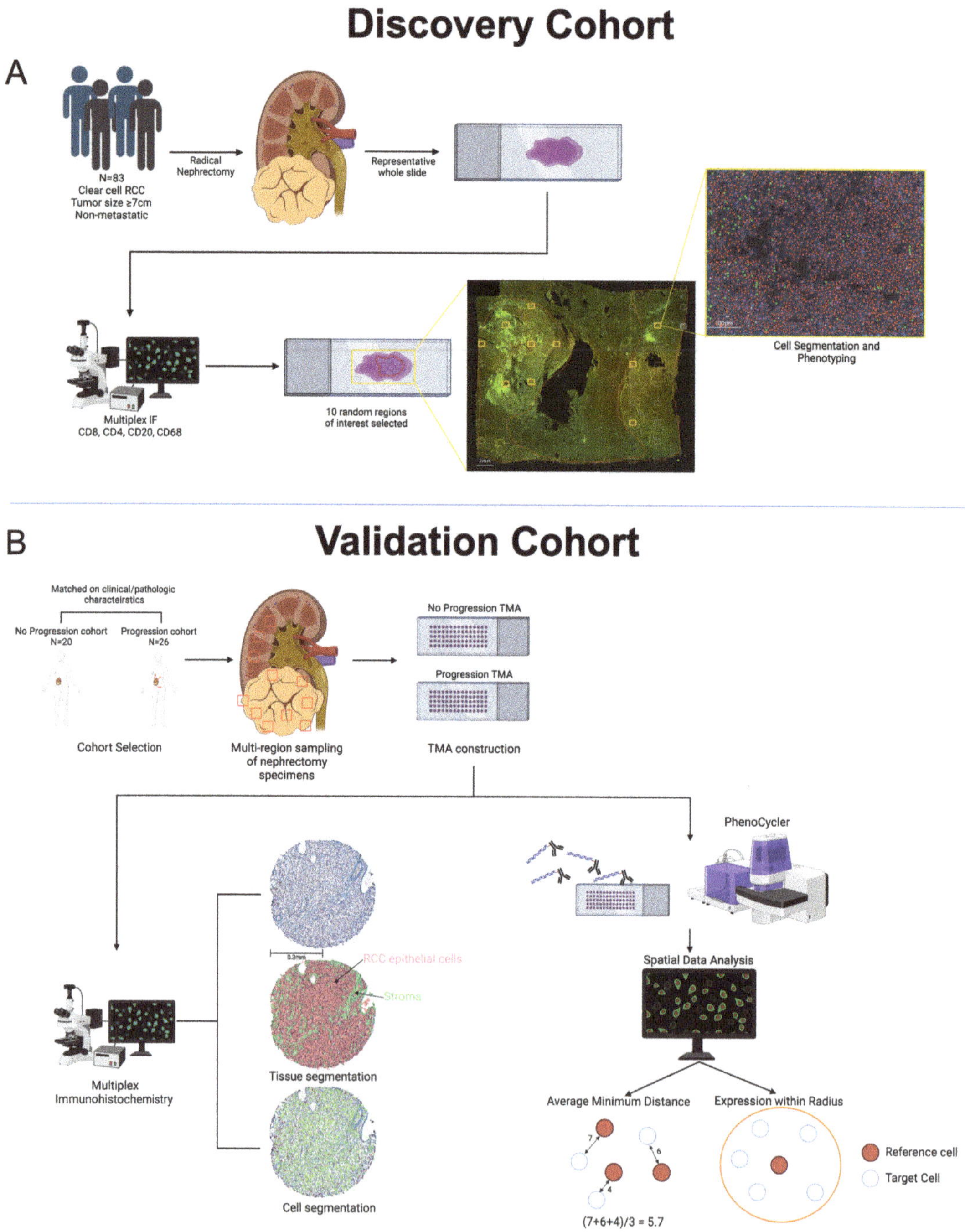

Figure 1. Evaluation of the immune microenvironment for surgically resected non-metastatic clear cell renal cell carcinoma. Two cohorts of patients were used to evaluate the immune microenvironment.

(**A**) A discovery cohort consisted of 83 patients with non-metastatic ccRCC with tumors ≥7 cm who underwent radical nephrectomy. Representative whole slides were selected for each case containing the greatest degree of tumor-infiltrating lymphocytes. These slides underwent multiplex immunofluorescence for immune cell markers. Ten regions of interest were selected randomly from each slide, and these regions underwent further cell segmentation, cell phenotyping, and subsequent quantification. (**B**) Findings from the discovery cohort were validated in a separate cohort of 46 patients with large (≥7 cm) non-metastatic ccRCC tumors. The patients used for the validation cohort were split into two groups (progression to metastatic disease versus no progression to metastatic disease), and these groups were matched on clinical characteristics that have previously been demonstrated to be prognostic for metastatic progression. Tissue microarrays (TMAs) were constructed from nephrectomy specimens. Each tumor underwent multi-region sampling, including 7–8 different tumor regions and 3 samples of non-adjacent normal renal parenchyma. These 10–11 cores from each tumor were used to construct the TMAs. The tissue microarrays were stained using multiplex immunohistochemistry for markers of immune cells and then quantified. The TMAs were additionally stained using the PhenoCycler multiplex immunofluorescence platform for further spatial quantification and measurements of immune cell exhaustion.

2.2. Discovery Cohort Histology and Immunostaining

After the selection of a representative slide from each tumor, additional slides were cut from archival FFPE tissue blocks. Slides were sectioned at 5 μm thickness and mounted on positively charged glass slides. Multiplex immunostaining was performed using the Opal method following the manufacturer's protocol (Akoya, Boston, MA, USA), and antibodies were applied in sequence as follows: CD4/Opal 540 (Ventana 790-4423, RTU, 10 min), CD68/Opal 570 (Ventana 790-2931, RTU, 5 min), CD8/Opal 620 (Ventana 790-4460, RTU, 30 min), CD20/Opal 650 (Ventana 790-4431, RTU, 15 min), and Pan-Cytokeratin (PCK)/Opal 570 (Abcam, ab234297, 1:1500, 15 min).

2.3. Discovery Cohort Slide Image Acquisition and Analysis

After staining, tissue section slides were scanned with Vectra 2 (Akoya, Boston, MA, USA) at 4× magnification, and then 10 fields of view were randomly selected per slide with PhenoChart software (Akoya, Boston, MA, USA) to be scanned with Vectra 2 at 20× magnification and analyzed with the InForm v.2.4 software (Akoya, Boston, MA, USA). A spectral library algorithm was created to unmix each individual signal, and the following pseudocolors were applied for image analysis: CD4 (yellow), CD68 (red), CD8 (green), CD20 (pink), and PCK (magenta). The InForm software was used to segment tissue compartments (epithelium vs. stroma) and subcellular compartments (nucleus, membrane, and cytoplasm). Individual cell segmentation was performed, and cell phenotypes were quantitated as cell density (cells/mm^2) (Figure 1A).

2.4. Validation Tissue Microarray Construction

To validate the findings of the discovery cohort as well as account for tissue heterogeneity, a validation cohort tissue microarray (TMA) was constructed. Tumors included in the TMA included two cohorts of patients who underwent surgery for non-metastatic ccRCC: those who progressed to metastatic disease (N = 26) and those who did not progress at last follow-up (N = 20) (Figure 1B). These patients were separate from the patients included in the discovery cohort. In order to account for potential other factors that may influence progression beyond the immune microenvironment, the patients in the progression and no progression cohorts were matched on age at the time of surgery, gender, performance status, preoperative serum C-reactive protein (CRP), neutrophil-lymphocyte ratio (NLR), pathologic stage, tumor size, tumor grade, and the presence of tumor thrombus. In order to account for differences in immune infiltration spatially throughout the tumor (i.e., spatial heterogeneity), the TMA was constructed to contain at least 7–8 tumor sample cores from each tumor, which were spatially distributed throughout the tumor. Additionally, 3 non-adjacent, non-tumor tissue cores of normal renal parenchyma were included for comparison. The TMA was constructed as previously described using a Manual Tissue Array (Beecher Instruments, Sun Prairie, WI, USA; model MTA-1), with 0.6 mm cores arranged 0.8 mm apart [19].

2.5. Validation of TMA-Multiplexed Immunohistochemistry

For this, 4 µm thick TMA slides were deparaffinized, and heat-induced epitope retrieval was carried out. Each slide underwent triple staining with two immune markers of interest and smooth muscle actin (SMA). For example, the progression and no progression TMA slides were incubated with the first primary antibody (CD4 Ventana #790-4423), then the slide was rinsed, and the Discovery OmniMap anti-Rabbit HRP (Ventana #760-4311) was applied. Slides were rinsed, and Reaction Buffer (Ventana #950-300) and Discovery ChromoMap DAB detection were applied (Ventana #760-159). Discovery Inhibitor (Ventana #760-4840) was applied, and then the slide was incubated with the second SMA antibody (Ventana #760-2833). Discovery OmniMap anti-mouse HQ (Ventana #740-4814) was applied with anti-HQ-HRP (Ventana #760-4820). The slide was rinsed, the Discovery Teal HRP detection kit (Ventana #760-247) was applied, and the denaturing agent Discovery Inhibitor (Ventana #760-4840) was applied. The slide was incubated with the primary anti-CD8 antibody (Ventana #790-4460), rinsed, and the Discovery UltraMap anti-rabbit AP (Ventana #760-4314) was applied. The slide was rinsed, and the Discovery Red HRP detection kit (Ventana #760-228) was applied. The slide was then washed, counterstained with hematoxylin, rinsed, dehydrated, and dipped in xylene. A similar protocol was used for our two additional markers of interest: CD20 (Ventana #760-2531) and CD68 (Ventana #790-2931).

2.6. Tissue Microarray Automated Image Acquisition and Analysis

Tissue microarray automated image acquisition and analysis followed a similar protocol as previously described [19]. Stained slides were loaded into the Vectra 2 slide scanner (Akoya, Boston, MA, USA), and an automated scanning protocol was created to acquire multi-spectral image cubes using the 20× objective. Control slides stained with only 1 chromogen were used to create a spectral library in Nuance v3.0.2 software (Akoya, Boston, MA, USA). Image cubes were opened in InForm v2.4 software, and images were chosen to set up an algorithm of differentiation for tissue and cell segmentation [19]. The algorithm was applied to the full set of TMA image cubes, and the expression of markers was quantified using segmentation settings sufficient to cover the cell membrane compartment. Protein expression data was exported, and cell density was calculated as cells/mm^2.

2.7. Tissue Microarray Staining and Image Acquisition Using the PhenoCycler Platform

We additionally performed high multiplex immunofluorescence using the PhenoCycler platform in order to understand in greater detail the spatial arrangement of cells within the tumor microenvironment using the validation TMAs [20]. Formalin-fixed paraffin-embedded validation TMA sections were analyzed using the PhenoCyclerTM-Open (formerly CODEX) platform (Akoya, Boston, MA, USA). Tissue was cut at 5 µm thickness and mounted onto superadhesive slides. The FFPE TMA tissue sections were dewaxed and rehydrated following standard histology methods. Epitope retrieval was performed using Tris-EDTA pH 9 for 20 min in a programmable pressure cooker (Instant PotTM). After allowing the pressure cooker to cool and depressurize naturally, the tissue was bleached by immersion in a solution of 4.5% (w/v) H202 and 20 mM NaOH in PBS under bright white LED light (A4-sized, Aibecy A4 Ultra Bright 25,000 Lux LED Light Box-Tracing Pads). The TMA was stained with a mixture of oligonucleotide-barcoded PhenoCycler antibodies (Supplementary Table S1) and post-fixed, according to the user manual. Imaging experiments were performed with the PhenoCyclerTM connected to a Keyence BZ-X800 epifluorescence microscope with a 20× objective (Nikon CFI Plan Apo 20×/0.75) (Figure 1B). The multiplex cycles were set up using Akoya's CODEX Instrument Manager (CIM), and the acquired images were then processed with the PhenoCyclerTM Processor to perform cycle alignment, background subtraction, deconvolution, extended depth of field, shading correction, tile registration, and stitching. The resulting QPTIFF image files were manually inspected for quality using the QuPath v0.5.0 software.

2.8. Cell Phenotype Labeling with PhenoCycler-Generated Images

After QPTIFF generation, QuPath v0.4 was used to process the 22-channel QPTIFF image generated from the PhenoCycler instrument. Cell segmentation is the first step of image analysis, which involves the identification of individual cells within the TMA cores and their corresponding 2-D x and y coordinates. Cell segmentation was performed with the StarDist (arXiv:1806.03535) nuclear segmentation algorithm using the DAPI channel and exported as a text file. Once cells were segmented, the CELESTA (cell-type identification with spatial information) algorithm was used to automate cell-type identification in our multiplexed image data [21]. CELESTA uses both protein expression and cell spatial neighborhood information from segmented imaging data for automated, unsupervised machine learning cell type identification. CELESTA requires two inputs, including the segmented imaging data as well as a cell-type signature matrix, which contains the cell types to be inferred from the markers used. For the purposes of this paper, we focused our matrix on Pan-Cytokeratin (PCK^+) malignant ccRCC cells, $CD45^+CD3e^+CD8^+$ T cells, and within $CD8^+$ T cells, and we defined $CD8^+$ T cells as exhausted if they co-expressed $PD1^+$ with $LAG3^+$ or $PD1^+$ with $TIM3^+$ [22–24]. CELESTA performs cell phenotype assignments and allows for cells to be plotted in two dimensions. Initially, quality control was conducted by excluding cells that exhibited either uniformly high or uniformly low expression across all markers, ensuring the removal of potential outliers or artifacts. Following the cell assignment outputs from the CELESTA algorithm, post-quality assessments were carried out by comparing these assignments with the original images obtained from the PhenoCycler platform. To further refine cell-type identification, we manually adjusted the 'high_expression_threshold' parameter for each cell type. This threshold, defining the minimum marker expression probability required for a marker to be considered as expressed, was determined by carefully comparing the expression probabilities against the corresponding PhenoCycler staining patterns for each marker.

2.9. Statistical Analysis and Spatial Analysis

Differences in clinical and pathologic characteristics were compared using the Wilcoxon rank-sum test and Fisher's exact test for continuous and categorical variables, respectively. The density of cell types was compared between patients who progressed versus those who did not progress using a mixed effects model, given that multiple samples (i.e., technical replicates) were taken from individual tumors from each patient. Survival analysis was performed using the Kaplan–Meier method, and differences in survival outcomes were estimated using the logrank test. When stratifying by immune cell density and comparing survival differences, the cell densities of each immune cell were averaged at the patient level. Patients were stratified into high- and low immune cell cohorts based on the median immune cell density. A Cox proportional hazard model was used to evaluate the association between survival and immune cell density. Logistic regression was used to evaluate the association between early progression and immune cell density. We used the PhenoCycler multiplex immunofluorescence data to evaluate the association of exhausted $CD8^+$ T cells. We evaluated cores with higher than the median percent of tumor cells and lower than the median $CD8^+$ T cell percentage as indicative of a "poor" immune response. For spatial analyses, TMA cores were categorized based on the median $CD8^+$ cell percentage within the progression and no progression cohorts. Subsequently, calculations were made to determine both the mean percentage of $CD8^+$ cells surrounding malignant cells and the average minimum distance between malignant cells and $CD8^+$ cells in the TMA cores with high $CD8^+$ T cell infiltration. These metrics were then compared between patients who progressed and those who did not, utilizing the Wilcoxon rank-sum test. The spatial analysis was conducted using SPIAT (v.1.2.3) [25]. The coefficient of variation (CV) was calculated for each tumor according to the following equation: $CV = \sigma/\mu$, where σ = the standard deviation of the cell density and μ = the mean cell density. The Wilcoxon rank-sum test was used to compare the CV between the progression versus no progression cohorts. Two-tailed p values < 0.05 were considered statistically significant. The statistical software used for analysis included STATA® SE v18 (StataCorp, College Station, TX, USA), GraphPad Prism v10.0.2 (GraphPad Software, Boston, MA, USA), and R v4.3.0.

3. Results

3.1. Discovery Cohort to Evaluate the Prognostic Impact of CD8⁺ T Cells

First, we evaluated whole slide images of non-metastatic ccRCC cases that were surgically resected at a single institution for patients with tumors ≥7 cm. Pathologic slides were evaluated by a trained GU pathologist, and representative slides were obtained from 83 cases. The cohort was divided based on metastatic progression at the last follow-up. Clinical and pathologic characteristics are listed in Table 1. Patients in the progression group were slightly older and tended to have higher grade disease. Median follow-up was similar in both groups, with a median follow-up of 37 months for those that did not progress versus 33 months for those that did progress (p = 0.9).

Table 1. Clinical and pathologic characteristics of the discovery cohort.

Discovery Cohort	No Progression	Progression	p Value
	N = 52	N = 31	
Median age, years (IQR)	58 (54–64)	64 (55–74)	0.03
Gender, no. of females (%)	15 (31)	6 (19)	0.3
ECOG Performance Status, no. (%)			1
0	43 (91)	29 (93)	
1	4 (9)	2 (7)	
Pathologic T-stage			1
T2	2 (4)	1 (3)	
T3–T4	50 (96)	30 (97)	
Median maximum pathologic tumor diameter, cm (IQR)	9.5 (8–11)	9.8 (8–12.5)	0.4
Grade, no. (%)			0.02
1–2	14 (27)	2 (6)	
3–4	38 (73)	29 (94)	
Thrombus	29 (56)	22 (71)	0.2
Died, no. (%)	11 (21)	17 (55)	0.004
Median follow-up, months (IQR)	37 (13–81)	33 (14–77)	0.9

ECOG = Eastern Cooperative Oncology Group, IQR = interquartile range.

Representative slides from each tumor were then sampled in 10 randomly selected regions to quantify the density of individual immune cells (Figure 1A). Slides were stained using multiplex immunofluorescence for $CD4^+$ T cells, $CD8^+$ T cells, $CD68^+$ macrophages, and $CD20^+$ B cells (Figure 2A). Based on other solid tumor types, we hypothesized that the $CD8^+$ T cell density would be higher among patients that did not progress. We indeed saw a higher mean $CD8^+$ T cell density among patients that did not progress (196.2 vs. 129.1 cells/mm^2, p = 0.005). We additionally found a higher density of $CD68^+$ macrophages and $CD20^+$ B cells among patients that progressed (Figure 2B). We then evaluated progression-free survival among patients in the discovery cohort. We calculated the average $CD8^+$ T cell density from each patient's tumor and stratified patients by the median $CD8^+$ T cell count from the entire discovery cohort. Patients stratified by high $CD8^+$ T cell density had improved progression-free survival (logrank p = 0.02) and a reduced risk of metastatic progression after surgery for localized ccRCC (HR 0.67, 95% CI 0.47–0.96; p = 0.03) (Figure 2C).

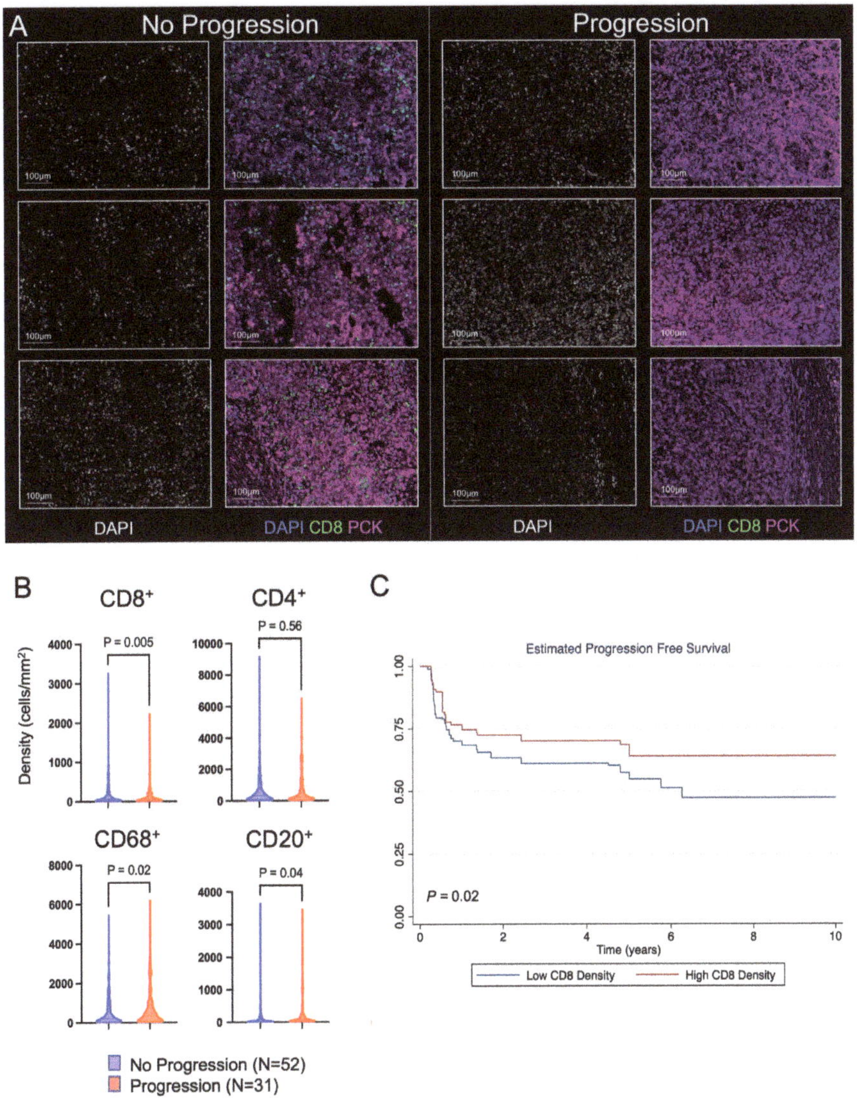

Figure 2. Discovery cohort immune microenvironment characteristics. (**A**) Representative multiplex immunofluorescence images of tumor tissue. The tissue images shown here were obtained by randomly selecting 3 patients in the no progression and progression cohorts. Within each cohort, the images on the left demonstrate the DAPI-stained individual cell nuclei, and the images on the right demonstrate the unmixed images, including the markers for DAPI, CD8 (effector T cell marker), and PCK (tumor cell marker). (**B**) Cell densities of individual immune cells for patients that either progressed to metastatic disease or did not. A mixed-effects model was used for comparison between the two groups. (**C**) Progression-free survival analysis. $CD8^+$ T cell densities were averaged at the individual patient level, and then patients were stratified by the overall median $CD8^+$ T cell density into high- and low-density cohorts. Differences in progression-free survival were calculated using the logrank test.

3.2. Validation Cohort to Evaluate the Prognostic Impact of CD8$^+$ T Cells

In order to validate the findings from the discovery cohort and reduce the risk of confounding factors that could influence prognosis and the degree of immune infiltration within the tumor microenvironment, we constructed TMAs of patients who either progressed or did not progress to metastatic disease after surgery for localized ccRCC. These TMAs consisted of patients with tumors ≥7 cm, and patients were matched on age at the time of surgery, gender, performance status, preoperative serum CRP, NLR, pathologic stage, tumor size, grade, and the presence of tumor thrombus. Multi-region sampling was performed, and tumors were sampled in seven to eight separate locations throughout the renal mass. Median follow-up for progressed and not progressed cohorts was 7 and 11 years from the date of surgery (Figure 1B). Table 2 demonstrates the validation cohorts' clinical and pathological characteristics. The progression and no progression cohorts had no significant differences between any of the matching criteria.

Table 2. Clinical and pathologic characteristics from the validation cohort.

Validation Cohort	No Progression	Progression	*p* Value
	N = 20	N = 26	
Median age, years (IQR)	61 (51–70)	57 (50–66)	0.3
Gender, no. of females (%)	8 (40)	9 (35)	0.8
ECOG Performance Status, no. (%)			0.6
0	19 (95)	23 (88)	
1	1 (5)	3 (12)	
Median preop NLR, (IQR)	4.4 (2.7–6.1)	2.9 (2.3–4.2)	0.2
Median preop CRP, (IQR)	2 (2–11)	1 (1–3)	0.2
Pathologic T-stage			0.2
T2	10 (50)	8 (31)	
T3–T4	10 (50)	18 (69)	
Median maximum pathologic tumor diameter, cm (IQR)	9 (7.2–9.4)	9.1 (8–13)	0.2
Grade, no. (%)			0.4
1–2	11 (55)	10 (38)	
3–4	9 (45)	16 (62)	
Thrombus	7 (35)	8 (31)	1
Died, no. (%)	5 (25)	14 (54)	0.07
Median follow-up, years (IQR)	11 (8–15)	7 (5–11)	0.08

ECOG = Eastern Cooperative Oncology Group, IQR = interquartile range, NLR = neutrophil lymphocyte ratio, CRP = C-reactive protein.

Immunohistochemistry staining was performed for markers of immune cells and quantified. As with the discovery cohort, the validation TMAs were stained for CD4$^+$ T cells, CD8$^+$ T cells, CD68$^+$ macrophages, and CD20$^+$ B cells (Figure 3A). Similar to the discovery cohort, an increased CD8$^+$ T cell density was demonstrated within the tumor immune microenvironment of localized ccRCC tumors that did not progress compared to those that did progress (median 330.6 cells/mm^2 vs. 181.5 cells/mm^2, $p = 0.004$). We did not find similar patterns among other immune cell markers that were found within the discovery cohort (Figure 3B). As with the discovery cohort, the CD8$^+$ T cell density was then averaged per patient, and patients were stratified by the median CD8$^+$ T cell density to create low- and high-density groups. Survival analysis again demonstrated that tumors highly infiltrated by CD8$^+$ T cells had a better prognosis (Figure 3C) and had a reduced risk of progression compared to tumors that had lower CD8$^+$ T cell infiltration (logrank $p = 0.02$).

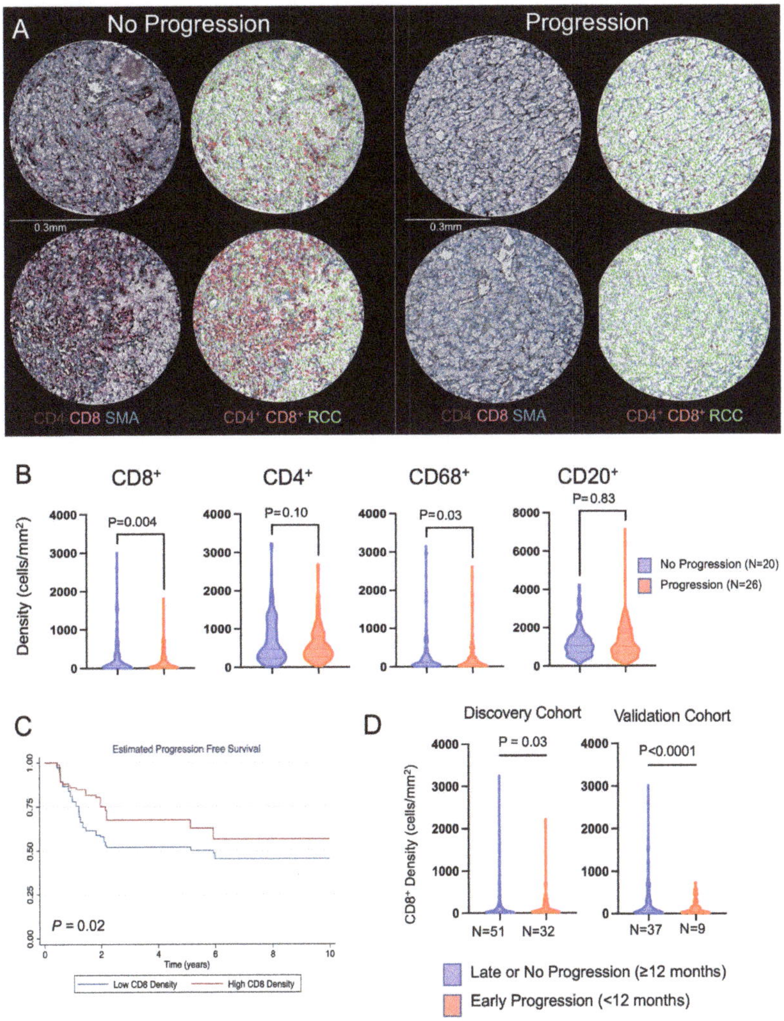

Figure 3. Validation cohort immune microenvironment characteristics. (**A**) Representative multiplex immunohistochemistry (IHC) images of tumor tissue. The tissue cores shown here were obtained by randomly selecting 2 patients from the no progression and progression cohorts. Within each cohort, the images on the left demonstrate brightfield IHC staining for CD4, CD8, and SMA (smooth muscle actin). The images on the right are the corresponding cell phenotypes as defined by the automated cell segmentation algorithm, with cell types being CD4$^+$ T cells in brown, CD8$^+$ T cells in red, and ccRCC cells in green. (**B**) Cell densities of individual immune cells for patients that either progressed to metastatic disease or did not. A mixed-effects model was used for comparison between the two groups. (**C**) CD8$^+$ T cell densities were averaged at the individual patient level, and then patients were stratified by the overall median CD8$^+$ T cell density into high- and low-density cohorts. Differences in progression-free survival were calculated using the logrank test. (**D**) For both the discovery and validation cohorts, patients were separated into those that progressed within 12 months of surgery and those that progressed after 12 months of surgery or did not progress at the last follow-up. A mixed-effects model was used to compare CD8$^+$ T cell densities between early and late/no progression groups.

We noted that a subset of patients in both the discovery (32/83, 39%) and validation (9/46, 20%) cohorts progressed early (<12 months after surgery). In our discovery cohort, we found that the mean $CD8^+$ T cell density was lower in patients that progressed early (<12 months) after surgery compared to patients that progressed ≥12 months after surgery or did not progress (137.2 cells/mm^2 vs. 189.5 cells/mm^2, $p = 0.03$). Patients with higher $CD8^+$ T cell density had reduced odds of early recurrence (OR 0.99, 95% CI 0.98–0.99; $p = 0.04$). The validation cohort had similar findings, with a mean $CD8^+$ T cell density of 128.9 cells/mm^2 in patients that progressed early compared to 275.1 cells/mm^2 among patients that progressed ≥12 months after surgery or did not progress (OR 0.99, 95% CI 0.98–0.99; $p = 0.02$) (Figure 3D).

We hypothesized that the specific $CD8^+$ T cell phenotype may be associated with progression to metastatic disease and that a higher percentage of exhausted $CD8^+$ T cells would be associated with a higher risk of progression. Using the PhenoCycler platform, we evaluated markers of T cell exhaustion. We found that a higher percentage of exhausted ($CD8^+PD1^+LAG3^+$) T cells were present in patients that recurred (Figure 4A). Additionally, we found that among patients who did not progress, the percent of exhausted $CD8^+$ T cells among all $CD8^+$ T cells was relatively stable regardless of how infiltrated the tumor tissue was. In the patients that progressed, however, there was a higher proportion of exhausted $CD8^+$ T cells among tumor cores that had lower CD8+ T cell infiltration, suggesting that tumors that progressed were more likely to have a weak immune response indicated by a lower $CD8^+$ T cell infiltration combined with a higher proportion of exhausted $CD8^+$ T cells (Figure 4B). Lastly, using logistic regression, we confirmed that an increasing percent of exhausted $PD1^+LAG3^+CD8^+$ T cells was associated with increased odds of progression to metastatic disease (OR 1.39, 95% CI 1.02–1.90, $p = 0.038$).

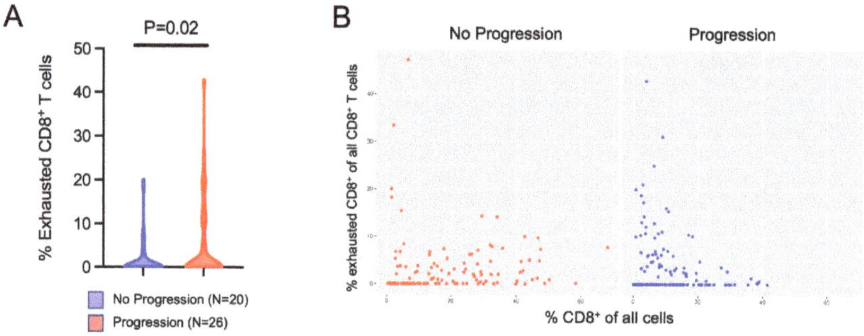

Figure 4. Evaluation of T cell exhaustion. Cell phenotyping was performed using the PhenoCycler multiplex immunofluorescence data. (**A**) The percentage of exhausted $CD8^+PD1^+LAG3^+$ T cells from all $CD8^+$ T cells was compared between patients that progressed versus those that did not progress. A mixed-effects model was used to compare the two cohorts. (**B**) The percentage of $CD8^+$ T cells was calculated from all cells present in individual tumor cores (x-axis). This was compared to the percentage of exhausted ($CD8^+PD1^+LAG3^+$) T cells from the total $CD8^+$ T cell population (y-axis).

3.3. Evaluation of Tissue Heterogeneity and Spatial Variation of Immune Cell Infiltration within Non-Metastatic ccRCC

Given that the validation TMAs were constructed using multi-regional sampling, we were able to use these TMAs to investigate the immune cell heterogeneity throughout individual RCC tumors. Figure 5 demonstrates the immune cell density within individual cores from each patient's ccRCC tumor. Substantial variation existed among individual cores. To quantify this variation, coefficients of variations (CVs) were calculated for each tumor, and the median CVs were compared between patients that progressed versus patients that did not progress. All immune cell markers had high CVs, with the highest being $CD8^+$ and $CD68^+$. $CD20^+$ B cell density had the lowest CV (Figure 5 and Table 3). No difference was found among the CVs for patients that progressed versus those that did not progress, indicating that all tumors, regardless of their capacity to progress, had substantial intratumoral immune cell density heterogeneity (Figure 5 and Table 3).

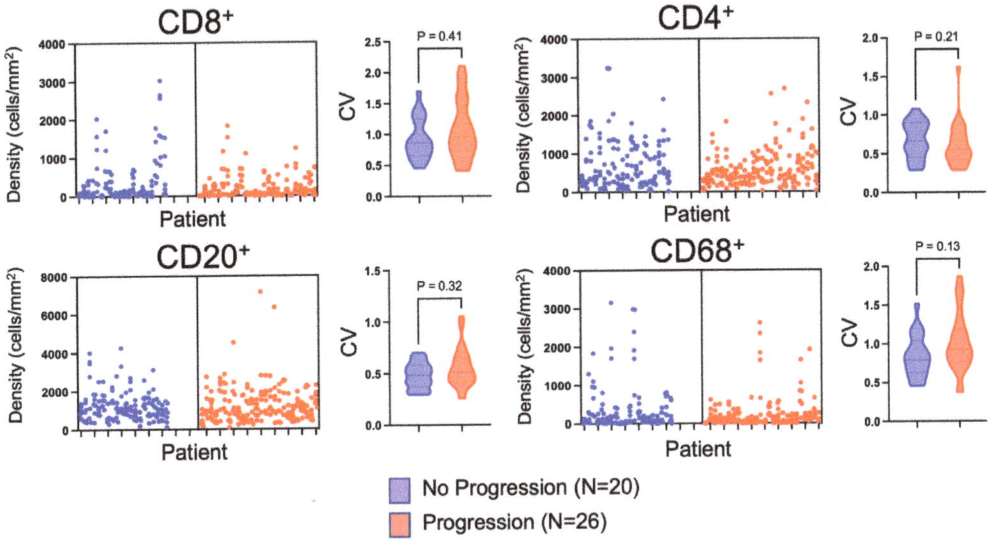

Figure 5. Evaluation of the heterogeneity. The cell densities of individual RCC cores were plotted for each patient, stratified by patients that progressed and those that did not progress, to visualize the variation in immune cell densities at the individual patient level. The x-axis of the cell density plots represents an individual patient. The cell density of each TMA core from that patient's tumor is represented by an individual dot stacked vertically. To quantify immune cell density heterogeneity, coefficients of variation (CV) were calculated between patients that progressed and did not progress, as represented by the violin plots.

Table 3. Coefficients of variation for immune infiltration among the validation cohort.

Median Coefficient of Variation	No Progression (N = 20)	Progression (N = 26)	p Value
CD8 CV	0.86	0.94	0.4
CD4 CV	0.66	0.54	0.2
CD20 CV	0.48	0.51	0.3
CD68 CV	0.79	0.92	0.1

CV = coefficient of variation.

To evaluate how immune cells infiltrated the kidney and RCC tumors, we assessed differences in benign versus tumor immune cell infiltration using the validation TMA, which captured both benign renal parenchyma and corresponding tumor tissue for comparison. As expected, we found tumor tissue contained a significantly higher degree of immune cell infiltration among all markers assessed (Figure 6A). We then evaluated immune cell penetration among the tumor epithelial cells versus within the surrounding stroma by segmenting each TMA core into epithelial and stromal compartments. Within the individual TMA RCC cores, immune cell infiltration was significantly less within the epithelial compartment, while the stromal compartment contained the majority of immune-infiltrating lymphocytes, except for $CD20^+$ cells, which were more prevalent within the epithelial compartment (Figure 6B). Overall, we demonstrate that the immune infiltration within the surrounding kidney is less than the tumor tissue. Additionally, within the RCC tissue, the majority of the immune infiltration occurs within the stromal tissue surrounding the tumor epithelial cells.

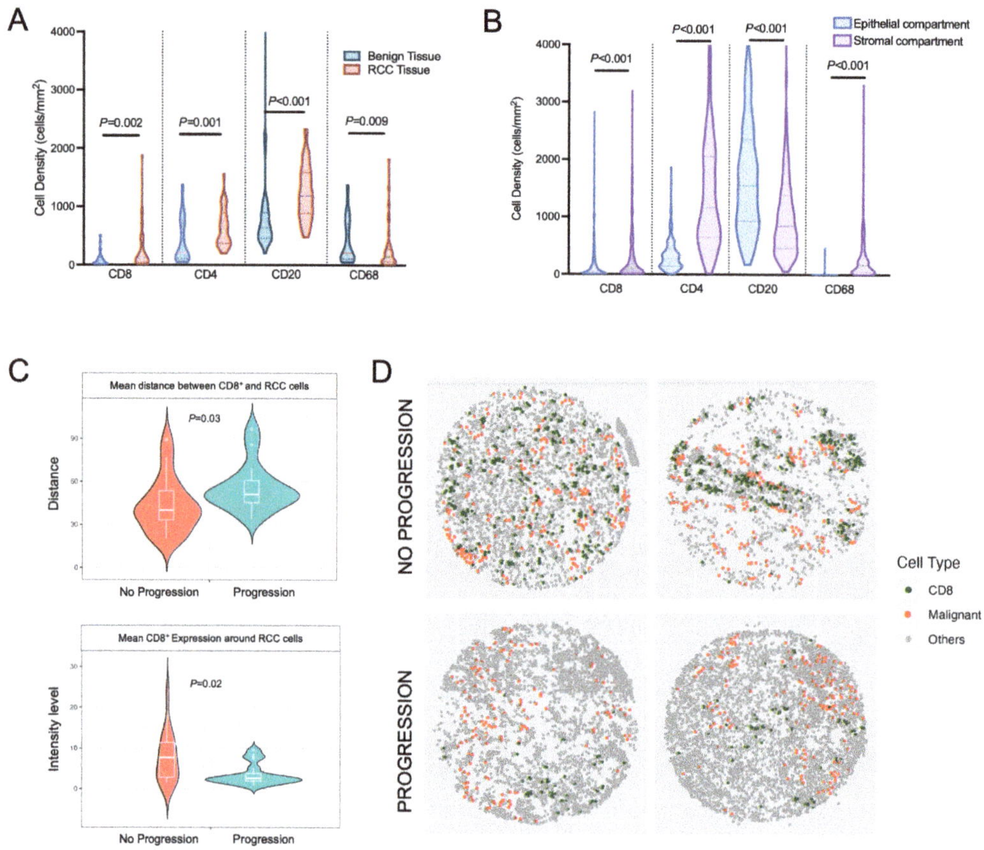

Figure 6. Evaluation of the spatial distribution of immune cell markers. (**A**) Patients within the validation cohort had immune cell markers quantified and compared between RCC tissue and non-adjacent normal renal parenchyma. (**B**) Using the validation cohort, tumor tissue cores were separated into stromal and epithelial compartments. The density of immune cells within each compartment was quantified and compared. Statistical comparisons of the mean cell densities between benign and RCC tissue as well as between stroma and epithelial compartments were made using the Wilcoxon paired signed rank test. (**C**) Validation TMAs were stained using the PhenoCycler platform and analyzed for spatial distribution of $CD8^+$ T cells in relation to malignant RCC cells (Pan-cytokeratin positive). The mean minimum distance between $CD8^+$ T cells and malignant cells was compared between the two cohorts. Also, the mean $CD8^+$ expression was calculated within a predefined radius of 150 pixels and compared between the progression and no progression cohorts. (**D**) Two representative TMA cores from the no progression and progression cohorts demonstrate differences in the spatial organization of $CD8^+$ T cells and ccRCC cells.

Using the PhenoCycler analysis of the validation TMAs, we were able to spatially resolve the distances between individual cell types within TMA tissue cores. We evaluated if the distance between $CD8^+$ T cells and RCC cells was different among patients who progressed versus those who did not progress. We demonstrated that patients who progressed to metastatic disease had smaller distances between $CD8^+$ T cells and RCC cells (mean distance of 20.72 µm for patients who progressed versus that of 15.92 µm for patients who did not progress, $p = 0.03$; Figure 6C). Additionally, the average $CD8^+$ infiltration was higher within a defined radius of 150 pixels around malignant cells for patients that did not progress compared to patients that did progress (Figure 6C). Taking all findings together, the inflammatory response appeared more robust, with a higher interaction between CD8$^+$ T cells and ccRCC cells among patients that did not progress to metastatic disease (Figure 6C,D).

4. Discussion

This study demonstrates that the composition of immune cells within the tumor microenvironment is spatially heterogeneous but has prognostic capacity in surgically resected, non-metastatic ccRCC. Using multi-region sampling, the degree of CD8$^+$ T cell infiltration was prognostic for metastatic progression, and a higher CD8$^+$ T cell density was associated with a better prognosis in a discovery cohort and validated in an independent cohort of patients. Furthermore, CD8$^+$ T cell infiltration was also associated with early progression to metastasis after surgery. Interestingly, patients who progressed to metastatic disease also had increased percentages of exhausted CD8$^+$ T cells. Finally, using spatial analysis, we demonstrated that patients without metastatic progression had CD8$^+$ T cells in closer proximity to ccRCC cells. Taken together, these findings demonstrate patterns of immune cell infiltration that are associated with metastatic progression in high-risk RCC, which could be used to identify patients for adjuvant therapy and clinical trials.

Prior studies have evaluated CD8$^+$ T cell infiltration as a biomarker but reported conflicting results regarding its prognostic capability [9,16–18,26,27]. Early data from Giraldo et al. suggested that increasing CD8$^+$ T cell density was associated with worse disease-free and overall survival for RCC patients [17]. This observation is unusual and differs from other solid tumors (e.g., glioma, melanoma, lung adenocarcinoma, and urothelial carcinoma), for which increased CD8$^+$ T cell infiltration is associated with a favorable prognosis [28]. Conversely, a later study by Jansen et al. found that CD8$^+$ T cell infiltration is associated with better outcomes [18]. There are multiple potential explanations for the conflicting data, which were investigated in this study. First, differences in study design or techniques used for evaluation of immune cell infiltration could significantly confound findings. Second, prior studies used patient cohorts that included both early and advanced-stage tumors that may have different quantities and phenotypes of immune cell infiltration. Finally, immune infiltration is heterogeneous throughout large tumors, as demonstrated in our study, which may skew findings depending on the quantity of immune infiltration in the portion of the tumor that was sampled.

This study was designed in a manner to address some possible differences in study design or techniques that might contribute to conflicting results from prior investigations. We used two independent cohorts and included only those patients who had non-metastatic disease in order to try and reduce the heterogeneity of the immune microenvironment phenotype (e.g., activated versus exhausted) [29]. Additionally, our validation cohort was matched on multiple clinical and pathologic variables known to be associated with progression; thus, differences in outcomes are less likely to be associated with these known confounding covariates. Also, unlike prior studies, our validation cohort had long-term follow-up, which is ideal for studies with non-metastatic patients. Given that over 90% of patients will progress within 10 years of surgery for localized ccRCC, the median follow-up for the validation cohort without progression was 11 years [3]. We focused our analysis on progression-free survival instead of overall survival, which is less likely to be confounded by changes over time in the systemic treatment availability for metastatic RCC (e.g., development-targeted therapy or immune checkpoint therapy). Lastly, we chose to utilize protein expression for the quantification of immune cells, compared to many prior studies that utilized gene expression. Protein expression provides a more direct measurement of the types and states (e.g., exhausted or activated) of immune cells present in the tumor microenvironment rather than inferring the presence of immune cells from gene expression signatures.

The immune microenvironment composition is dynamic during progression from early to advanced tumors, and studies investigating the immune microenvironment as a prognostic biomarker must account for this in their design. Contradictory results regarding the prognostic value of $CD8^+$ T cells may result when different-stage tumors are analyzed in aggregate. In the study by Giraldo et al., which suggested that immune infiltration was associated with poor outcomes, over half (54%) of the ccRCC tumors were locally advanced or metastatic (stage III or IV) tumors. In the metastatic setting, other studies have suggested that $CD8^+$ T cell infiltration is associated with a worse prognosis, likely attributable to an immune-exhausted state [29–31]. Similar to Jansen et al. [18], we found that increasing $CD8^+$ T cell infiltration was associated with improved progression-free survival among ccRCC tumors that were non-metastatic. By focusing our analysis for biomarker development on only non-metastatic disease, these data may avoid the confounding influence of different clinical tumor stages on the quantity and type of immune infiltration. Additionally, we demonstrated that increased exhaustion of $CD8^+$ T cells, defined by co-expression of $PD1^+$ and $LAG3^+$, was associated with increased odds of progression. This supports the concept that the immune microenvironment evolves to a more exhausted phenotype in patients who develop advanced disease.

A key objective of this study was to evaluate immune microenvironment heterogeneity within high-risk non-metastatic ccRCC tumors because this subpopulation would benefit from a prognostic biomarker. This study demonstrated that there is substantial variability in immune cell infiltration within large tumors, creating a potential for sampling error in prior studies when only one area is used for biomarker development. The coefficients of variation for all of the immune cells evaluated were high, and the majority of immune cells were located outside the epithelial compartment within the stromal compartment, suggesting the immune cells have limited capacity to penetrate the tumor beyond the invasive margin. Given the high degree of heterogeneity within ccRCC tumors, our study is strengthened in that our validation cohort sampled seven to eight different locations throughout the primary tumor to adequately address issues of tumor heterogeneity and accurately reflect the degree of immune infiltration. While the optimal sampling strategy is not known, a study by the TRACERx Renal Consortium evaluated the number of biopsies that are required to adequately capture the genetic drives of ccRCC [32]. The study suggested that for larger tumors, between four and eight biopsies are needed to capture the majority of genetic driver events [32]. We applied this rationale to our study design by selecting seven to eight different tumor regions for TMA construction in an effort to evaluate tumor heterogeneity. Our findings also have implications for using immune cells as biomarkers. We demonstrate that using only a single or few biopsy locations is unlikely to completely reflect the tumor immune infiltration. Future efforts to evaluate immune infiltration as a prognostic biomarker should be focused on defining an ideal tissue sampling strategy.

We demonstrated that the spatial organization of immune and malignant cells within the primary tumor differs between patients who progress to metastatic disease versus those who do not. Among patients who did not progress, $CD8^+$ T cells were more closely associated spatially with tumor cells. It appears that not only the degree of $CD8^+$ T cell inflammation but also the organization of these cells around malignant cells is greater among patients who do not progress to metastatic disease. While similar findings have been demonstrated in colorectal cancer, the prognostic value of the spatial organization of immune cells within the non-metastatic ccRCC microenvironment has not been well characterized [33,34].

We showed that $CD8^+$ T cell infiltration can help identify patients likely to rapidly progress (within 12 months) after surgery. Both the discovery and validation cohorts demonstrated that higher $CD8^+$ T cell infiltration was associated with a reduced risk of rapid progression. Identifying patients at risk of rapid progression is critically important, particularly in light of the newly approved adjuvant immunotherapy, pembrolizumab [8]. With further validation, using $CD8^+$ T cell infiltration as a prognostic biomarker could identify patients who may benefit from adjuvant immunotherapy. Currently, adjuvant immunotherapy is approved for a broad, heterogeneous group of patients. Identifying the patients most likely to benefit (i.e., those at highest risk of progression) will reduce the number of patients receiving unnecessary immunotherapy, decreasing both the cost of therapy and exposure to immune-related toxicity [35]. A future study will focus on not only the prognostic value of the immune microenvironment but also its ability to predict response to adjuvant immunotherapy in a non-metastatic setting.

This study has limitations. While we did attempt to address tumor heterogeneity by sampling multiple tumor locations, heterogeneity likely still impacts our results. The techniques used to quantify immune cells have inherent limitations, including the ability of antibodies to bind to cell surface proteins, which may lead to overestimation or underestimation of the number of immune cells present. We attempted to address these issues with multiple quality control measures, but errors are still possible. The findings from this study are primarily associative, and causal mechanisms defining why higher $CD8^+$ T cell infiltration is associated with reduced progression cannot be determined from this study alone, requiring further investigation.

5. Conclusions

In conclusion, the immune microenvironment in non-metastatic ccRCC is highly heterogeneous. If using components of the immune microenvironment as a potential clinically applicable prognostic biomarker, multiple samples should be obtained, with the optimal strategy to be determined in future studies. Our study demonstrates that $CD8^+$ T cells are associated with prognosis. Increased infiltration with non-exhausted $CD8^+$ T cells correlated with reduced rates of rapid progression after surgery and improved progression-free survival.

Supplementary Materials: The following supporting information can be downloaded at: https://www.mdpi.com/article/10.3390/cancers16030478/s1, Table S1: PhenoCycler panel used for staining validation tissue microarray.

Author Contributions: Conceptualization, D.D.S., T.L. and E.J.A.; methodology, D.D.S., E.X., L.C., C.K. and E.J.A.; formal analysis, D.D.S., T.L., E.X., L.C. and C.K.; resources, D.D.S. and E.J.A.; data curation, D.D.S., T.L., L.C., E.X., C.K. and E.J.A.; writing—original draft preparation, D.D.S., T.L. and E.J.A.; writing—review and editing, D.D.S., T.L., L.C., E.X., I.L., M.H.L., W.H., D.F.J., C.M.C., G.O.A., R.H., T.K., K.E., P.F.L., C.K. and E.J.A.; visualization, D.D.S., L.C. and E.X.; supervision, D.D.S. and E.J.A.; funding acquisition, D.D.S. and E.J.A. All authors have read and agreed to the published version of the manuscript.

Funding: This study was supported in part by a Wisconsin Urological Society Pilot Grant (DDS) and by NCI/NIH P30 CA014520 to the University of Wisconsin Carbone Cancer Center (UWCCC).

Institutional Review Board Statement: The study was conducted according to the guidelines of the Declaration of Helsinki and approved by the Institutional Review Board of the University of Wisconsin—Madison (protocol 2018-0018).

Informed Consent Statement: Not applicable.

Data Availability Statement: All data will be made available upon reasonable request to the corresponding author.

Acknowledgments: The authors thank the University of Wisconsin Translational Research Initiatives in Pathology Laboratory (TRIP), supported by the UW Department of Pathology and Laboratory Medicine and UWCCC (P30 CA014520), for the use of its facilities and services. Figures are created using BioRender.com and exported under a paid subscription. The contents of this article do not necessarily reflect the views or policies of the Department of Health and Human Services, nor does the mention of trade names, commercial products, or organizations imply endorsement by the US government.

Conflicts of Interest: CMC reports honorarium from Bayer, Elephas, Nektar Therapeutics, Novartis, and WiCell Research Institute, who had no input in the study design, analysis, manuscript preparation, or decision to submit for publication. The authors declare that no other competing interests exist.

References

1. Siegel, R.L.; Miller, K.D.; Wagle, N.S.; Jemal, A. Cancer statistics, 2023. *CA Cancer J. Clin.* **2023**, *73*, 17–48. [CrossRef] [PubMed]
2. Campbell, S.C.; Clark, P.E.; Chang, S.S.; Karam, J.A.; Souter, L.; Uzzo, R.G. Renal mass and localized renal cancer: Evaluation, management, and follow-up: Aua guideline: Part I. *J. Urol.* **2021**, *206*, 199–208. [CrossRef]
3. Campbell, S.C.; Uzzo, R.G.; Karam, J.A.; Chang, S.S.; Clark, P.E.; Souter, L. Renal mass and localized renal cancer: Evaluation, management, and follow-up: Aua guideline: Part II. *J. Urol.* **2021**, *206*, 209–218. [CrossRef] [PubMed]
4. Dabestani, S.; Beisland, C.; Stewart, G.D.; Bensalah, K.; Gudmundsson, E.; Lam, T.B.; Gietzmann, W.; Zakikhani, P.; Marconi, L.; Fernandéz-Pello, S.; et al. Long-term outcomes of follow-up for initially localised clear cell renal cell carcinoma: Recur database analysis. *Eur. Urol. Focus* **2019**, *5*, 857–866. [CrossRef]
5. Abel, E.J.; Margulis, V.; Bauman, T.M.; Karam, J.A.; Christensen, W.P.; Krabbe, L.M.; Haddad, A.; Golla, V.; Wood, C.G. Risk factors for recurrence after surgery in non-metastatic rcc with thrombus: A contemporary multicentre analysis. *BJU Int.* **2016**, *117*, E87–E94. [CrossRef]
6. Correa, A.F.; Jegede, O.; Haas, N.B.; Flaherty, K.T.; Pins, M.R.; Messing, E.M.; Manola, J.; Wood, C.G.; Kane, C.J.; Jewett, M.A.S.; et al. Predicting renal cancer recurrence: Defining limitations of existing prognostic models with prospective trial-based validation. *J. Clin. Oncol.* **2019**, *37*, 2062–2071. [CrossRef] [PubMed]
7. Powles, T.; Tomczak, P.; Park, S.H.; Venugopal, B.; Ferguson, T.; Symeonides, S.N.; Hajek, J.; Gurney, H.; Chang, Y.H.; Lee, J.L.; et al. Pembrolizumab versus placebo as post-nephrectomy adjuvant therapy for clear cell renal cell carcinoma (keynote-564): 30-month follow-up analysis of a multicentre, randomised, double-blind, placebo-controlled, phase 3 trial. *Lancet Oncol.* **2022**, *23*, 1133–1144. [CrossRef]
8. Choueiri, T.K.; Tomczak, P.; Park, S.H.; Venugopal, B.; Ferguson, T.; Chang, Y.-H.; Hajek, J.; Symeonides, S.N.; Lee, J.L.; Sarwar, N.; et al. Adjuvant pembrolizumab after nephrectomy in renal-cell carcinoma. *N. Engl. J. Med.* **2021**, *385*, 683–694. [CrossRef]
9. Şenbabaoğlu, Y.; Gejman, R.S.; Winer, A.G.; Liu, M.; Allen, E.M.V.; Velasco, G.d.; Miao, D.; Ostrovnaya, I.; Drill, E.; Luna, A.; et al. Tumor immune microenvironment characterization in clear cell renal cell carcinoma identifies prognostic and immunotherapeutically relevant messenger rna signatures. *Genome Biol.* **2016**, *17*, 231. [CrossRef]
10. Lasorsa, F.; Rutigliano, M.; Milella, M.; Ferro, M.; Pandolfo, S.D.; Crocetto, F.; Tataru, O.S.; Autorino, R.; Battaglia, M.; Ditonno, P.; et al. Cellular and molecular players in the tumor microenvironment of renal cell carcinoma. *J. Clin. Med.* **2023**, *12*, 3888. [CrossRef]
11. Motzer, R.J.; Tannir, N.M.; McDermott, D.F.; Frontera, O.A.; Melichar, B.; Choueiri, T.K.; Plimack, E.R.; Barthélémy, P.; Porta, C.; George, S.; et al. Nivolumab plus ipilimumab versus sunitinib in advanced renal-cell carcinoma. *N. Engl. J. Med.* **2018**, *378*, 1277–1290. [CrossRef]
12. Motzer, R.J.; Penkov, K.; Haanen, J.; Rini, B.; Albiges, L.; Campbell, M.T.; Venugopal, B.; Kollmannsberger, C.; Negrier, S.; Uemura, M.; et al. Avelumab plus axitinib versus sunitinib for advanced renal-cell carcinoma. *N. Engl. J. Med.* **2019**, *380*, 1103–1115. [CrossRef] [PubMed]
13. Motzer, R.; Alekseev, B.; Rha, S.Y.; Porta, C.; Eto, M.; Powles, T.; Grünwald, V.; Hutson, T.E.; Kopyltsov, E.; Méndez-Vidal, M.J.; et al. Lenvatinib plus pembrolizumab or everolimus for advanced renal cell carcinoma. *N. Engl. J. Med.* **2021**, *384*, 1289–1300. [CrossRef] [PubMed]
14. Rini, B.I.; Plimack, E.R.; Stus, V.; Gafanov, R.; Hawkins, R.; Nosov, D.; Pouliot, F.; Alekseev, B.; Soulieres, D.; Melichar, B.; et al. Pembrolizumab plus axitinib versus sunitinib for advanced renal-cell carcinoma. *N. Engl. J. Med.* **2019**, *380*, 1116–1127. [CrossRef] [PubMed]
15. Lasorsa, F.; di Meo, N.A.; Rutigliano, M.; Milella, M.; Ferro, M.; Pandolfo, S.D.; Crocetto, F.; Tataru, O.S.; Autorino, R.; Battaglia, M.; et al. Immune checkpoint inhibitors in renal cell carcinoma: Molecular basis and rationale for their use in clinical practice. *Biomedicines* **2023**, *11*, 1071. [CrossRef] [PubMed]
16. Shapiro, D.D.; Dolan, B.; Laklouk, I.A.; Rassi, S.; Lozar, T.; Emamekhoo, H.; Wentland, A.L.; Lubner, M.G.; Abel, E.J. Understanding the tumor immune microenvironment in renal cell carcinoma. *Cancers* **2023**, *15*, 2500. [CrossRef] [PubMed]
17. Giraldo, N.A.; Becht, E.; Pagès, F.; Skliris, G.; Verkarre, V.; Vano, Y.; Mejean, A.; Saint-Aubert, N.; Lacroix, L.; Natario, I.; et al. Orchestration and prognostic significance of immune checkpoints in the microenvironment of primary and metastatic renal cell cancer. *Clin. Cancer Res.* **2015**, *21*, 3031–3040. [CrossRef]
18. Jansen, C.S.; Prokhnevska, N.; Master, V.A.; Sanda, M.G.; Carlisle, J.W.; Bilen, M.A.; Cardenas, M.; Wilkinson, S.; Lake, R.; Sowalsky, A.G.; et al. An intra-tumoral niche maintains and differentiates stem-like cd8 t cells. *Nature* **2019**, *576*, 465–470. [CrossRef]
19. Bauman, T.M.; Huang, W.; Lee, M.H.; Abel, E.J. Neovascularity as a prognostic marker in renal cell carcinoma. *Hum. Pathol.* **2016**, *57*, 98–105. [CrossRef]
20. Black, S.; Phillips, D.; Hickey, J.W.; Kennedy-Darling, J.; Venkataraaman, V.G.; Samusik, N.; Goltsev, Y.; Schürch, C.M.; Nolan, G.P. Codex multiplexed tissue imaging with dna-conjugated antibodies. *Nat. Protoc.* **2021**, *16*, 3802–3835. [CrossRef]
21. Zhang, W.; Li, I.; Reticker-Flynn, N.E.; Good, Z.; Chang, S.; Samusik, N.; Saumyaa, S.; Li, Y.; Zhou, X.; Liang, R.; et al. Identification of cell types in multiplexed in situ images by combining protein expression and spatial information using celesta. *Nat. Methods* **2022**, *19*, 759–769. [CrossRef] [PubMed]

22. Baitsch, L.; Baumgaertner, P.; Devêvre, E.; Raghav, S.K.; Legat, A.; Barba, L.; Wieckowski, S.; Bouzourene, H.; Deplancke, B.; Romero, P.; et al. Exhaustion of tumor-specific cd8+ t cells in metastases from melanoma patients. *J. Clin. Investig.* **2011**, *121*, 2350–2360. [CrossRef] [PubMed]
23. Ahmadzadeh, M.; Johnson, L.A.; Heemskerk, B.; Wunderlich, J.R.; Dudley, M.E.; White, D.E.; Rosenberg, S.A. Tumor antigen-specific cd8 t cells infiltrating the tumor express high levels of pd-1 and are functionally impaired. *Blood* **2009**, *114*, 1537–1544. [CrossRef] [PubMed]
24. Giles, J.R.; Globig, A.-M.; Kaech, S.M.; Wherry, E.J. Cd8+ t cells in the cancer-immunity cycle. *Immunity* **2023**, *56*, 2231–2253. [CrossRef]
25. Feng, Y.; Yang, T.; Zhu, J.; Li, M.; Doyle, M.; Ozcoban, V.; Bass, G.T.; Pizzolla, A.; Cain, L.; Weng, S.; et al. Spatial analysis with spiat and spasim to characterize and simulate tissue microenvironments. *Nat. Commun.* **2023**, *14*, 2697. [CrossRef]
26. Giraldo, N.A.; Becht, E.; Vano, Y.; Petitprez, F.; Lacroix, L.; Validire, P.; Sanchez-Salas, R.; Ingels, A.; Oudard, S.; Moatti, A.; et al. Tumor-infiltrating and peripheral blood t-cell immunophenotypes predict early relapse in localized clear cell renal cell carcinoma. *Clin. Cancer Res.* **2017**, *23*, 4416–4428. [CrossRef]
27. Braun, D.A.; Hou, Y.; Bakouny, Z.; Ficial, M.; Angelo, M.S.; Forman, J.; Ross-Macdonald, P.; Berger, A.C.; Jegede, O.A.; Elagina, L.; et al. Interplay of somatic alterations and immune infiltration modulates response to pd-1 blockade in advanced clear cell renal cell carcinoma. *Nat. Med.* **2020**, *26*, 909–918. [CrossRef]
28. Varn, F.S.; Wang, Y.; Mullins, D.W.; Fiering, S.; Cheng, C. Systematic pan-cancer analysis reveals immune cell interactions in the tumor microenvironment. *Cancer Res.* **2017**, *77*, 1271–1282. [CrossRef]
29. Braun, D.A.; Street, K.; Burke, K.P.; Cookmeyer, D.L.; Denize, T.; Pedersen, C.B.; Gohil, S.H.; Schindler, N.; Pomerance, L.; Hirsch, L.; et al. Progressive immune dysfunction with advancing disease stage in renal cell carcinoma. *Cancer Cell* **2021**, *39*, 632–648.e638. [CrossRef]
30. Clark, D.J.; Dhanasekaran, S.M.; Petralia, F.; Pan, J.; Song, X.; Hu, Y.; Leprevost, F.d.V.; Reva, B.; Lih, T.-S.M.; Chang, H.-Y.; et al. Integrated proteogenomic characterization of clear cell renal cell carcinoma. *Cell* **2019**, *179*, 964–983.e931. [CrossRef]
31. Hakimi, A.A.; Voss, M.H.; Kuo, F.; Sanchez, A.; Liu, M.; Nixon, B.G.; Vuong, L.; Ostrovnaya, I.; Chen, Y.-B.; Reuter, V.; et al. Transcriptomic profiling of the tumor microenvironment reveals distinct subgroups of clear cell renal cell cancer—Data from a randomized phase iii trial. *Cancer Discov.* **2019**, *9*, CD-18-0957. [CrossRef] [PubMed]
32. Turajlic, S.; Xu, H.; Litchfield, K.; Rowan, A.; Horswell, S.; Chambers, T.; O'Brien, T.; Lopez, J.I.; Watkins, T.B.K.; Nicol, D.; et al. Deterministic evolutionary trajectories influence primary tumor growth: Tracerx renal. *Cell* **2018**, *173*, 595–610.e11. [CrossRef] [PubMed]
33. Schürch, C.M.; Bhate, S.S.; Barlow, G.L.; Phillips, D.J.; Noti, L.; Zlobec, I.; Chu, P.; Black, S.; Demeter, J.; McIlwain, D.R.; et al. Coordinated cellular neighborhoods orchestrate antitumoral immunity at the colorectal cancer invasive front. *Cell* **2020**, *182*, 1341–1359.e19. [CrossRef] [PubMed]
34. Bhate, S.S.; Barlow, G.L.; Schürch, C.M.; Nolan, G.P. Tissue schematics map the specialization of immune tissue motifs and their appropriation by tumors. *Cell Syst.* **2021**, *13*, 109–130.e6. [CrossRef]
35. Sharma, V.; Wymer, K.M.; Joyce, D.D.; Moriarty, J.; Khanna, A.; Borah, B.J.; Thompson, R.H.; Costello, B.A.; Leibovich, B.C.; Boorjian, S.A. Cost-effectiveness of adjuvant pembrolizumab after nephrectomy for high-risk renal cell carcinoma: Insights for patient selection from a markov model. *J. Urol.* **2023**, *209*, 89–98. [CrossRef]

cancers

MDPI

Article

Targeting NPC1 in Renal Cell Carcinoma

Rushaniya Fazliyeva [1], Peter Makhov [2], Robert G. Uzzo [3] and Vladimir M. Kolenko [1,*]

1 Nuclear Dynamics and Cancer Program, Fox Chase Cancer Center, Philadelphia, PA 19111, USA; rushaniya.fazliyeva@fccc.edu
2 Cancer Signaling and Microenvironment Program, Fox Chase Cancer Center, Philadelphia, PA 19111, USA; petr.makhov@fccc.edu
3 Department of Urology, Fox Chase Cancer Center, Philadelphia, PA 19111, USA; robert.uzzo@fccc.edu
* Correspondence: vladimir.kolenko@fccc.edu

Simple Summary: The development of multi-targeted tyrosine kinase inhibitors (TKIs) and immunotherapeutic agents notably changed the treatment paradigm of advanced kidney cancer. However, despite the therapeutic progress, complete and durable responses have been noted in only a few cases. Our studies demonstrate that all major lipoproteins have a comparable ability to supply cholesterol to tumor cells and compromise the antitumor activity of TKIs. Endolysosomal cholesterol transport regulated by NPC1 protein is an attractive therapeutic target based on the fact that this is a point where LDL-, HDL-, and VLDL-derived cholesterol trafficking routes converge and therefore may be simultaneously targeted. Our studies elucidated the role of NPC1 as a potential therapeutic target in clear cell renal cell carcinoma (ccRCC).

Abstract: Rapidly proliferating cancer cells have a greater requirement for cholesterol than normal cells. Tumor cells are largely dependent on exogenous lipids given that their growth requirements are not fully met by endogenous pathways. Our current study shows that ccRCC cells have redundant mechanisms of cholesterol acquisition. We demonstrate that all major lipoproteins (i.e., LDL, HDL, and VLDL) have a comparable ability to support the growth of ccRCC cells and are equally effective in counteracting the antitumor activities of TKIs. The intracellular trafficking of exogenous lipoprotein-derived cholesterol appears to be distinct from the movement of endogenously synthesized cholesterol. De novo synthetized cholesterol is transported from the endoplasmic reticulum directly to the plasma membrane and to the acyl-CoA: cholesterol acyltransferase, whereas lipoprotein-derived cholesterol is distributed through the NPC1-dependent endosomal trafficking system. Expression of NPC1 is increased in ccRCC at mRNA and protein levels, and high expression of NPC1 is associated with poor prognosis. Our current findings show that ccRCC cells are particularly sensitive to the inhibition of endolysosomal cholesterol export and underline the therapeutic potential of targeting NPC1 in ccRCC.

Keywords: cancer; cholesterol; LDL; HDL; VLDL; NPC1; ccRCC; TKI

Citation: Fazliyeva, R.; Makhov, P.; Uzzo, R.G.; Kolenko, V.M. Targeting NPC1 in Renal Cell Carcinoma. *Cancers* **2024**, *16*, 517. https://doi.org/10.3390/cancers16030517

Academic Editors: José I. López and Claudia Manini

Received: 11 December 2023
Revised: 5 January 2024
Accepted: 20 January 2024
Published: 25 January 2024

1. Introduction

The incidence of kidney cancer has risen steadily over several decades and continues to increase. Renal cell carcinoma (RCC) is the most common form of kidney cancer, whereas clear cell RCC (ccRCC) is the most frequent (75–80%) and the best-studied subtype of RCC. Papillary RCC and chromophobe RCC represent the most common remaining histologic subtypes with an incidence of 7% to 14% and 6% to 11%, respectively [1]. Traditional chemotherapy and radiation therapy are largely ineffective in the treatment of all RCC subtypes [2]. Recently, significant progress in the treatment of advanced ccRCC was achieved with the introduction of targeted agents and checkpoint inhibitors. However, despite the therapeutic progress, complete and durable responses have been noted in only a few cases [3–5]. We and others have demonstrated that, in addition to the inhibition of angiogenesis, TKIs also manifest a direct cytotoxic effect on tumor cells [6–9]. Importantly, TKIs may selectively accumulate in tumor tissue at high concentrations [6,8,10]. Indeed, intra-tumor TKI levels are much higher than peak serum levels ($\geq$10 µM vs. $\leq$1 µM, respectively) [6,8,10].

ccRCC is a highly lipogenic tumor. The term "clear cell" itself originates from the clear (empty) microscopic appearance of the cytoplasm after the lipids are removed in the process of fixation. ccRCC tissue contains five to eight-fold more total cholesterol than normal kidney tissue [11]. Due to the absolute requirement of cholesterol for the synthesis of cell membranes, rapidly proliferating cancer cells have a greater requirement for cholesterol than normal cells [12,13]. Tumor cells are dependent on exogenous lipids given that their growth requirements are not fully met by endogenous pathways [14]. Also, increased uptake of exogenous cholesterol is preferential for cancer cells compared with time- and energy-consuming de novo cholesterol synthesis. This concept is supported by several studies showing low activity of HMG-CoA reductase (the rate-limiting enzyme in cholesterol synthesis) and reduced cholesterol synthesis in ccRCC cells [11,15].

There are several levels of control in the regulation of cholesterol homeostasis, i.e., cholesterol synthesis, uptake, intracellular trafficking, and efflux. Cholesterol synthesis can be effectively blocked by statins, competitive inhibitors of HMG-CoA reductase. However, cancer cells can bypass the effects of statins by unrestrained cholesterol importation via the LDL receptor (LDLR) pathway [16,17]. These findings provide an explanation for why many tumor cells are resistant to statin treatment. Cholesterol uptake is mediated by a protein transport mechanism. The LDLR supports the efficient uptake of LDL and VLDL [18]. Cholesterol uptake may be also regulated by the VLDL receptor (VLDLR), which shows considerable similarity to the LDLR [15]. The scavenger receptor class B type I (SR-BI) mediates HDL cholesterol uptake [19]. SR-BI can also bind LDL and VLDL, however, less efficiently than LDLR [20]. Up-regulation of SR-BI promotes tumor progression in ccRCC [21]. Lipoprotein-derived cholesterol is delivered to early endosomes. Importantly, the trafficking of exogenous lipoprotein-derived cholesterol appears to be distinct from the movement of endogenously synthesized cholesterol. Newly synthesized cholesterol is transported from the endoplasmic reticulum (ER) directly to the plasma membrane and to acyl-CoA:cholesterol acyltransferase [22], whereas exogenous cholesterol is distributed through the Niemann–Pick type C1 protein (NPC1)-regulated endosomal trafficking system [23–27]. Inhibition of NPC1 causes accumulation of cholesterol in the endolysosomes, a phenotype similar to that observed in Niemann–Pick disease [26,28]. NPC1 plays a critical role in maintaining adequate cholesterol supply in cells that cannot produce endogenous cholesterol [29]. Importantly, the trafficking of de novo synthetized cholesterol in normal cells is not affected by pharmacological or genetic inhibition of NPC1, in contrast to the trafficking of exogenously derived cholesterol [22]. Cellular cholesterol efflux is carried out mainly by ABCA1 and ABCG1 (ATP-binding cassette transporters) [30]. Recent findings enlightened the role of Liver X receptors (LXRs) in cholesterol homeostasis. Activation of LXRs reduces cholesterol uptake by decreasing expression of LDLR and VLDLR [17,31–33]. In addition, activation of LXRs stimulates ABCA1-dependent cholesterol efflux [34,35].

Our current studies demonstrate that ccRCC cells depend on the uptake of exogenous cholesterol for their growth and survival and have redundant mechanisms of cholesterol acquisition. We reveal herein that HDL, LDL, and VLDL are equally effective in supplying cholesterol to ccRCC cells and in compromising the antitumor activity of TKIs. Based on our findings, it was anticipated that only concomitant targeting of all sources of cholesterol acquisition or common routes of intracellular cholesterol trafficking would deprive tumor cells of cholesterol supply. We addressed this issue by targeting NPC1-dependent endolysosomal cholesterol transport based on the fact that this is a point where trafficking routes of different lipoproteins converge. Our data show that pharmacological or genetic inhibition of NPC1 reduces viability and sensitizes ccRCC cells to TKIs.

2. Materials and Methods

2.1. The Cells and Culture Conditions

The 786-O (human ccRCC cell line), RWPE-1, and PZ-HPV-7 (normal prostate epithelial cell lines) were obtained from ATCC (Manassas, VA, USA). PNX0010 cell line was established from a lung metastatic lesion of a ccRCC patient undergoing nephron-sparing surgery at our institution and represents an aggressive TKI-resistant VHL-negative variant of ccRCC [36,37]. SK-RC-45 (human ccRCC cell line), NK680, NK686, NKE (normal kidney epithelial cell lines), and HUVEC (endothelial cell line) were obtained from the Cell Culture Facility (Fox Chase Cancer Center, Philadelphia, PA, USA). Initial stocks were cryopreserved, and at every 6-month interval, a fresh aliquot of frozen cells was used for the experiments. Cells were cultured in RPMI 1640 (Bio-Whittaker, Walkersville, MD, USA) supplemented with 10% FCS (Hyclone, Logan, UT, USA), gentamicin (50 mg/L), sodium pyruvate (1 mM), and non-essential amino acids (0.1 mM) under conditions indicated in the figure legends.

2.2. Antibodies and Reagents

LDL, HDL, and VLDL were obtained from Lee Biosolutions (Maryland Heights, MO, USA). Cabozantinib, sunitinib, axitinib, pazopanib, U18666A, and posaconazole were obtained from Cayman Chemical Company (Ann Arbor, MI, USA). Antibody against AR was obtained from Cell Signaling Technology (Danvers, MA, USA).

2.3. Western Blot Analysis

Western blot analysis was performed as described previously [38].

2.4. Cell Viability and Drug Interaction Analysis

Cell viability was analyzed by CellTiter Blue cell viability assay (Promega, Madison, WI, USA) as described previously [37]. Effective doses (EDs) were calculated using XLfit 2.0, a Microsoft Excel add-in. The synergistic interaction between pharmacological agents was evaluated by the combination index (CI) using CalcuSyn 2.0 software [39]. CI 0.85–0.9: slight synergism; CI 0.7–0.85: moderate synergism; CI 0.3–0.7: synergism; CI 0.1–0.3: strong synergism; CI < 0.1: very strong synergism.

2.5. siRNA Transfection

786-O and PNX0010 cells were transfected using SMARTpool siRNA targeting NPC1 (Horizon Discovery, Cambridge, UK, Cat# L-003486-00-0005) essentially as described in our previous report [40].

3. Results

3.1. LDL, HDL, and VLDL Are Equally Effective in Supporting Viability of ccRCC Cells

Studies by Parinaud et al. demonstrate that all major lipoproteins, (i.e., LDL, HDL, and VLDL) have a similar ability to supply exogenous cholesterol [41]. Based on these findings, we examined the ability of different lipoproteins to support the viability of ccRCC cells. SK-RC-45 [8,42] and PNX0010 [43,44] ccRCC cell lines were cultured in RPMI-1640 medium supplemented with lipid-depleted fetal bovine serum (FBS) in the presence or absence of LDL, HDL, or VLDL. As demonstrated in Figure 1, LDL, HDL, and VLDL were equally effective in maintaining the viability of SK-RC-45 and PNX0010 ccRCC cells cultured under lipid-depleted conditions. The results of this experiment also show that ccRCC cells are largely dependent on exogenous cholesterol supply. In contrast, normal kidney epithelial NK680 cells were less dependent on exogenous cholesterol as indicated by their capacity to maintain viability under lipid-depleted conditions (Figure 1).

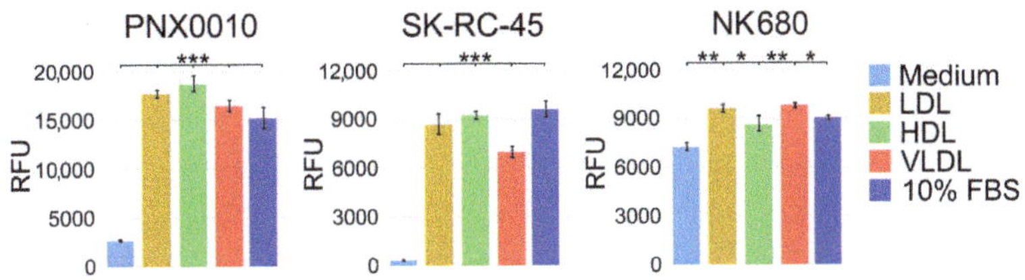

Figure 1. LDL, HDL, and VLDL are equally effective in supporting viability of ccRCC cells. PNX0010 and SK-45 ccRCC cells and NK680 normal kidney epithelial cells were cultured in RPMI 1640 medium supplemented with lipid-depleted fetal bovine serum in the presence or absence of LDL, HDL, or VLDL (all at 100 µg/mL) for 96 h. Cell viability was analyzed by CellTiter Blue assay (Promega). Results are expressed as the mean (n = 3) ± s.e.m. * $p < 0.01$; ** $p < 0.001$; *** $p < 0.0001$. RFU-relative fluorescence units.

3.2. LDL, HDL, and VLDL Compromise the Antitumor Activity of TKIs

Our previous findings demonstrate that LDL cholesterol compromises the efficacy of TKIs against ccRCC and endothelial cells [37]. Therefore, we examined whether treatment with HDL and VLDL also affects the antitumor activity of TKIs. Our previous work has shown that TKIs may differ substantially with respect to the mechanism of their antitumor activity [42]. To demonstrate that our observations are not limited to a specific pharmacological agent, we treated SK-RC-45 and PNX0010 ccRCC cells with several clinically relevant TKIs, such as cabozantinib, sunitinib, axitinib, and pazopanib. As shown in Figure 2, LDL, HDL, and VLDL effectively rescued the viability of SK-RC-45 and PNX0010 cells treated with all tested TKIs.

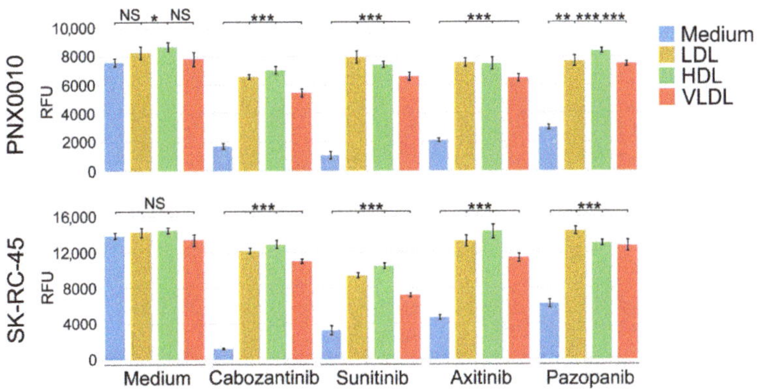

Figure 2. LDL, HDL, and VLDL compromise the antitumor activity of TKIs. SK-45 and PNX0010 ccRCC cells were cultured with TKIs (all at 5 µM) with or without LDL, HDL, or VLDL (all at 100 µg/mL) in RPMI 1640 medium supplemented with regular 10% FBS for 96 h. Cell viability was analyzed by CellTiter Blue assay. Results are expressed as the mean (n = 3) ± s.e.m. * $p < 0.01$; ** $p < 0.001$; *** $p < 0.0001$; NS—Not significant.

3.3. Inhibition of Endosomal Cholesterol Trafficking Sensitizes ccRCC Cells to TKIs

Expression of NPC1 is increased in ccRCC at mRNA and protein levels (Figure 3A,B), and high expression of NPC1 is associated with poor prognosis based on TCGA data analysis (Figure 3C). Given that trafficking of de novo synthetized cholesterol in normal cells is not affected by pharmacological or genetic inhibition of NPC1 [22], we anticipated that ccRCC cells would be particularly sensitive to the inhibition of NPC-1-regulated endosomal cholesterol trafficking. Indeed, ccRCC cells manifested much higher sensitivity to the NPC1 inhibitor U18666A (90) compared with normal cells of various origins (Table 1).

Of note, the effect of U1866A and posaconazole, an approved antifungal agent, which directly binds NPC1 and blocks endosomal cholesterol trafficking [45], on tumor cell viability was significantly increased under hypoxia, a central event in renal tumorigenesis (Table 2).

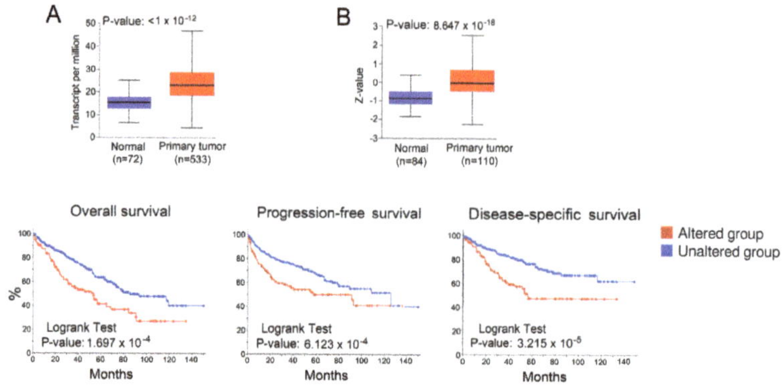

Figure 3. Relationship between NPC1 expression and clinical outcomes in ccRCC. (**A**) Expression of NPC1 is increased in ccRCC at (**A**) mRNA (TCGA) and (**B**) protein (CPTAC) levels according to the University of Alabama at Birmingham cancer data analysis portal (UALCAN). (**C**) NPC1 is an unfavorable prognostic marker in ccRCC based on TCGA analysis via cBioPortal.

Table 1. U18666A preferentially inhibits viability of ccRCC cells. ccRCC (786-O, SK-RC-45, and PNX0010), normal kidney epithelial (NKE and NK686), normal prostate epithelial (RWPE-1 and PZ-HPV-7), and human umbilical vein endothelial (HUVEC) cells were treated with escalating concentrations of U18666A for 72 h. Cell viability and effective doses (ED, μM) were evaluated as described in Materials and Methods.

Cell Line	ED25	IC50	IC75	IC90
786-O	2.1	2.8	3.7	4.9
SK-RC-45	0.09	0.4	1.8	8.2
PNX0010	5.2	10.4	15.9	31.3
NK686	30.7	54.8	97.9	174.7
NKE	26.5	52.9	105.9	211
RWPE-1	21.8	30.3	42.1	58.5
PZ-HPV-7	13.7	28.0	57	116.2
HUVEC	19.7	34.1	59.4	103.6

Unobstructed endolysosomal cholesterol trafficking is required to sustain the activity of several signaling pathways that confer drug resistance (i.e., Akt/mTOR, NF-κB, and Erk1/2) [37,46–49]. Given that, we tested whether NPC1 inhibition enhances the antitumor effect of TKIs. Our experiments reveal that administration of sunitinib, cabozantinib, or pazopanib in combination with U18666A showed a clear synergistic inhibitory effect on the viability of ccRCC cells at all effective dose levels (Table 3).

The results of these experiments were further validated using siRNA-mediated knockdown of NPC1. Genetic depletion of NPC1 reduced viability and sensitized ccRCC cells to TKI treatment (Figure 4 and Table 4).

Table 2. The effectiveness of NPC1 inhibitors is enhanced under hypoxic conditions. 786-O and PNX0010 ccRCC cells were cultured under normoxic (N) (20% O_2) or hypoxic (H) (2% O_2) conditions in RPMI1640 medium supplemented with 10% FBS and treated with escalating concentrations of either U18666A or posaconazole for 72 h. Cell viability and effective doses (ED, μM) were evaluated as described in Materials and Methods.

Drug	Cell Line	ED25	ED50	ED75	ED90
U18666A	786-O (N)	1.5	2.3	3.5	5.4
	786-O (H)	0.7	1.2	2.1	3.6
	PNX0010 (N)	2.2	3.3	4.9	7.2
	PNX0010 (H)	0.4	0.9	1.9	3.8
Posaconazole	786-O (N)	4.7	10.1	21.6	46.2
	786-O (H)	1.3	3.0	6.7	15.3
	PNX0010 (N)	4.6	8.0	14.2	24.9
	PNX0010 (H)	0.9	1.9	4.2	9.2

Table 3. The synergistic effect of combined treatment with U18666A and TKIs on the viability of ccRCC cells. The cells were treated with escalating concentrations of U18666A and TKIs for 72 h. Analysis of the synergistic interaction between U18666A and TKIs was performed as described in Materials and Methods. ED: effective doses. Combination index (CI) 0.85–0.9: slight synergism; CI 0.7–0.85: moderate synergism; CI 0.3–0.7: synergism; CI 0.1–0.3: strong synergism; CI < 0.1: very strong synergism.

Drug	Cell Line	ED25	ED50	ED75	ED90
Sunitinib	786-O	0.56	0.52	0.47	0.41
	SK-RC-45	0.45	0.41	0.37	0.32
	PNX0010	0.50	0.46	0.42	0.39
Cabozantinib	786-O	0.44	0.40	0.35	0.30
	SK-RC-45	0.39	0.33	0.27	0.22
	PNX0010	0.52	0.45	0.41	0.38
Pazopanib	786-O	0.37	0.31	0.25	0.20
	SK-RC-45	0.42	0.36	0.29	0.23
	PNX0010	0.48	0.40	0.33	0.26

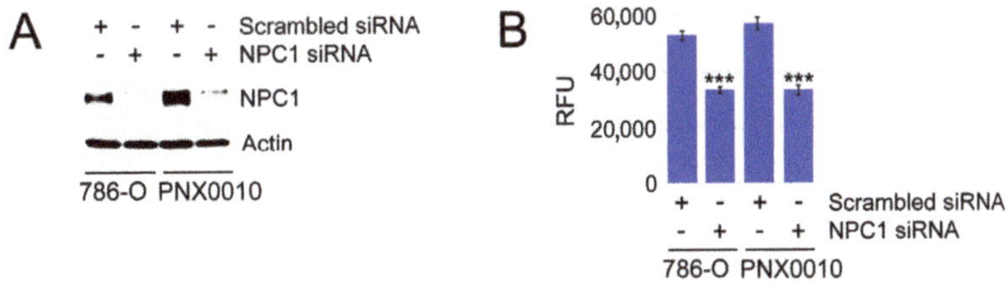

Figure 4. NPC1 depletion reduces viability of ccRCC cells. (**A**) Western blot analysis of NPC1 expression in 786-O and PNX00100 cells treated with control siRNA and NPC1 SmartPool siRNA. The uncropped blots are shown in the supplementary materials. (**B**) Aliquots of the cells used in panel A were cultured in RPMI1640 medium supplemented with 10% FBS for 72 h. Cell viability was analyzed as described in legend to Figure 2. *** $p < 0.0001$.

Table 4. Depletion of NPC1 sensitizes ccRCC cells to TKIs. Control and NPC1-depleted cells described in Figure 4 were treated with escalating concentrations of axitinib and cabozantinib for 72 h. Viability and EDs (μM) were calculated as described in the legend in Table 1.

Drug	Cell Line	ED25	ED50	ED75	ED90
Axitinib	786-O	5.4	7.4	10.1	13.9
	786-O-NPC1^{KD}	1.3	2.5	5.0	9.9
	PNX0010	4.1	6.9	11.5	19.3
	PNX0010-NPC1^{KD}	1.5	2.6	4.7	8.5
Cabozantinib	786-O	5.6	8.5	12.8	19.3
	786-O-NPC1^{KD}	2.5	4.0	6.4	10.4
	PNX0010	4.6	6.8	10.0	14.9
	PNX0010-NPC1^{KD}	2.7	3.9	5.7	8.2

4. Discussion

The lack of sensitivity of ccRCC to chemotherapy and radiation therapy prompted early research efforts into the development of new treatment options. The introduction of TKIs and immune checkpoint inhibitors (ICIs) notably changed the treatment paradigm of ccRCC. Recent clinical studies suggest that combining or sequencing TKIs with ICIs may provide effective treatments that reduce or delay disease progression [50–57]. However, despite the introduction of novel therapeutic approaches in the past decade, advanced ccRCC continues to be a treatment-resistant malignancy [3–5].

Multiple studies demonstrate that disruption of cholesterol homeostasis suppresses tumor growth and suggest that targeting cholesterol metabolism may be a promising strategy for antitumor therapy. Indeed, due to the absolute requirement of cholesterol for the synthesis of cell membranes, rapidly proliferating cancer cells have a greater requirement for cholesterol than normal cells (1,2). Our studies demonstrate that ccRCC cells can utilize lipoprotein-derived cholesterol irrespective of the particular class of cholesterol. donor and that all major classes of lipoproteins are capable of compromising the antitumor activity of TKIs. Thus, only simultaneous targeting of different lipoproteins is postulated to have therapeutic benefits. There are several potential approaches for targeting cholesterol homeostasis in cancer cells such as inhibition of cholesterol biosynthesis, blockade of cholesterol uptake, and modulation of intracellular cholesterol trafficking. Increased serum levels of cholesterol in humans can be effectively lowered by PCSK9 inhibitors and statins [58–61]. PCSK9 is a proprotein convertase that is involved in the degradation of LDLR in the liver [62]. PCSK9 inhibition increases LDLR expression by hepatocytes, which causes increased uptake of circulating LDL, thereby reducing plasma LDL-cholesterol. Statins are competitive inhibitors of HMG-CoA reductase, the rate-limiting enzyme of cholesterol biosynthesis [63,64]. However, these drugs up-regulate HDL levels [60,65–68], which can serve as a source of cholesterol for tumor cells. Also, statins demonstrate effects on a number of signaling pathways including protein isoprenylation [69]. Anticancer effects of statins have been largely attributed to the inhibition of this post-translational mechanism [69,70]. Moreover, cancer cells can bypass the effects of statins by unrestrained cholesterol importation [16,17]. Given that cholesterol uptake receptors have promiscuous ligand-binding properties, simultaneous targeting of different receptors may be necessary to reduce cellular cholesterol import, which may not be a clinically feasible approach.

As discussed above, tumor cells are largely dependent on exogenous lipids. Endolysosomal cholesterol transport is an attractive therapeutic target based on the fact that this is a point where LDL-, HDL, and VLDL-derived cholesterol trafficking routes converge and therefore may be simultaneously targeted. Furthermore, the trafficking of exogenous lipoprotein-derived cholesterol, in contrast to endogenously produced cholesterol, occurs through an NPC1-mediated endosomal trafficking system [23–26]. Therefore, the trafficking of de novo synthetized cholesterol in normal cells is not affected by NPC1 inhibition [22]. Indeed, our studies demonstrate that ccRCC cells are particularly sensitive to the inhibition of endosomal cholesterol trafficking by NPC1-specific inhibitor U18666A compared with normal cells of various origins. Importantly, the effect of U1866A and posaconazole on tumor cell viability was significantly increased under hypoxia, a central event in renal tumorigenesis. This observation could be potentially explained by the fact that hypoxia results in the depletion of cholesterol from the plasma membrane [71], increasing the demand for cholesterol transport from an endosomal compartment to the plasma membrane. These findings may widen the therapeutic window of NPC1 inhibitors for more selective targeting of malignant cells.

5. Conclusions

Our studies demonstrate that ccRCC cells are highly dependent on the uptake of exogenous cholesterol for their growth and survival and that all major lipoproteins have a comparable ability to supply cholesterol to tumor cells and compromise the antitumor activity of TKIs. Thus, concomitant targeting of all sources of cholesterol acquisition or common routes of cholesterol trafficking may be required to deprive tumor cells of cholesterol supply. Our findings indicate that ccRCC cells are particularly sensitive to the inhibition of NPC-1-dependent endosomal cholesterol trafficking. Importantly, NPC1 expression is increased in ccRCC, and high expression of NPC1 is associated with poor prognosis. Taken together, our work suggests that NPC1 may serve as a potential therapeutic target in ccRCC.

Supplementary Materials: The following supporting information can be downloaded at: https://www.mdpi.com/article/10.3390/cancers16030517/s1, Figure S1: full size blot of Figure 4A.

Author Contributions: Conceptualization, V.M.K.; methodology, V.M.K. and P.M.; investigation, P.M. and R.F.; writing—original draft preparation, V.M.K. and P.M.; writing—review and editing, V.M.K. and R.G.U.; supervision, V.M.K.; project administration, V.M.K.; funding acquisition, V.M.K. and P.M. All authors have read and agreed to the published version of the manuscript.

Funding: This work was supported in part by the National Institutes of Health Grants R21 CA263362 (to P.M.); the Department of Defense Idea Development Awards KC170127, KC220055, and PC210308 (to V.M.K.); MPI FCCC pilot application (to V.M.K.); and the NCI Core Grant NCI P30 CA006927 (to Fox Chase Cancer Center).

Institutional Review Board Statement: Not applicable.

Informed Consent Statement: Not applicable.

Data Availability Statement: The data presented in this study are available from the corresponding author (V.M.K.) upon request.

Acknowledgments: The authors acknowledge the support from the Cell Culture Facility at Fox Chase Cancer Center. PNX0010 ccRCC cell line was a kind gift from Vladimir Khazak.

Conflicts of Interest: The authors declare no conflicts of interest.

References

1. Shuch, B.; Amin, A.; Armstrong, A.J.; Eble, J.N.; Ficarra, V.; Lopez-Beltran, A.; Martignoni, G.; Rini, B.I.; Kutikov, A. Understanding pathologic variants of renal cell carcinoma: Distilling therapeutic opportunities from biologic complexity. *Eur. Urol.* **2015**, *67*, 85–97. [CrossRef] [PubMed]
2. Maines, F.; Caffo, O.; Veccia, A.; Trentin, C.; Tortora, G.; Galligioni, E.; Bria, E. Sequencing new agents after docetaxel in patients with metastatic castration-resistant prostate cancer. *Crit. Rev. Oncol. Hematol.* **2015**, *96*, 498–506. [CrossRef]
3. Albiges, L.; Oudard, S.; Negrier, S.; Caty, A.; Gravis, G.; Joly, F.; Duclos, B.; Geoffrois, L.; Rolland, F.; Guillot, A.; et al. Complete remission with tyrosine kinase inhibitors in renal cell carcinoma. *J. Clin. Oncol.* **2012**, *30*, 482–487. [CrossRef] [PubMed]
4. Park, K.; Lee, J.L.; Park, I.; Park, S.; Ahn, Y.; Ahn, J.H.; Ahn, S.; Song, C.; Hong, J.H.; Kim, C.S.; et al. Comparative efficacy of vascular endothelial growth factor (VEGF) tyrosine kinase inhibitor (TKI) and mammalian target of rapamycin (mTOR) inhibitor as second-line therapy in patients with metastatic renal cell carcinoma after the failure of first-line VEGF TKI. *Med. Oncol.* **2012**, *29*, 3291–3297. [CrossRef] [PubMed]

5. Weinstock, M.; McDermott, D. Targeting PD-1/PD-L1 in the treatment of metastatic renal cell carcinoma. *Ther. Adv. Urol.* **2015**, *7*, 365–377. [CrossRef] [PubMed]
6. Gotink, K.J.; Broxterman, H.J.; Labots, M.; de Haas, R.R.; Dekker, H.; Honeywell, R.J.; Rudek, M.A.; Beerepoot, L.V.; Musters, R.J.; Jansen, G.; et al. Lysosomal sequestration of sunitinib: A novel mechanism of drug resistance. *Clin. Cancer Res.* **2011**, *17*, 7337–7346. [CrossRef]
7. Kutikov, A.; Makhov, P.; Golovine, K.; Canter, D.J.; Sirohi, M.; Street, R.; Simhan, J.; Uzzo, R.G.; Kolenko, V.M. Interleukin-6: A Potential Biomarker of Resistance to Multitargeted Receptor Tyrosine Kinase Inhibitors in Castration-Resistant Prostate Cancer. *Urology* **2011**, *8*, 968.e7–968.e11.
8. Makhov, P.B.; Golovine, K.; Kutikov, A.; Teper, E.; Canter, D.J.; Simhan, J.; Uzzo, R.G.; Kolenko, V.M. Modulation of Akt/mTOR Signaling Overcomes Sunitinib Resistance in Renal and Prostate Cancer Cells. *Mol. Cancer Ther.* **2012**, *11*, 1510–1517. [CrossRef]
9. Xin, H.; Zhang, C.; Herrmann, A.; Du, Y.; Figlin, R.; Yu, H. Sunitinib inhibition of Stat3 induces renal cell carcinoma tumor cell apoptosis and reduces immunosuppressive cells. *Cancer Res.* **2009**, *69*, 2506–2513. [CrossRef]
10. Adelaiye, R.; Ciamporcero, E.; Miles, K.M.; Sotomayor, P.; Bard, J.; Tsompana, M.; Conroy, D.; Shen, L.; Ramakrishnan, S.; Ku, S.Y.; et al. Sunitinib dose escalation overcomes transient resistance in clear cell renal cell carcinoma and is associated with epigenetic modifications. *Mol. Cancer Ther.* **2015**, *14*, 513–522. [CrossRef]
11. Gebhard, R.L.; Clayman, R.V.; Prigge, W.F.; Figenshau, R.; Staley, N.A.; Reesey, C.; Bear, A. Abnormal cholesterol metabolism in renal clear cell carcinoma. *J. Lipid Res.* **1987**, *28*, 1177–1184. [CrossRef] [PubMed]
12. Llaverias, G.; Danilo, C.; Mercier, I.; Daumer, K.; Capozza, F.; Williams, T.M.; Sotgia, F.; Lisanti, M.P.; Frank, P.G. Role of cholesterol in the development and progression of breast cancer. *Am. J. Pathol.* **2011**, *178*, 402–412. [CrossRef] [PubMed]
13. Freeman, M.R.; Di Vizio, D.; Solomon, K.R. The Rafts of the Medusa: Cholesterol targeting in cancer therapy. *Oncogene* **2010**, *29*, 3745–3747. [CrossRef] [PubMed]
14. Riscal, R.; Skuli, N.; Simon, M.C. Even Cancer Cells Watch Their Cholesterol! *Mol. Cell* **2019**, *76*, 220–231. [CrossRef] [PubMed]
15. Sundelin, J.P.; Stahlman, M.; Lundqvist, A.; Levin, M.; Parini, P.; Johansson, M.E.; Boren, J. Increased expression of the very low-density lipoprotein receptor mediates lipid accumulation in clear-cell renal cell carcinoma. *PLoS ONE* **2012**, *7*, e48694. [CrossRef] [PubMed]
16. Gabitova, L.; Gorin, A.; Astsaturov, I. Molecular pathways: Sterols and receptor signaling in cancer. *Clin. Cancer Res.* **2014**, *20*, 28–34. [CrossRef] [PubMed]
17. Guo, D.; Reinitz, F.; Youssef, M.; Hong, C.; Nathanson, D.; Akhavan, D.; Kuga, D.; Amzajerdi, A.N.; Soto, H.; Zhu, S.; et al. An LXR agonist promotes glioblastoma cell death through inhibition of an EGFR/AKT/SREBP-1/LDLR-dependent pathway. *Cancer Discov.* **2011**, *1*, 442–456. [CrossRef]
18. Pompey, S.; Zhao, Z.; Luby-Phelps, K.; Michaely, P. Quantitative fluorescence imaging reveals point of release for lipoproteins during LDLR-dependent uptake. *J. Lipid Res.* **2013**, *54*, 744–753. [CrossRef]
19. Trigatti, B.L.; Krieger, M.; Rigotti, A. Influence of the HDL receptor SR-BI on lipoprotein metabolism and atherosclerosis. *Arterioscler. Thromb. Vasc. Biol.* **2003**, *23*, 1732–1738. [CrossRef]
20. Rigotti, A.; Miettinen, H.E.; Krieger, M. The role of the high-density lipoprotein receptor SR-BI in the lipid metabolism of endocrine and other tissues. *Endocr. Rev.* **2003**, *24*, 357–387. [CrossRef]
21. Xu, G.H.; Lou, N.; Shi, H.C.; Xu, Y.C.; Ruan, H.L.; Xiao, W.; Liu, L.; Li, X.; Xiao, H.B.; Qiu, B.; et al. Up-regulation of SR-BI promotes progression and serves as a prognostic biomarker in clear cell renal cell carcinoma. *BMC Cancer* **2018**, *18*, 88. [CrossRef]
22. Liscum, L.; Dahl, N.K. Intracellular cholesterol transport. *J. Lipid Res.* **1992**, *33*, 1239–1254. [CrossRef]
23. Wojtanik, K.M.; Liscum, L. The transport of low density lipoprotein-derived cholesterol to the plasma membrane is defective in NPC1 cells. *J. Biol. Chem.* **2003**, *278*, 14850–14856. [CrossRef] [PubMed]
24. Pfisterer, S.G.; Peranen, J.; Ikonen, E. LDL-cholesterol transport to the endoplasmic reticulum: Current concepts. *Curr. Opin. Lipidol.* **2016**, *27*, 282–287. [CrossRef] [PubMed]
25. Kwon, H.J.; Abi-Mosleh, L.; Wang, M.L.; Deisenhofer, J.; Goldstein, J.L.; Brown, M.S.; Infante, R.E. Structure of N-terminal domain of NPC1 reveals distinct subdomains for binding and transfer of cholesterol. *Cell* **2009**, *137*, 1213–1224. [CrossRef] [PubMed]
26. Peake, K.B.; Vance, J.E. Defective cholesterol trafficking in Niemann-Pick C-deficient cells. *FEBS Lett.* **2010**, *584*, 2731–2739. [CrossRef] [PubMed]
27. Soffientini, U.; Graham, A. Intracellular cholesterol transport proteins: Roles in health and disease. *Clin. Sci.* **2016**, *130*, 1843–1859. [CrossRef] [PubMed]
28. Lange, Y.; Ye, J.; Rigney, M.; Steck, T. Cholesterol movement in Niemann-Pick type C cells and in cells treated with amphiphiles. *J. Biol. Chem.* **2000**, *275*, 17468–17475. [CrossRef] [PubMed]
29. O'Neill, K.I.; Kuo, L.W.; Williams, M.M.; Lind, H.; Crump, L.S.; Hammond, N.G.; Spoelstra, N.S.; Caino, M.C.; Richer, J.K. NPC1 Confers Metabolic Flexibility in Triple Negative Breast Cancer. *Cancers* **2022**, *14*, 3543. [CrossRef] [PubMed]
30. Burns, V.E.; Kerppola, T.K. ATR-101 inhibits cholesterol efflux and cortisol secretion by ATP-binding cassette transporters, causing cytotoxic cholesterol accumulation in adrenocortical carcinoma cells. *Br. J. Pharmacol.* **2017**, *174*, 3315–3332. [CrossRef]
31. Do, H.T.; Bruelle, C.; Tselykh, T.; Jalonen, P.; Korhonen, L.; Lindholm, D. Reciprocal regulation of very low density lipoprotein receptors (VLDLRs) in neurons by brain-derived neurotrophic factor (BDNF) and Reelin: Involvement of the E3 ligase Mylip/Idol. *J. Biol. Chem.* **2013**, *288*, 29613–29620. [CrossRef]

32. Hong, C.; Duit, S.; Jalonen, P.; Out, R.; Scheer, L.; Sorrentino, V.; Boyadjian, R.; Rodenburg, K.W.; Foley, E.; Korhonen, L.; et al. The E3 ubiquitin ligase IDOL induces the degradation of the low density lipoprotein receptor family members VLDLR and ApoER2. *J. Biol. Chem.* **2010**, *285*, 19720–19726. [CrossRef] [PubMed]
33. Ren, K.; Jiang, T.; Zhao, G.J. Quercetin induces the selective uptake of HDL-cholesterol via promoting SR-BI expression and the activation of the PPARgamma/LXRalpha pathway. *Food Funct.* **2018**, *9*, 624–635. [CrossRef] [PubMed]
34. Venkateswaran, A.; Laffitte, B.A.; Joseph, S.B.; Mak, P.A.; Wilpitz, D.C.; Edwards, P.A.; Tontonoz, P. Control of cellular cholesterol efflux by the nuclear oxysterol receptor LXR alpha. *Proc. Natl. Acad. Sci. USA* **2000**, *97*, 12097–12102. [CrossRef] [PubMed]
35. Hu, Y.W.; Wang, Q.; Ma, X.; Li, X.X.; Liu, X.H.; Xiao, J.; Liao, D.F.; Xiang, J.; Tang, C.K. TGF-beta1 up-regulates expression of ABCA1, ABCG1 and SR-BI through liver X receptor alpha signaling pathway in THP-1 macrophage-derived foam cells. *J. Atheroscler. Thromb.* **2010**, *17*, 493–502. [CrossRef] [PubMed]
36. Golovine, K.; Makhov, P.; Naito, S.; Raiyani, H.; Tomaszewski, J.; Mehrazin, R.; Tulin, A.; Kutikov, A.; Uzzo, R.G.; Kolenko, V.M. Piperlongumine and its analogs down-regulate expression of c-Met in renal cell carcinoma. *Cancer Biol. Ther.* **2015**, *16*, 743–749. [CrossRef] [PubMed]
37. Naito, S.; Makhov, P.; Astsaturov, I.; Golovine, K.; Tulin, A.; Kutikov, A.; Uzzo, R.G.; Kolenko, V.M. LDL cholesterol counteracts the antitumour effect of tyrosine kinase inhibitors against renal cell carcinoma. *Br. J. Cancer* **2017**, *116*, 1203–1207. [CrossRef] [PubMed]
38. Golovine, K.; Makhov, P.; Uzzo, R.G.; Shaw, T.; Kunkle, D.; Kolenko, V.M. Overexpression of the zinc uptake transporter hZIP1 inhibits nuclear factor-kappaB and reduces the malignant potential of prostate cancer cells in vitro and in vivo. *Clin. Cancer Res.* **2008**, *14*, 5376–5384. [CrossRef]
39. Chou, T.C. Theoretical basis, experimental design, and computerized simulation of synergism and antagonism in drug combination studies. *Pharmacol. Rev.* **2006**, *58*, 621–681. [CrossRef]
40. Makhov, P.; Sohn, J.A.; Serebriiskii, I.G.; Fazliyeva, R.; Khazak, V.; Boumber, Y.; Uzzo, R.G.; Kolenko, V.M. CRISPR/Cas9 genome-wide loss-of-function screening identifies druggable cellular factors involved in sunitinib resistance in renal cell carcinoma. *Br. J. Cancer* **2020**, *123*, 1749–1756. [CrossRef]
41. Parinaud, J.; Perret, B.; Ribbes, H.; Chap, H.; Pontonnier, G.; Douste-Blazy, L. High density lipoprotein and low density lipoprotein utilization by human granulosa cells for progesterone synthesis in serum-free culture: Respective contributions of free and esterified cholesterol. *J. Clin. Endocrinol. Metab.* **1987**, *64*, 409–417. [CrossRef]
42. Canter, D.; Kutikov, A.; Golovine, K.; Makhov, P.; Simhan, J.; Uzzo, R.G.; Kolenko, V.M. Are all multi-targeted tyrosine kinase inhibitors created equal? An in vitro study of sunitinib and pazopanib in renal cell carcinoma cell lines. *Can. J. Urol.* **2011**, *18*, 5819–5825.
43. Kirsanov, K.I.; Kotova, E.; Makhov, P.; Golovine, K.; Lesovaya, E.A.; Kolenko, V.M.; Yakubovskaya, M.G.; Tulin, A.V. Minor grove binding ligands disrupt PARP-1 activation pathways. *Oncotarget* **2014**, *5*, 428–437. [CrossRef] [PubMed]
44. Thomas, C.; Ji, Y.; Lodhi, N.; Kotova, E.; Pinnola, A.D.; Golovine, K.; Makhov, P.; Pechenkina, K.; Kolenko, V.; Tulin, A.V. Non-NAD-Like poly(ADP-Ribose) Polymerase-1 Inhibitors effectively Eliminate Cancer in vivo. *EBioMedicine* **2016**, *13*, 90–98. [CrossRef] [PubMed]
45. Trinh, M.N.; Lu, F.; Li, X.; Das, A.; Liang, Q.; De Brabander, J.K.; Brown, M.S.; Goldstein, J.L. Triazoles inhibit cholesterol export from lysosomes by binding to NPC1. *Proc. Natl. Acad. Sci. USA* **2017**, *114*, 89–94. [CrossRef]
46. Li, Y.; Schwabe, R.F.; DeVries-Seimon, T.; Yao, P.M.; Gerbod-Giannone, M.C.; Tall, A.R.; Davis, R.J.; Flavell, R.; Brenner, D.A.; Tabas, I. Free cholesterol-loaded macrophages are an abundant source of tumor necrosis factor-alpha and interleukin-6: Model of NF-kappaB- and map kinase-dependent inflammation in advanced atherosclerosis. *J. Biol. Chem.* **2005**, *280*, 21763–21772. [CrossRef] [PubMed]
47. Xu, J.; Dang, Y.J.; Ren, Y.R.Z.; Liu, J.O. Cholesterol trafficking is required for mTOR activation in endothelial cells. *Proc. Natl. Acad. Sci. USA* **2010**, *107*, 4764–4769. [CrossRef] [PubMed]
48. Morales, M.L.; Arenas, A.; Ortiz-Ruiz, A.; Leivas, A.; Rapado, I.; Rodriguez-Garcia, A.; Castro, N.; Zagorac, I.; Quintela-Fandino, M.; Gomez-Lopez, G.; et al. MEK inhibition enhances the response to tyrosine kinase inhibitors in acute myeloid leukemia. *Sci. Rep.* **2019**, *9*, 18630. [CrossRef] [PubMed]
49. Blakely, C.M.; Pazarentzos, E.; Olivas, V.; Asthana, S.; Yan, J.J.; Tan, I.; Hrustanovic, G.; Chan, E.; Lin, L.; Neel, D.S.; et al. NF-kappaB-activating complex engaged in response to EGFR oncogene inhibition drives tumor cell survival and residual disease in lung cancer. *Cell Rep.* **2015**, *11*, 98–110. [CrossRef]
50. Bedke, J.; Stuhler, V.; Stenzl, A.; Brehmer, B. Immunotherapy for kidney cancer: Status quo and the future. *Curr. Opin. Urol.* **2018**, *28*, 8–14. [CrossRef]
51. Mosillo, C.; Ciccarese, C.; Bimbatti, D.; Fantinel, E.; Volta, A.D.; Bisogno, I.; Zampiva, I.; Santoni, M.; Massar, F.; Brunelli, M.; et al. Renal cell carcinoma in one year: Going inside the news of 2017—A report of the main advances in RCC cancer research. *Cancer Treat. Rev.* **2018**, *67*, 29–33. [CrossRef] [PubMed]
52. Makhov, P.; Joshi, S.; Ghatalia, P.; Kutikov, A.; Uzzo, R.G.; Kolenko, V.M. Resistance to Systemic Therapies in Clear Cell Renal Cell Carcinoma: Mechanisms and Management Strategies. *Mol. Cancer Ther.* **2018**, *17*, 1355–1364. [CrossRef] [PubMed]
53. Atkins, M.B.; Plimack, E.R.; Puzanov, I.; Fishman, M.N.; McDermott, D.F.; Cho, D.C.; Vaishampayan, U.; George, S.; Olencki, T.E.; Tarazi, J.C.; et al. Axitinib in combination with pembrolizumab in patients with advanced renal cell cancer: A non-randomised, open-label, dose-finding, and dose-expansion phase 1b trial. *Lancet Oncol.* **2018**, *19*, 405–415. [CrossRef] [PubMed]

54. Calvo, E.; Porta, C.; Grunwald, V.; Escudier, B. The Current and Evolving Landscape of First-Line Treatments for Advanced Renal Cell Carcinoma. *Oncologist* **2019**, *24*, 338–348. [CrossRef] [PubMed]
55. Fuca, G.; de Braud, F.; Di Nicola, M. Immunotherapy-based combinations: An update. *Curr. Opin. Oncol.* **2018**, *30*, 345–351. [CrossRef] [PubMed]
56. Vano, Y.A.; Ladoire, S.; Elaidi, R.; Dermeche, S.; Eymard, J.C.; Falkowski, S.; Gross-Goupil, M.; Malouf, G.; Narciso, B.; Sajous, C.; et al. First-Line Treatment of Metastatic Clear Cell Renal Cell Carcinoma: What Are the Most Appropriate Combination Therapies? *Cancers* **2021**, *13*, 5548. [CrossRef] [PubMed]
57. Massari, F.; Rizzo, A.; Mollica, V.; Rosellini, M.; Marchetti, A.; Ardizzoni, A.; Santoni, M. Immune-based combinations for the treatment of metastatic renal cell carcinoma: A meta-analysis of randomised clinical trials. *Eur. J. Cancer* **2021**, *154*, 120–127. [CrossRef] [PubMed]
58. Colhoun, H.M.; Leiter, L.A.; Muller-Wieland, D.; Cariou, B.; Ray, K.K.; Tinahones, F.J.; Domenger, C.; Letierce, A.; Israel, M.; Samuel, R.; et al. Effect of alirocumab on individuals with type 2 diabetes, high triglycerides, and low high-density lipoprotein cholesterol. *Cardiovasc. Diabetol.* **2020**, *19*, 14. [CrossRef]
59. Mombelli, G.; Castelnuovo, S.; Pavanello, C. Potential of PCSK9 as a new target for the management of LDL cholesterol. *Res. Rep. Clin. Cardiol.* **2015**, *6*, 73–86. [CrossRef]
60. Sabatine, M.S.; Giugliano, R.P.; Wiviott, S.D.; Raal, F.J.; Blom, D.J.; Robinson, J.; Ballantyne, C.M.; Somaratne, R.; Legg, J.; Wasserman, S.M.; et al. Efficacy and safety of evolocumab in reducing lipids and cardiovascular events. *N. Engl. J. Med.* **2015**, *372*, 1500–1509. [CrossRef]
61. Moride, Y.; Hegele, R.A.; Langer, A.; McPherson, R.; Miller, D.B.; Rinfret, S. Clinical and public health assessment of benefits and risks of statins in primary prevention of coronary events: Resolved and unresolved issues. *Can. J. Cardiol.* **2008**, *24*, 293–300. [CrossRef] [PubMed]
62. Hajar, R. PCSK 9 Inhibitors: A Short History and a New Era of Lipid-lowering Therapy. *Heart Views* **2019**, *20*, 74–75. [CrossRef] [PubMed]
63. McFarlane, S.I.; Muniyappa, R.; Francisco, R.; Sowers, J.R. Clinical review 145: Pleiotropic effects of statins: Lipid reduction and beyond. *J. Clin. Endocrinol. Metab.* **2002**, *87*, 1451–1458. [CrossRef] [PubMed]
64. Schaefer, E.J.; Asztalos, B.F. The effects of statins on high-density lipoproteins. *Curr. Atheroscler. Rep.* **2006**, *8*, 41–49. [CrossRef]
65. Barter, P.J.; Brandrup-Wognsen, G.; Palmer, M.K.; Nicholls, S.J. Effect of statins on HDL-C: A complex process unrelated to changes in LDL-C: Analysis of the VOYAGER Database. *J. Lipid Res.* **2010**, *51*, 1546–1553. [CrossRef] [PubMed]
66. McTaggart, F.; Jones, P. Effects of statins on high-density lipoproteins: A potential contribution to cardiovascular benefit. *Cardiovasc. Drugs Ther.* **2008**, *22*, 321–338. [CrossRef] [PubMed]
67. Yamashita, S.; Tsubakio-Yamamoto, K.; Ohama, T.; Nakagawa-Toyama, Y.; Nishida, M. Molecular mechanisms of HDL-cholesterol elevation by statins and its effects on HDL functions. *J. Atheroscler. Thromb.* **2010**, *17*, 436–451. [CrossRef] [PubMed]
68. Teramoto, T.; Kobayashi, M.; Tasaki, H.; Yagyu, H.; Higashikata, T.; Takagi, Y.; Uno, K.; Baccara-Dinet, M.T.; Nohara, A. Efficacy and Safety of Alirocumab in Japanese Patients With Heterozygous Familial Hypercholesterolemia or at High Cardiovascular Risk with Hypercholesterolemia Not Adequately Controlled With Statins- ODYSSEY JAPAN Randomized Controlled Trial. *Circ. J.* **2016**, *80*, 1980–1987. [CrossRef]
69. Brusselmans, K.; Timmermans, L.; Van de Sande, T.; Van Veldhoven, P.P.; Guan, G.; Shechter, I.; Claessens, F.; Verhoeven, G.; Swinnen, J.V. Squalene synthase, a determinant of Raft-associated cholesterol and modulator of cancer cell proliferation. *J. Biol. Chem.* **2007**, *282*, 18777–18785. [CrossRef]
70. Ribas, V.; Garcia-Ruiz, C.; Fernandez-Checa, J.C. Mitochondria, cholesterol and cancer cell metabolism. *Clin. Transl. Med.* **2016**, *5*, 22. [CrossRef]
71. Huetsch, J.C.; Suresh, K.; Shimoda, L.A. When higher cholesterol is better: Membrane cholesterol loss and endothelial Ca^{2+} signaling. *Am. J. Physiol. Heart Circ. Physiol.* **2018**, *314*, H780–H783. [CrossRef]

cancers

MDPI

Article

Efficacy of Immune Checkpoint Inhibitors vs. Tyrosine Kinase Inhibitors/Everolimus in Adjuvant Renal Cell Carcinoma: Indirect Comparison of Disease-Free Survival

Andrea Ossato [1,†], Lorenzo Gasperoni [2,†], Luna Del Bono [3], Andrea Messori [4,*] and Vera Damuzzo [5,6]

1 Department of Pharmaceutical and Pharmacological Sciences, University of Padua, 35131 Padova, Italy; andrea.ossato@studenti.unipd.it
2 Oncological Pharmacy Unit, IRCCS Istituto Romagnolo per lo Studio dei Tumori (IRST) "Dino Amadori", 47014 Meldola, Italy; lorenzo.gasperoni@irst.emr.it
3 Azienda Ospedaliera Universitaria Pisana, 56100 Pisa, Italy; lunadelbono@gmail.com
4 HTA Unit, Regional Health Service, 50139 Florence, Italy
5 Hospital Pharmacy, Vittorio Veneto Hospital, 31029 Vittorio Veneto, Italy
6 Italian Society of Clinical Pharmacy and Therapeutics (SIFaCT), 10123 Turin, Italy
* Correspondence: andrea.messorit@regione.toscana.it
† These authors contributed equally to this work.

Simple Summary: The proven efficacy of mTOR inhibitor (mTORI), tyrosine kinase inhibitor (TKI) or immune checkpoint inhibitor (ICI) therapies in metastatic renal cell carcinoma (RCC) suggests that these agents should be investigated as adjuvant therapy with the aim of eliminating undetectable microscopic residual disease after curative resection. Our study aimed to compare the efficacy of these treatments using an innovative method that reconstructs individual patient data from Kaplan–Meier (KM) curves. Nine phase III trials describing different treatment options for adjuvant RCC were selected. Individual patient data were reconstructed from KM curves of disease-free survival (DFS) using the IPDfromKM method. DFS was then compared between the combination treatments and the control arm (placebo). The results were summarized as multi-treatment KM curves. Standard statistical tests were used, including the hazard ratio for superiority and the likelihood ratio test for heterogeneity. In the population of these nine trials, our study showed that two ICIs (nivolumab plus ipilimumab and pembrolizumab) and one TKI (sunitinib) were superior to the placebo, whereas the remaining TKIs and mTORIs were not. As we assessed DFS as the primary endpoint for the adjuvant comparison, the overall survival benefit remains unknown. This novel approach to studying survival has allowed us to make all of the indirect head-to-head comparisons between these agents in a context where no "real" comparative trials have been conducted.

Abstract: Background: The proven efficacy of mTOR inhibitors (mTORIs), tyrosine kinase inhibitors (TKIs) or immune checkpoint inhibitors (ICIs) in metastatic renal cell carcinoma (RCC) suggests that these agents should be investigated as adjuvant therapy with the aim of eliminating undetectable microscopic residual disease after curative resection. The aim of our study was to compare the efficacy of these treatments using an innovative method of reconstructing individual patient data. Methods: Nine phase III trials describing adjuvant RCC treatments were selected. The IPDfromKM method was used to reconstruct individual patient data from Kaplan–Meier (KM) curves. The combination treatments were compared with the control arm (placebo) for disease-free survival (DFS). Multi-treatment KM curves were used to summarize the results. Standard statistical tests were performed. These included hazard ratio and likelihood ratio tests for heterogeneity. Results: In the overall population, the study showed that two ICIs (nivolumab plus ipilimumab and pembrolizumab) and one TKI (sunitinib) were superior to the placebo, whereas both TKIs and mTORIs were inferior. As we assessed DFS as the primary endpoint for the adjuvant comparison, the overall survival benefit remains unknown. Conclusions: This novel approach to investigating survival has allowed us to conduct all indirect head-to-head comparisons between these agents in a context where no "real" comparative trials have been conducted.

Citation: Ossato, A.; Gasperoni, L.; Del Bono, L.; Messori, A.; Damuzzo, V. Efficacy of Immune Checkpoint Inhibitors vs. Tyrosine Kinase Inhibitors/Everolimus in Adjuvant Renal Cell Carcinoma: Indirect Comparison of Disease-Free Survival. *Cancers* **2024**, *16*, 557. https://doi.org/10.3390/cancers16030557

Academic Editors: José I. López and Claudia Manini

Received: 18 December 2023
Revised: 19 January 2024
Accepted: 25 January 2024
Published: 28 January 2024

Keywords: indirect comparison; Shiny method; reconstructed individual patient data; disease-free survival; renal cell carcinoma; adjuvant setting

1. Introduction

Renal cell carcinoma (RCC) is the most common type of kidney cancer, accounting for about 2% of all cancer diagnoses worldwide. Approximately 80% of RCCs are clear cell tumors [1].

Over the last 20 years, the prognosis for patients with metastatic RCC has improved thanks to the results of clinical trials with mTOR inhibitors (mTORIs), tyrosine kinase inhibitors (TKIs) or immune checkpoint inhibitors (ICIs). The most recent revolution is the treatment of metastatic RCC with ICI combinations or ICI–TKI combinations [2–5], which has resulted in a significant improvement in survival, although not a cure.

Prevention of metastatic disease remains a priority in the curative setting of early stage RCC. For patients with locoregional RCC, partial or radical nephrectomy is the standard of care, and adjuvant treatment is an option to reduce the risk of recurrence, considering that 40% of surgically resected patients with stage II–III disease will relapse [6–10].

The proven efficacy of ICI, TKI and mTORI therapies in metastatic RCC suggested that these agents should be investigated as adjuvant therapy with the aim of eliminating any residual undetectable microscopic disease after curative resection.

Numerous randomized phase III trials in the adjuvant treatment of patients with RCC have ended with conflicting results, but overall, they seem to show a greater benefit of treatment with ICIs compared to TKIs and mTORIs [11–19]. As a result of this uncertainty, adjuvant therapies have also had different regulatory pathways. In fact, sunitinib, the first adjuvant therapy approved by the Food and Drug Administration (FDA) for the adjuvant treatment of RCC, was not approved by the European Medicines Agency (EMA).

In this scenario, given the available results and the lack of head-to-head comparisons between the adjuvant treatment options studied to date, we conducted an analysis using a new artificial intelligence technique (called the "IPDfromKM" method or Shiny method) to compare disease-free survival (DFS) in patients with resected primary RCC at risk of recurrence. Only results from phase III randomized clinical trials (RCTs) with adjuvant mTORI or ICI or TKI were included in the analysis.

The IPDfromKM method is a new artificial intelligence tool that reconstructs individual patient data from the graph of Kaplan–Meier (KM) curves and allows cross-study comparisons to be made based on reconstructed patients [20,21]. This is a relatively new method for generating new original clinical evidence and is particularly suitable for indirect comparison of time-to-event endpoints, especially those with a long follow-up, because it takes into consideration the time at which events occurred, while a standard binary meta-analysis ignores this information. In addition, the IPDfromKM method presents an easy-to-understand summary of results generating a unique KM chart containing the curves based on reconstructed patients and pooling all patients who received the same treatment, regardless of the clinical trial. In other words, the Forest plot typical of standard binary meta-analysis is replaced by a survival plot with as many KM curves as the number of treatments compared. The main treatment regimens can be further compared with each other using standard statistical methods such as the hazard ratio. Thanks to this method, in the present report we provide a comparative overview of the main adjuvant treatments available for RCC patients and determine their place in therapy and their relative efficacy.

2. Materials and Methods

2.1. Study Design

In accordance with recent publications [5,21,22], we conducted a comprehensive literature review to identify primary treatment options for adjuvant treatment of RCC. Following the selection of relevant studies, we utilized the IPDfromKM method to reconstruct individual patient data from KM graphs and consequently perform head-to-head indirect comparisons between treatments [20]. Our analysis focused on disease-free survival (DFS) as the primary endpoint. Results were shown through multi-treatment KM curves.

2.2. Literature Search

We searched the ClinicalTrials.gov database to identify randomized controlled trials (RCTs) that were eligible for our analysis (last search on 15 November 2023). The following search terms were used: "renal cell carcinoma" OR "RCC". Only phase III randomized interventional clinical trials were selected applying the filter option. By eliminating duplicates, 146 RCTs emerged. The main inclusion criteria were: (a) RCC treatment (not diagnostic, imaging or surgical trials, not other diseases or solid tumors of different origin); (b) adjuvant treatment (not neoadjuvant/adjuvant treatment, not perioperative treatment, not locally advanced or metastatic treatment); (c) ICI or TKI or mTORI treatment; (d) inclusion of non-metastatic patients; (e) DFS endpoint; and (f) publication of results as a KM curve. The selection of articles from our literature search was based on the PRISMA algorithm, which recorded the reasons for inclusion and exclusion of each trial [23]; the final list of included trials was determined in the last step of the PRISMA flow.

For each trial included, we recorded the number of patients enrolled and the number of events (defined as first documented local or distant recurrence of RCC, secondary systemic malignancy, or death from any cause, whichever occurred first). To avoid duplicate inclusion of patients from the same trial, only the most recent publication was included.

2.3. Reconstruction of Individual Patient Data

Patient-level data were reconstructed from KM curves using the IPDfromKM method, as previously described [20,21]. The curves were digitized with Webplot digitizer (version 4.5 online; URL https://apps.automeris.io/wpd/, accessed on 10 January 2024) and subsequently entered into the individual patient data reconstruction tool of the Shiny software (version 1.2.3.0).

Reconstructed individual patient data included observation time (defined as difference between enrolment and last follow-up) and patient outcome at last follow-up (alive, dead, censored).

2.4. Statistical Analysis

Restricted mean survival and DFS were estimated for each experimental treatment in comparison to placebo using Cox statistics for time-to-event endpoints; hazard ratio (HR) with 95% confidence interval (CI) and medians with 95% CI were also calculated. The likelihood ratio test was used to assess heterogeneity in outcomes between control groups of different RCTs. Moreover, indirect comparisons between treatments were assessed using the Cox model through four specific R-platform (version 4.2.1) packages for statistical analyses: survival, survRM2, survminer and ggsurvplot (2020; https://www.R-project.org/, accessed on 18 December 2023).

3. Results

Nine trials met the criteria for inclusion in our analysis (see Figure 1 for the PRISMA flowchart and Table 1 for RCT characteristics). Three trials involved an experimental treatment with parenteral ICIs while six trials were based on oral therapies such as TKIs (n = 5) or mTORI (n = 1). In all nine trials, the control arm was the placebo. In the application of the IPDfromKM method, we reconstructed 20 patient cohorts which represented the clinical material to perform our indirect comparisons.

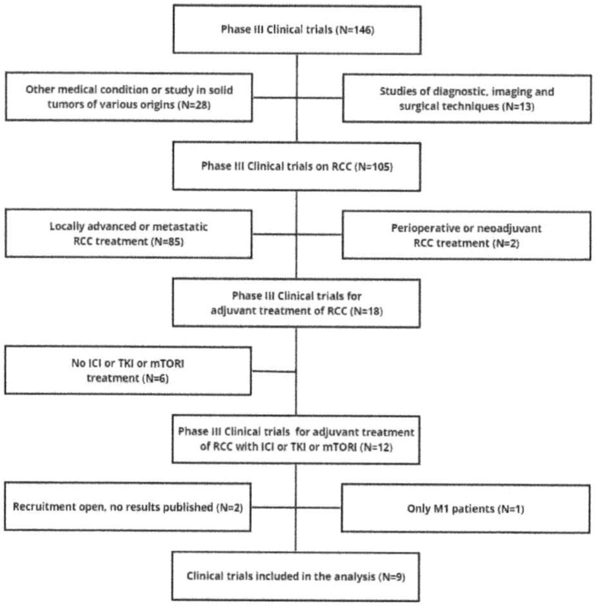

Figure 1. PRISMA flowchart of the process of trial selection. Abbreviations: RCC, renal cell carcinoma; DFS, disease-free survival.

Table 1. Main information about the nine RCTs included in the analysis.

#	Trial	Reference	Treatments under Comparison	DFS	
				Treatment Group (Events/Patients)	Controls (Events/Patients)
ARCC01	KEYNOTE-564 (two-arm)	Powles et al. [11]	Pembrolizumab (200 mg) Q3W (max 17 cycles); Placebo Q3W.	114/496	169/498
ARCC02	IMmotion010 (two-arm)	Pal et al. [12]	Atezolizumab (1200 mg) Q3W (max 16 cycles); Placebo Q3W.	164/390	168/388
ARCC03	CheckMate914 (two-arm)	Motzer et al. [13]	Nivolumab (240 mg) Q2W (max 12 cycles) + Ipilimumab (1 mg/kg) Q6W (max 4 cycles); Placebo Q2W + Q6W.	110/405	118/411
TKI01	ATLAS (two-arm)	Gross-Goupil et al. [14]	Axitinib (5 mg) BID (max 3 years); Placebo BID (max 3 years).	96/363	107/361
TKI02	SORCE (three-arm)	Eisen et al. [15]	Sorafenib (400 mg) BID for 3 years	245/639	167/430
			Sorafenib (400 mg) BID for 1 year plus placebo for 2 years; Placebo BID for 3 years.	242/642	
TKI03	ASSURE (three-arm)	Haas et al. [16]	Sunitinib (50 mg) (4 weeks on/2 off) for 54 weeks;	284/647	287/647
			Sorafenib (400 mg) BID for 54 weeks; Placebo for 54 weeks.	284/649	
TKI04	PROTECT (two-arm)	Motzer et al. [17]	Pazopanib (600 mg) for 1 year; Placebo for 1 year.	194/571	202/564
TKI05	S-TRAC (two-arm)	Ravaud et al. [18]	Sunitinib (50 mg) (4 weeks on/2 off) for 52 weeks; Placebo for 52 weeks.	113/309	144/306
mTORI01	EVEREST (two-arm)	Ryan et al. [19]	Everolimus (10 mg) for 54 weeks; Placebo for 54 weeks.	262/755	294/744

To conduct the indirect comparisons between the three ICI treatments, the DFS KM curves from the reconstructed patients of the ICI trials were plotted individually and reported in a single multi-treatment graph, with the three placebo cohorts pooled into a single graph. In this way, a total of four curves were generated (Figure 2A). Our DFS analysis on these reconstructed patients demonstrated superiority, in comparison with the placebo, as adjuvant therapy for both pembrolizumab (HR 0.62; 95% CI 0.50–0.78; $p < 0.001$) and nivolumab plus ipilimumab (HR 0.74; 95% CI 0.60–0.93; $p = 0.008$). Atezolizumab showed no advantage in DFS compared to the placebo (HR 1.04; 95% CI 0.86–1.26; p = NS).

The likelihood ratio test carried out on control arms showed no heterogeneity between these cohorts (likelihood ratio test, 5.01 on 2 df, $p < 0.08$; Figure 2B).

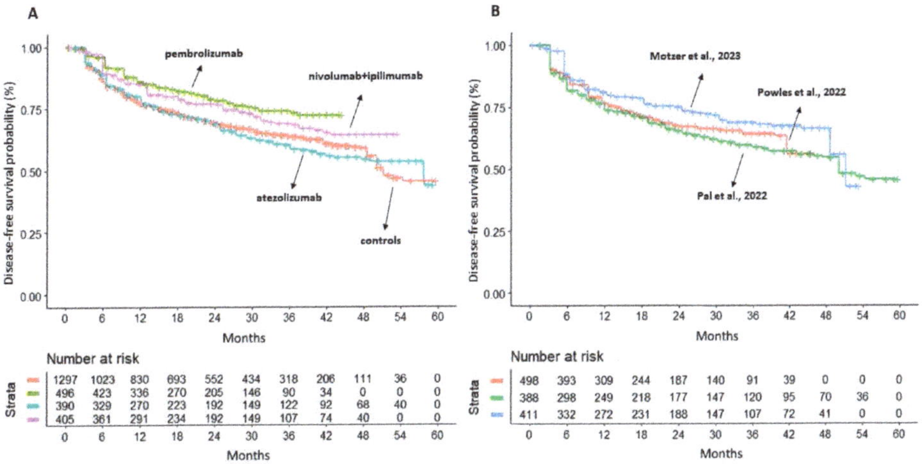

Figure 2. Panel (**A**) Shiny technique applied to placebo (n = 1297 from three trials; in red) and to three combination treatments: (a) pembrolizumab 200 mg (n = 496; in green); (b) atezolizumab 12,000 mg (n = 390; in light blue); (c) nivolumab 240 mg + ipilimumab 1 mg/kg (n = 405; in purple). Panel (**B**) Kaplan–Meier curves generated after reconstructing patient-level data from the three control arms of the included trials (placebo treated): Powles et al., 2022 (n = 498; in red [11]); Pal et al., 2022 (n = 388; in green [12]); Motzer et al., 2023 (n = 411; in blue [13]). Endpoint: disease-free survival (DFS), time in months. Abbreviations: n, number of patients.

Detailed results of indirect comparisons of the three ICIs treatments in all head-to-head combinations are reported as Forest plots of HRs with 95% CI (Figure 3 and Table S1). This analysis shows that pembrolizumab (HR = 0.60; 95% CI 0.45–0.81; $p < 0.001$) and nivolumab plus ipilimumab (HR = 0.72; 95% CI 0.54–0.96; $p = 0.024$) were superior to atezolizumab; on the other hand, pembrolizumab was not significantly superior to nivolumab plus ipilimumab (HR = 0.83; 95% CI 0.61–1.14; p = NS).

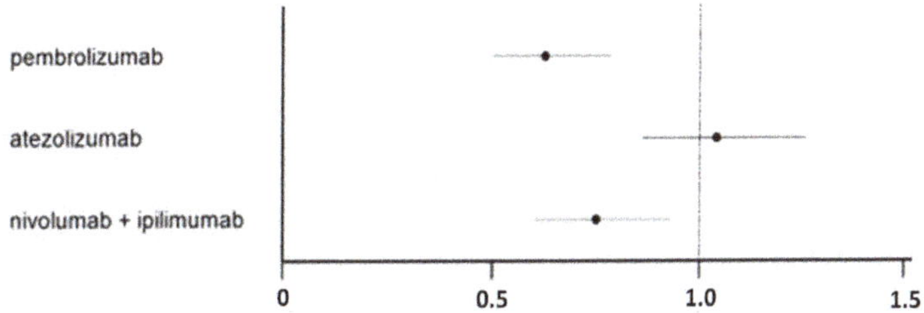

Figure 3. Forest plot showing the disease-free survival of adjuvant renal cell carcinoma patients treated with ICI treatment. Values are reported as HR of disease-free survival compared with controls.

To conduct indirect comparisons across oral treatments, the DFS KM curves from reconstructed patients of TKI and mTORI trials were plotted separately and reported as a single multi-treatment graph. The six control arms of oral therapy treatments demonstrated significant heterogeneity (Figure S1). Our head-to-head indirect comparisons were designed as follows. Firstly, we identified two non-heterogeneous subgroups among these trials because patients in the ATLAS trial were comparable to those in SORCE and EVEREST trials (subgroup #1) while patients in the ASSURE trial were comparable to those in PROTECT and S-TRAC trials (subgroup #2). Likewise, patients treated with the placebo in each of these two subgroups were combined to form two separate control cohorts. Finally, the DFS of the three active arms within each subgroup was compared to that of the respective pooled controls (Figure 4).

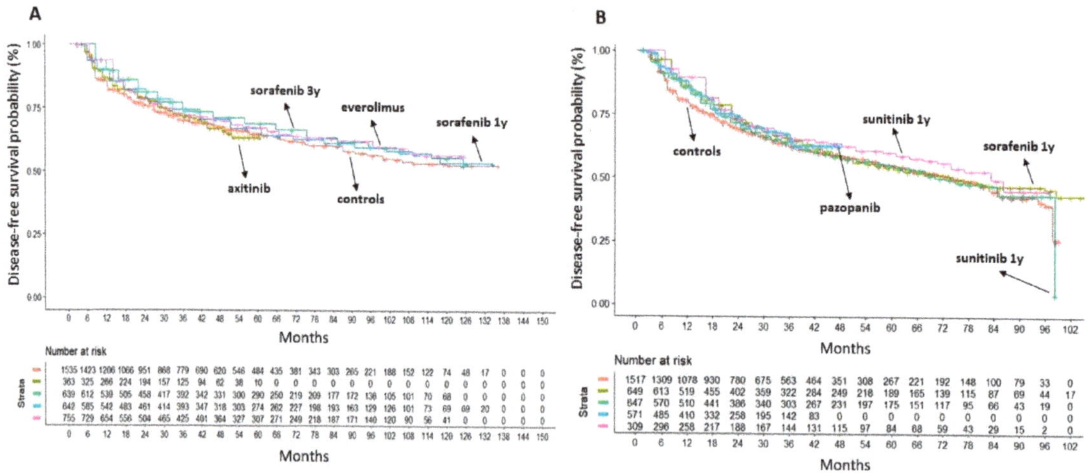

Figure 4. Application of the Shiny method in our subgroup analysis. Panel (**A**): comparison of DFS of the placebo (n = 1535 from ATLAS, SORCE and EVEREST trials; in red) with treatment arms of ATLAS ((axitinib n = 363; in gold); [14]), SORCE ((sorafenib for 3 years; n = 639; in turquoise) and (sorafenib for one year; n = 642; in blue); [15]) and EVEREST RCT ((everolimus; n = 755; in purple); [19]). Panel (**B**): comparison of DFS of placebo (n = 1517 from ASSURE, PROTECT and S-TRAC trials; in red) with treatment arms of ASSURE (sorafenib for 1 year (n = 649; in gold) and (sunitinib; n = 647; in light green); [16]), PROTECT (pazopanib (n = 571; in blue); [17]) and S-TRAC RCT ((sunitinib; n = 309; in purple); [18]). End-point: disease-free survival (DFS), time in months. Abbreviations: n, number of patients.

Most oral drug treatments did not show any DFS benefit over the placebo. Only two studies with sorafenib or sunitinib demonstrated a DFS advantage, but the results were somewhat contradictory. In fact, in the SORCE trial, patients receiving sorafenib 400 mg BID for 3 years showed a significantly longer DFS than the controls (HR = 0.83; 95% CI 0.71–0.98; p = 0.028) but not those treated for one year (HR = 0.89; 95% CI 0.76–1.04; p = NS). Conversely, in the ASSURE study, patients treated with sorafenib for approximately one year (54 weeks) showed a significantly longer DFS than placebo (HR = 0.85; 95% CI 0.74–0.99; p = 0.036). In the S-TRAC study, sunitinib determined a significant advantage over the placebo (HR = 0.81; 95% CI 0.66–0.99; p = 0.041), but in the ASSURE study the sunitinib treatment arm did not confirm this result, even though the dosage was the same.

Based on these intermediate results of our analysis, in the trials in which the active arm reported a DFS benefit over placebo (namely, KEYNOTE-564, CheckMate-914, S-TRAC, SORCE, based on the advantage at 3 years in sorafenib arm, and ASSURE, based on the advantage at 1 year in sorafenib arm), the treatment arms were indirectly compared with one another in all combinations (Table S2). At the same time, we performed the heterogeneity test on the control arms (Figure S2). With a likelihood ratio test of 34.42 on 4 df, $p < 0.001$, the SORCE trial proved to be an outlier compared with the other control arms and, therefore, was excluded from the indirect comparison. After this exclusion of the SORCE study, the analysis of the control arms showed no heterogeneity (likelihood ratio test = 3.68 on 3 df, $p = 0.3$).

Figure 5 shows the results of our main analysis in which pembrolizumab (HR 0.67; 95% CI 0.54–0.83; $p < 0.001$), nivolumab plus ipilimumab (HR 0.81; 95% CI 0.66–0.99; $p = 0.05$) and sunitinib (HR 0.82; 95% CI 0.67–0.99; $p = 0.05$) demonstrated superiority compared to the placebo. These three treatments were indirectly compared with one another, and these comparisons showed that pembrolizumab was significantly superior to 1-year treatment with sorafenib (HR = 0.76; 95% CI 0.60–0.98; $p = 0.038$); no significant difference was observed in the remaining comparisons (Figure 6, Table S3).

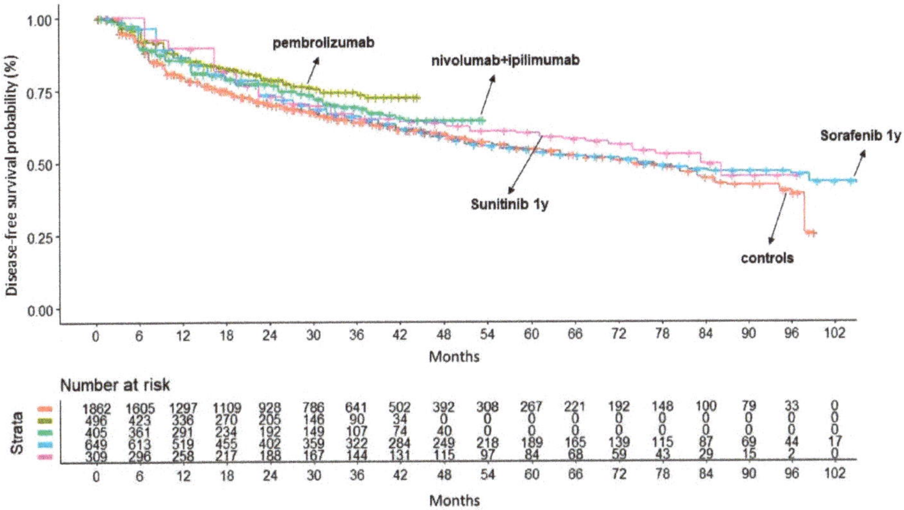

Figure 5. Comparison between DFS of the placebo (n = 1862 from 4 trials; in red) and of four treatments that showed survival benefit compared to control: (a) pembrolizumab (n = 496; in gold); (b) nivolumab + ipilimumab (n = 405; in light green); (c) sorafenib for 1 year (n = 649; in blue) and (d) sunitinib (n = 309; in purple). Endpoint: disease-free survival (DFS), time in months. Abbreviations: n, number of patients.

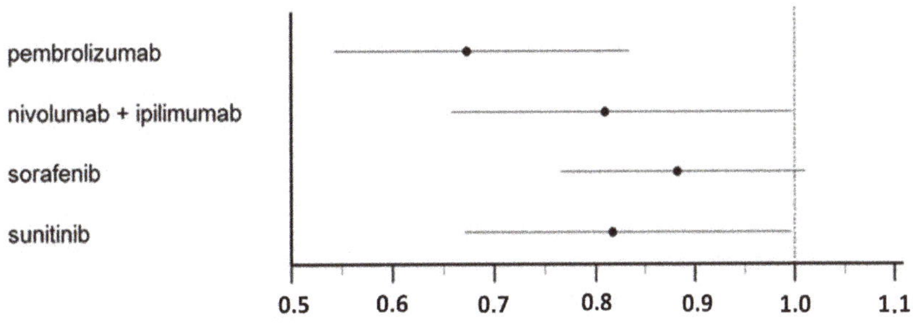

Figure 6. Forest plot showing the disease-free survival of adjuvant renal cell carcinoma patients treated with ICI and TKI regimens. Values are reported as HR of disease-free survival compared with controls.

Overall, the efficacy of adjuvant treatments for RCC ranks as follows: (1) pembrolizumab; (2) nivolumab + ipilimumab; (3) sunitinib.

In our estimation of absolute outcome parameters, only the S-TRAC trial had a sufficient follow-up to reach the median DFS; hence, we decided to calculate the Restricted Mean Survival Times (RMSTs) as an alternative to medians. Our RMST analysis was truncated at 44 months; its results are reported in Table S3. All the five active treatments showed a DFS benefit compared to the placebo. In terms of clinical relevance, pembrolizumab produced a three-month advantage over the placebo (RMST of 36.09 vs. 32.9 months, respectively), while it was only slightly superior to the other treatments.

4. Discussion

The present study investigated the main treatments for adjuvant RCC using an innovative tool to reconstruct individual patient data, known as the "IPDfromKM" or "Shiny method". This technique, applied to indirect comparisons based on standard statistics, is a valid alternative to network meta-analysis, mainly because of its ability to adjust for the different lengths of follow-up in the included trials. For this reason, the IPDfromKM method is particularly suitable for studies in the field of oncology and hemato-oncology [22], and its use has recently been extended to cardiology, where long follow-up is also common [24,25].

Additional advantages of the IPDfromKM method are the straightforward fitting procedure and the summary of results expressed as a KM curves of immediate visual interpretation. The technique also comes with some limitations like the great dependence on availability of KM curves in the original publication of the trial, which often impairs subgroup analysis, a certain dependence on the accuracy of graph processing and the absence of statistical techniques for analyzing variability such as sensitivity analysis. Nevertheless, validation studies document that, in using the Shiny method, the agreement between KM curves based on the real patients and KM curves based on reconstructed patients is excellent [26,27].

Our study is the first to indirectly compare the DFS of different drug classes in the adjuvant treatment of RCC. Our results show that two ICIs (pembrolizumab and nivolumab plus ipilimumab) and one TKI (sunitinib) showed superiority over the placebo, whereas both TKI and mTORI did not.

The pattern of heterogeneity estimated in the included trials is an interesting finding of our study. For the ICI trials, despite some differences in patient inclusion criteria (e.g., the KEYNOTE-564 and IMmotion010 trials enrolled 6% and 14% of M1 patients, respectively), the heterogeneity test showed substantial comparability between these populations. In contrast, the analysis of the TKI or mTORI trials showed significant heterogeneity in the control arms, which negatively affected the reliability of our indirect comparisons. This heterogeneity was likely due to differences in patient selection criteria, such as the presence of a high proportion of patients with early-stage tumors, sarcomatoid features or specific factors that increase the risk of relapse.

Patients at high risk of relapse should be the focus of adjuvant trials, as currently recommended [28,29], and indeed the benefit of adjuvant TKI treatment may be greater in patients at higher risk, as demonstrated in the S-TRAC trial [18] with sunitinib. In addition to eligibility criteria, the efficacy of TKIs and mTORIs may be related to other factors that influence drug exposure, such as patient adherence, discontinuation due to adverse drug reactions, different dose reduction schedules and duration of treatment. For example, in the sunitinib arm of the ASSURE trial, the dose was allowed to be reduced to 25 mg, and midway through the trial the starting dose was changed from 50 mg to 37.5 mg (a dose level not allowed in the S-TRAC trial). In addition, the duration of oral treatment was quite heterogeneous, as ATLAS trial and the sorafenib arm in the SORCE trial proposed 3 years of treatment, compared with 1 year in the other trials. Furthermore, grade 3–4 adverse events occurred in 46% of patients receiving everolimus and 49 to 72% in the TKI trials, which is more frequent than in the adjuvant ICI trials. The occurrence of adverse events may require discontinuation of treatment due to intolerance, but this may also be related to patient choice. Indeed, patients who have undergone nephrectomy are considered disease-free and may be less willing to accept serious adverse events and reduced quality of life [30].

Finally, it is possible that the vascular endothelial growth factor pathway, which is primarily targeted by TKIs, is less involved in the growth of early stage RCC, while remaining a hallmark of metastatic disease.

Overall, heterogeneity may be intrinsic and related to eligibility criteria or it may be related to less measurable variables related to the active treatment (e.g., adherence to therapy, duration of treatment, dose reduction schedule, occurrence of ADRs). In our study, we mitigated the first source of heterogeneity by performing the likelihood ratio test and comparing only studies whose placebo arms did not show heterogeneity.

Given that life expectancy after nephrectomy is nearly 40% at 10 years [29,30], a limitation of the current analysis is that OS was not investigated. Indeed, most of the studies included in the analysis had DFS as primary endpoint and OS as secondary endpoint. The availability of OS results is certainly an advantage as it is a simple endpoint, reliable to measure, easy to interpret and of great clinical utility. However, OS analyses require long follow-up times, and results may be affected by non-cancer deaths and subsequent lines of therapy. Further follow-up and maturation of OS data might influence the conclusions drawn from the study. However, even if DFS benefit may not translate into an OS benefit, a longer DFS will certainly postpone the initiation of therapies for the advanced setting which require greater intensity of care and lead to greater patient involvement and reduced quality of life. We accepted DFS as the primary endpoint for the adjuvant comparison, but, as the benefit in OS remains unknown, it is an open question whether the DFS advantage is sufficient to support the financial burden of these therapies. In addition to these economic issues, further follow-up of these trials will show whether there are long-term survivors and whether the superiority of pembrolizumab over other TKIs is confirmed. Patient perception can contribute to a more complete picture of the patient experience of adjuvant therapy, although the heterogeneity of patient selection remains a limitation. The patient-reported outcome (PRO) analysis in the ASSURE trial highlighted that sunitinib was associated with significant fatigue compared to sorafenib treatment or the placebo [31]. The PRO analysis in the S-TRAC trial showed that adjuvant sunitinib therapy was not associated with clinically meaningful deterioration in most quality-of-life measures, with the exceptions of PRO scores for diarrhea and loss of appetite, which reached a clinically meaningful difference [32]. Adjuvant pembrolizumab therapy in KEYNOTE-564 did not compromise patient health-related quality of life [33]. Finally, the combination of clinical findings from the adjuvant RCC treatment trials with artificial intelligence tools (e.g., IPDfromKM) and molecular insights from the machine learning study may lead to a more comprehensive understanding of RCC biology, its impact on treatment outcomes and treatment efficacy. These techniques may lead to new hypotheses for future research and contribute to the development of more effective and individualized treatment strategies [34].

5. Conclusions

In conclusion, while waiting for data with longer follow-up, most adjuvant treatments for RCC are currently approved, and the oncologist is faced with difficult clinical decisions. Our study, despite the limitations described above, mostly related to heterogeneity issues, allowed head-to-head comparisons between different adjuvant regimens for RCC, ranking their efficacy and showing pembrolizumab as the most effective option.

Supplementary Materials: The following supporting information can be downloaded at: https://www.mdpi.com/article/10.3390/cancers16030557/s1, Figure S1: Visual representation of indirect treatment comparisons of six control arms of oral therapy treatments; Figure S2: Visual representation of indirect treatment comparisons for DFS of control arms of KEYNOTE-564, CheckMate-914, S-TRAC, SORCE and ASSURE Trials. Table S1: HR of inter-treatment comparison between the three ICIs treatments; Table S2: HR of inter-treatment comparison between adjuvant treatments for RCC; Table S3: Restricted mean survival time (RMST) of the five active treatments that showed a DFS benefit compared to the placebo. References [11,13–19] are cited in the Supplementary Materials.

Author Contributions: Conceptualization, A.O. and V.D.; data curation, L.D.B. and V.D.; formal analysis, A.O., L.D.B., A.M. and V.D.; investigation, L.G.; methodology, A.O., L.G., A.M. and V.D.; software, A.M.; writing—original draft, A.O. and V.D.; writing—review and editing, L.G. and A.M. All authors have read and agreed to the published version of the manuscript.

Funding: This research received no external funding.

Institutional Review Board Statement: Not applicable.

Informed Consent Statement: Not applicable.

Data Availability Statement: The data presented in this study are available in the article and in the Supplementary Materials.

Acknowledgments: Italian Society for Clinical Pharmacy and Therapeutics (SIFaCT), Turin, Italy.

Conflicts of Interest: The authors declare no conflicts of interest.

References

1. Padala, S.A.; Barsouk, A.; Thandra, K.C.; Saginala, K.; Mohammed, A.; Vakiti, A.; Rawla, P.; Barsouk, A. Epidemiology of Renal Cell Carcinoma. *World J. Oncol.* **2020**, *11*, 79–87. [CrossRef] [PubMed]
2. Calvo, E.; Escudier, B.; Motzer, R.J.; Oudard, S.; Hutson, T.E.; Porta, C.; Bracarda, S.; Grünwald, V.; Thompson, J.A.; Ravaud, A.; et al. Everolimus in metastatic renal cell carcinoma: Subgroup analysis of patients with 1 or 2 previous vascular endothelial growth factor receptor-tyrosine kinase inhibitor therapies enrolled in the phase III RECORD-1 study. *Eur. J. Cancer* **2012**, *48*, 333–339. [CrossRef]
3. Xu, J.X.; Maher, V.E.; Zhang, L.; Tang, S.; Sridhara, R.; Ibrahim, A.; Kim, G.; Pazdur, R. FDA Approval Summary: Nivolumab in Advanced Renal Cell Carcinoma After Anti-Angiogenic Therapy and Exploratory Predictive Biomarker Analysis. *Oncologist* **2017**, *22*, 311–317. [CrossRef]
4. Lalani, A.A.; Heng, D.Y.C.; Basappa, N.S.; Wood, L.; Iqbal, N.; McLeod, D.; Soulières, D.; Kollmannsberger, C. Evolving landscape of first-line combination therapy in advanced renal cancer: A systematic review. *Ther. Adv. Med. Oncol.* **2022**, *14*, 17588359221108685. [CrossRef] [PubMed]
5. Ossato, A.; Mengato, D.; Chiumente, M.; Messori, A.; Damuzzo, V. Progression-Free and Overall Survival of First-Line Treatments for Advanced Renal Cell Carcinoma: Indirect Comparison of Six Combination Regimens. *Cancers* **2023**, *15*, 2029. [CrossRef]
6. Powles, T.; Albiges, L.; Bex, A.; Grünwald, V.; Porta, C.; Procopio, G.; Schmidinger, M.; Suárez, C.; de Velasco, G.; ESMO Guidelines Committee. Electronic address: Clinicalguidelines@esmo.org. ESMO Clinical Practice Guideline update on the use of immunotherapy in early stage and advanced renal cell carcinoma. *Ann. Oncol.* **2021**, *32*, 1511–1519. [CrossRef]
7. Ljungberg, B.; Albiges, L.; Abu-Ghanem, Y.; Bedke, J.; Capitanio, U.; Dabestani, S.; Fernández-Pello, S.; Giles, R.H.; Hofmann, F.; Hora, M.; et al. European Association of Urology Guidelines on Renal Cell Carcinoma: The 2022 Update. *Eur. Urol.* **2022**, *82*, 399–410. [CrossRef] [PubMed]
8. Martinez Chanza, N.; Tripathi, A.; Harshman, L.C. Adjuvant Therapy Options in Renal Cell Carcinoma: Where Do We Stand? *Curr. Treat. Options Oncol.* **2019**, *20*, 44. [CrossRef]
9. Motzer, R.J.; Ravaud, A.; Patard, J.J.; Pandha, H.S.; George, D.J.; Patel, A.; Chang, Y.H.; Escudier, B.; Donskov, F.; Magheli, A.; et al. Adjuvant Sunitinib for High-risk Renal Cell Carcinoma After Nephrectomy: Subgroup Analyses and Updated Overall Survival Results. *Eur. Urol.* **2018**, *73*, 62–68. [CrossRef]
10. Harshman, L.C.; Xie, W.; Moreira, R.B.; Bossé, D.; Ruiz Ares, G.J.; Sweeney, C.J.; Choueiri, T.K. Evaluation of disease-free survival as an intermediate metric of overall survival in patients with localized renal cell carcinoma: A trial-level meta-analysis. *Cancer* **2018**, *124*, 925–933. [CrossRef]
11. Powles, T.; Tomczak, P.; Park, S.H.; Venugopal, B.; Ferguson, T.; Symeonides, S.N.; Hajek, J.; Gurney, H.; Chang, Y.H.; Lee, J.L.; et al. Investigators. Pembrolizumab versus placebo as post-nephrectomy adjuvant therapy for clear cell renal cell carcinoma (KEYNOTE-564): 30-month follow-up analysis of a multicentre, randomised, double-blind, placebo-controlled, phase 3 trial. *Lancet Oncol.* **2022**, *23*, 1133–1144. [CrossRef]
12. Pal, S.K.; Uzzo, R.; Karam, J.A.; Master, V.A.; Donskov, F.; Suarez, C.; Albiges, L.; Rini, B.; Tomita, Y.; Kann, A.G.; et al. Adjuvant atezolizumab versus placebo for patients with renal cell carcinoma at increased risk of recurrence following resection (IMmotion010): A multicentre, randomised, double-blind, phase 3 trial. *Lancet* **2022**, *400*, 1103–1116. [CrossRef]
13. Motzer, R.J.; Russo, P.; Grünwald, V.; Tomita, Y.; Zurawski, B.; Parikh, O.; Buti, S.; Barthélémy, P.; Goh, J.C.; Ye, D.; et al. Adjuvant nivolumab plus ipilimumab versus placebo for localised renal cell carcinoma after nephrectomy (CheckMate 914): A double-blind, randomised, phase 3 trial. *Lancet* **2023**, *401*, 821–832. [CrossRef]
14. Gross-Goupil, M.; Kwon, T.G.; Eto, M.; Ye, D.; Miyake, H.; Seo, S.I.; Byun, S.S.; Lee, J.L.; Master, V.; Jin, J.; et al. Axitinib versus placebo as an adjuvant treatment of renal cell carcinoma: Results from the phase III, randomized ATLAS trial. *Ann. Oncol.* **2018**, *29*, 2371–2378. [CrossRef]

15. Eisen, T.; Frangou, E.; Oza, B.; Ritchie, A.W.S.; Smith, B.; Kaplan, R.; Davis, I.D.; Stockler, M.R.; Albiges, L.; Escudier, B.; et al. Adjuvant Sorafenib for Renal Cell Carcinoma at Intermediate or High Risk of Relapse: Results From the SORCE Randomized Phase III Intergroup Trial. *J. Clin. Oncol.* **2020**, *38*, 4064–4075. [CrossRef]
16. Haas, N.B.; Manola, J.; Uzzo, R.G.; Flaherty, K.T.; Wood, C.G.; Kane, C.; Jewett, M.; Dutcher, J.P.; Atkins, M.B.; Pins, M.; et al. Adjuvant sunitinib or sorafenib for high-risk, non-metastatic renal-cell carcinoma (ECOG-ACRIN E2805): A double-blind, placebo-controlled, randomised, phase 3 trial. *Lancet* **2016**, *387*, 2008–2016. [CrossRef] [PubMed]
17. Motzer, R.J.; Haas, N.B.; Donskov, F.; Gross-Goupil, M.; Varlamov, S.; Kopyltsov, E.; Lee, J.L.; Melichar, B.; Rini, B.I.; Choueiri, T.K.; et al. Randomized Phase III Trial of Adjuvant Pazopanib Versus Placebo After Nephrectomy in Patients With Localized or Locally Advanced Renal Cell Carcinoma. *J. Clin. Oncol.* **2017**, *35*, 3916–3923. [CrossRef] [PubMed]
18. Ravaud, A.; Motzer, R.J.; Pandha, H.S.; George, D.J.; Pantuck, A.J.; Patel, A.; Chang, Y.H.; Escudier, B.; Donskov, F.; Magheli, A.; et al. Adjuvant Sunitinib in High-Risk Renal-Cell Carcinoma after Nephrectomy. *N. Engl. J. Med.* **2016**, *375*, 2246–2254. [CrossRef] [PubMed]
19. Ryan, C.W.; Tangen, C.M.; Heath, E.I.; Stein, M.N.; Meng, M.V.; Alva, A.S.; Pal, S.K.; Puzanov, I.; Clark, J.I.; Choueiri, T.K.; et al. Adjuvant everolimus after surgery for renal cell carcinoma (EVEREST): A double-blind, placebo-controlled, randomised, phase 3 trial. *Lancet* **2023**, *402*, 1043–1051. [CrossRef]
20. Liu, N.; Zhou, Y.; Lee, J.J. IPDfromKM: Reconstruct individual patient data from published Kaplan-Meier survival curves. *BMC Med. Res. Methodol.* **2021**, *21*, 111. [CrossRef]
21. Messori, A. Synthetizing Published Evidence on Survival by Reconstruction of Patient-Level Data and Generation of a Multi-Trial Kaplan-Meier Curve. *Cureus* **2021**, *13*, e19422. [CrossRef]
22. Messori, A.; Damuzzo, V.; Rivano, M.; Cancanelli, L.; Di Spazio, L.; Ossato, A.; Chiumente, M.; Mengato, D. Application of the IPDfromKM-Shiny Method to Compare the Efficacy of Novel Treatments Aimed at the Same Disease Condition: A Report of 14 Analyses. *Cancers* **2023**, *15*, 1633. [CrossRef]
23. Shamseer, L.; Moher, D.; Clarke, M.; Ghersi, D.; Liberati, A.; Petticrew, M.; Shekelle, P.; Stewart, L.A.; PRISMA-P Group. Preferred reporting items for systematic review and meta-analysis protocols (PRISMA-P) 2015: Elaboration and explanation. *BMJ* **2015**, *350*, g7647. [CrossRef]
24. Messori, A. Reconstruction of individual-patient data from the analysis of Kaplan-Meier curves: The use of this method has extended from oncology to cardiology. *Open Sci. Framew.* 2023, *preprint*. Available online: https://osf.io/qejus(accessed on 15 November 2023).
25. de Sá Marchi, M.F.; Calomeni, P.; Gauza, M.M.; Kanhouche, G.; Ravani, L.V.; Rodrigues, C.V.F.; Tarasoutchi, F.; de Brito, F.S., Jr.; Rodés-Cabau, J.; Van Mieghem, N.M.; et al. Impact of periprocedural myocardial injury after transcatheter aortic valve implantation on long-term mortality: A meta-analysis of Kaplan-Meier derived individual patient data. *Front. Cardiovasc. Med.* **2023**, *10*, 1228305. [CrossRef]
26. Saluja, R.; Cheng, S.; Delos Santos, K.A.; Chan, K.K.W. Estimating hazard ratios from published Kaplan-Meier survival curves: A methods validation study. *Res. Synth. Methods* **2019**, *10*, 465–475. [CrossRef]
27. Everest, L.; Blommaert, S.; Tu, D.; Pater, J.L.; Hay, A.; Cheung, M.C.; Chan, K.K.W. Validating restricted mean survival time estimates from reconstructed kaplan-meier data against original trial individual patient data from trials conducted by the canadian cancer trials group. *Value Health* **2022**, *25*, 1157–1164. [CrossRef] [PubMed]
28. Agrawal, S.; Haas, N.B.; Bagheri, M.; Lane, B.R.; Coleman, J.; Hammers, H.; Bratslavsky, G.; Chauhan, C.; Kim, L.; Krishnasamy, V.P.; et al. Eligibility and Radiologic Assessment for Adjuvant Clinical Trials in Kidney Cancer. *JAMA Oncol.* **2020**, *6*, 133–141. [CrossRef] [PubMed]
29. Janowitz, T.; Welsh, S.J.; Zaki, K.; Mulders, P.; Eisen, T. Adjuvant therapy in renal cell carcinoma-past, present, and future. *Semin. Oncol.* **2013**, *40*, 482–491. [CrossRef] [PubMed]
30. Giberti, C.; Oneto, F.; Martorana, G.; Rovida, S.; Carmignani, G. Radical nephrectomy for renal cell carcinoma: Long-term results and prognostic factors on a series of 328 cases. *Eur. Urol.* **1997**, *31*, 40–48. [CrossRef] [PubMed]
31. Zhao, F.; Cella, D.; Manola, J.; DiPaola, R.S.; Wagner, L.I.; Haas, N.S.B. Fatigue among patients with renal cell carcinoma receiving adjuvant sunitinib or sorafenib: Patient-reported outcomes of ECOG-ACRIN E2805 trial. *Support. Care Cancer* **2018**, *26*, 1889–1895. [CrossRef] [PubMed]
32. Staehler, M.; Motzer, R.J.; George, D.J.; Pandha, H.S.; Donskov, F.; Escudier, B.; Pantuck, A.J.; Patel, A.; DeAnnuntis, L.; Bhattacharyya, H.; et al. Adjuvant sunitinib in patients with high-risk renal cell carcinoma: Safety, therapy management, and patient-reported outcomes in the S-TRAC trial. *Ann. Oncol.* **2018**, *29*, 2098–2104. [CrossRef] [PubMed]
33. Choueiri, T.K.; Tomczak, P.; Park, S.H.; Venugopal, B.; Symeonides, S.; Hajek, J.; Ferguson, T.; Chang, Y.H.; Lee, J.L.; Haas, N.; et al. Patient-Reported Outcomes in KEYNOTE-564: Adjuvant Pembrolizumab Versus Placebo for Renal Cell Carcinoma. *Oncologist* **2023**, *17*, oyad231. [CrossRef] [PubMed]
34. Marquardt, A.; Solimando, A.G.; Kerscher, A.; Bittrich, M.; Kalogirou, C.; Kübler, H.; Rosenwald, A. Subgroup-Independent Mapping of Renal Cell Carcinoma-Machine Learning Reveals Prognostic Mitochondrial Gene Signature Beyond Histopathologic Boundaries. *Front. Oncol.* **2021**, *11*, 621278. [CrossRef] [PubMed]

MDPI
St. Alban-Anlage 66
4052 Basel
Switzerland
www.mdpi.com

Cancers Editorial Office
E-mail: cancers@mdpi.com
www.mdpi.com/journal/cancers

www.ingramcontent.com/pod-product-compliance
Lightning Source LLC
Chambersburg PA
CBHW061440210926
51850CB00008B/1562
* 9 7 8 3 7 2 5 8 0 4 7 5 7 *